One Health for Dog-mediated Rabies Elimination in Asia

A Collection of Local Experiences

One Health for Dog-mediated Rabies Elimination in Asia

A Collection of Local Experiences

Edited by

Vanessa Slack

Texas Department of State Health Services, Zoonosis Control Branch, USA

Deborah Nadal

Ca' Foscari University of Venice, Italy; University of Glasgow, UK; World Health Organization, Geneva, Switzerland

Sandul Yasobant

Centre for One Health Education, Research & Development, Indian Institute of Public Health Gandhinagar, India

Florence Cliquet

WOAH/WHO Reference Laboratory for Rabies, France

Waqas Ahmad

Department of Clinical Sciences, University of Veterinary and Animal Sciences, Lahore, Narowal Campus, Narowal, Pakistan

Nihal Pushpakumara

Department of Wildlife Conservation, Sri Lanka

Sumon Ghosh

Department of Public Health, University of Tennessee, Knoxville, USA

CABI

CABI is a trading name of CAB International

CABI
Nosworthy Way
Wallingford
Oxfordshire OX10 8DE
UK

Tel: +44 (0)1491 832111
E-mail: info@cabi.org
Website: www.cabi.org

CABI
200 Portland Street
Boston
MA 02114
USA

T: +1 (617)682-9015
E-mail: cabi-nao@cabi.org

A catalogue record for this book is available from the British Library, London, UK.

A catalogue record for this book is available from the British Library, London, UK.

ISBN-13: 9781800622951 (hardback)
9781800622968 (OA ePDF)
9781800622975 (OA ePub)

DOI: 10.1079/ 9781800622975.0000

Commissioning Editor: Alexandra Lainsbury
Editorial Assistant: Emma McCann
Production Editor: James Bishop

Typeset by Exeter Premedia Services Pvt Ltd, Chennai, India

Contents

Contributors

Abila, Ronello – World Organisation for Animal Health (WOAH), Sub-Regional Representation for South-east Asia, Bangkok, Thailand. Email: r.abila@woah.org

Ahmad, Fiaz – Department of Clinical Sciences, University of Veterinary and Animal Sciences, Lahore, Narowal Campus, Narowal, Pakistan. Email: fiaz.ahmad@uvas.edu.pk

Ahmad, Manzoor – Department of Clinical Sciences, University of Veterinary and Animal Sciences, Lahore, Narowal Campus, Narowal, Pakistan. Email: manzoor.ahmad@uvas.edu.pk

Ahmad, Waqas – Department of Clinical Sciences, University of Veterinary and Animal Sciences, Lahore, Narowal Campus, Narowal, Pakistan. Email: waqas.hussain@uvas.edu.pk

Arikawa, Genki – Public Welfare and Health Department, Miyazaki Prefecture, Miyazaki, Japan. Email: genki_arikawa@med.miyazaki-u.ac.jp

Baig-Ansari, Naila – Indus Hospital Research Center, Indus Hospital and Health Network, Karachi, Pakistan. Email: naila.baigansari@tih.org.pk

Barber, Suzanne – Perdue University, West Lafayette, Indiana, USA. Email: sbarber4851@gmail.com

Bharti, Omesh Kumar – State Institute of Health and Family Welfare, Department of Health & Family Welfare, Shimla, India. Email: bhartiomesh@yahoo.com

Brookes, Victoria J. – Sydney School of Veterinary Science, Faculty of Science, The University of Sydney, Camperdown, New South Wales, Australia. Email: victoria.brookes@sydney.edu.au

Brum, Eric – Food and Agriculture Organization of the United Nations, Regional Office for Asia and the Pacific (FAORAP), Bangkok, Thailand; Emergency Centre for Transboundary Animal Diseases (ECTAD), Food and Agriculture Organization of the United Nations (FAO), Dhaka, Bangladesh. Email: Eric.Brum@fao.org

Chanachai, Karoon – USAID Regional Development Mission for Asia, Bangkok, Thailand. Email: kchanachai@usaid.gov

Masud, M.A. – Natural Product Informatics Research Center, KIST Gangneung Institute of Natural Products, Gangneung-si, Gangwon-do, South Korea. Email: masud@kist.re.kr

Miranda, Mary Elizabeth Gutierrez – Field Epidemiology Training Program Alumni Foundation Inc, Manila, Philippines. Email: megmiranda@fetpafi.org

Miranda, Noel Lee Jose – Independent Regional Consultant for Asia. Email: nljmiranda@yahoo.com

Misawa, Naoaki – Center for Animal Disease Control, University of Miyazaki, Miyazaki, Japan; Department of Veterinary Sciences, Faculty of Agriculture, University of Miyazaki, Miyazaki, Japan. Email: a0d901u@cc.miyazaki-u.ac.jp

Nadal, Deborah – Department of Humanities, Ca' Foscari University of Venice, Italy; School of Biodiversity, One Health and Veterinary Medicine, University of Glasgow, UK; Department of Control of Neglected Tropical Diseases, World Health Organization, Geneva, Switzerland. Email: nadal.deborah@gmail.com

Newman, Scott – Food and Agriculture Organization of the United Nations, Regional Office for Asia and the Pacific (FAORAP), Bangkok, Thailand. Email: scott.newman@fao.org

Oh, Yooni – Food and Agriculture Organization of the United Nations, Regional Office for Asia and the Pacific (FAORAP), Bangkok, Thailand. Email: Yooni.Oh@fao.org

Perera, K.A.D.N. – Formerly of Department of Rabies Research and Vaccine Quality Control, Medical Research Institute, Colombo, Sri Lanka. Email: devikanandani@yahoo.com

Perera, Oswin – Department of Veterinary Clinical Sciences, Faculty of Veterinary Medicine and Animal Science, University of Peradeniya, Peradeniya, Sri Lanka. Email: oswinperera@yahoo.com

Petsophonsakul, Wilaiwan – Lanna Dog Welfare, Chiang Mai, Thailand; Laboratory for Rabies Vaccine Evaluation, AMS Clinical Service Centre, Faculty of Associated Sciences, Chiang Mai University, Chiang Mai, Thailand. Email: wilaiwan_pet@yahoo.com

Phuentshok, Yoenten – Food and Agriculture Organization of the United Nations, Regional Office for Asia and the Pacific (FAORAP), Bangkok, Thailand. Email: y.phuentshok@woah.org

Pimburage, Ruwini M.S. – Public Health Veterinary Services, Ministry of Health, Colombo, Sri Lanka. Email: ruwinipimburage@gmail.com

Prabsriphum, Supalak – World Organisation for Animal Health (WOAH), Sub-Regional Representation for South-east Asia, Bangkok, Thailand. Email: s.prabsriphum@woah.org

Pushpakumara, Nihal – Department of Wildlife Conservation, Koduwatuwana, Ampara, Sri Lanka. Email: npkumara29@gmail.com

Quiambao, Beatriz Puzon – Clinical Research Division, Research Institute for Tropical Medicine, Muntinlupa, Philippines. Email: QuiambaoBeych2019@gmail.com

Radhakrishnan, Sreejith – School of Biodiversity, One Health and Veterinary Medicine of the University of Glasgow, UK. Email: sreejith.radhakrishnan@glasgow.ac.uk

Sagarasaeranee, Onpawee – Department of Livestock Development, Ministry of Agriculture, Bangkok, Thailand. Email: ammieloveu@hotmail.com

Salahuddin, Naseem – Infectious Diseases, Indus Hospital and Health Network, Karachi, Pakistan. Email: naseem.salahuddin@tih.org.pk

Schoonman, Luuk – Food and Agriculture Organization of the United Nations, Regional Office for Asia and the Pacific (FAORAP), Bangkok, Thailand. Email: luuk.schoonman@fao.org

Slack, Vanessa – Formerly of Texas Department of State Health Services, Zoonosis Control Branch, Texas, USA. Email: vanessa.m.slack@gmail.com

Sutherland, David – World Health Organization, Regional Office for South-East Asia (WHO-SEARO), New Delhi, India. Email: david.sutherland@fao.org

Telmo, Shynie – Regional Animal Disease Diagnostic Laboratory IV-B, Naujan, Oriental Mindoro, Philippines. Email: shynie.telmo.dvm@gmail.com

Thompson, Lesa – World Organisation for Animal Health (WOAH), Regional Representation for Asia and the Pacific, Tokyo, Japan. Email: l.thompson@woah.org

Tshering, Pasang – World Organisation for Animal Health (WOAH), Regional Representation for Asia and the Pacific, Tokyo, Japan. Email: p.tshering@woah.org

Ullah, Muhammad Rafi – Department of Clinical Sciences, University of Veterinary and Animal Sciences, Lahore, Narowal Campus, Narowal, Pakistan. Email: muhammad.rafi@uvas.edu.pk

Vanak, Abi Tamim – Ashoka Trust for Research in Ecology and the Environment, Bengaluru, India; School of Life Sciences, University of KwaZulu-Natal, Pietermaritzburg, South Africa; DBT-Wellcome Trust India Alliance, Hyderabad, India. Email: avanak@atree.org

Vos, Ad – Ceva Innovation Center GmbH, Am Pharmapark, Dessau-Rosslau, Germany. Email: ad.vos@ceva.com

Warembourg, Charlotte – Veterinary Public Health Institute, Vetsuisse Faculty, University of Bern, Bern, Switzerland. Email: charlotte.warembourg59@gmail.com

Wera, Ewaldus – Kupang State Agricultural Polytechnic (Politeknik Pertanian Negeri Kupang), West Timor, Indonesia. Email: ewaldus.wera@yahoo.com

Wijayathilaka, Tikiri – World Organisation for Animal Health (WOAH), Sub-Regional Representation for South-east Asia, Bangkok, Thailand. Email: t.wijayathilaka@woah.org

Wimalaratne, Omala – Formerly of Department of Rabies Research and Vaccine Quality Control, Medical Research Institute, Colombo, Sri Lanka. Email: omalaw@hotmail.com

Wongphruksasoong, Vilaiporn – Department of Livestock Development, Ministry of Agriculture, Bangkok, Thailand. Email: lhin_001@hotmail.com

Yajima, Aya – World Health Organization, Regional Office for the Western Pacific (WHO-WPRO), Manila, Philippines. Email: Yajimaa@who.int

Yasobant, Sandul – Centre for One Health Education, Research & Development, Indian Institute of Public Health Gandhinagar, Gandhinagar, Gujarat, India. Email: yasobant@iiphg.org

Yoshida, Takahiko – Department of Social Medicine, School of Medicine, Asahikawa Medical University, Asahikawa, Japan. Email: tyoshida@asahikawa-med.ac.jp

Yoshii, Kentaro – Laboratory of Public Health, Faculty of Veterinary Medicine, Hokkaido University, Sapporo, Japan; National Research Center for the Control and Prevention of Infectious Diseases, Nagasaki University, Nagasaki, Japan. Email: kyoshii@nagasaki-u.ac.jp

Yuson, Mirava – Field Epidemiology Training Program Alumni Foundation, Inc., Quezon City, Metro Manila, Philippines. Email: myuson@fetpafi.org

1 Global Dog and Human Rabies Control Efforts from Ancient Times to 2030 and Beyond

Deborah Nadal[1,2,3]* and Sreejith Radhakrishnan[2]
[1]Ca' Foscari University of Venice, Italy; [2]University of Glasgow, UK; [3]World Health Organization, Geneva, Switzerland

Abstract
Nowadays, rabies is mainly present in Africa and Asia, where every year it causes an estimated 59,000 human deaths and costs US$8.6 billion. A key date in the history of rabies control is 6 July 1885, when the first dose of rabies vaccine was successfully inoculated to an exposed individual in Paris. Yet, long before and after this event, many attempts at stopping rabies transmission, managing dog bites and preventing rabies symptoms have occurred throughout the world. Each step forwards – and backwards too – has been crucial to advance the scientific knowledge of rabies and how to control this disease at the interface of challenges of ecological, political and social nature. As the world starts to recover from the coronavirus 2019 (COVID-19) pandemic and move towards the 2030 goal of eliminating dog-mediated human rabies, learning from the past is vital for achieving a world with reduced rabies risk.

1.1 Current Epidemiological Situation

The rabies virus (RABV) belongs to the order *Mononegavirales* (which are viruses with non-segmented, negative-stranded RNA genomes), the family *Rhabdoviridae* (which groups bullet-shaped viruses), and the genus *Lyssavirus* (Fooks and Jackson, 2020). Rabies is a viral zoonotic disease most often transmitted through the bite of a rabid animal, but transmission can also occur through licks, scratches, or the contamination of mucosa or open skin wounds with infectious saliva. RABV affects mammals, including humans, by infecting their central nervous system and ultimately causing inflammation of the brain and death. In humans, the incubation period of the disease can vary from 1 week to 1 year, though it is typically 2–3 months. In dogs, the incubation period is usually shorter.

Clinically, rabies has two forms: (i) furious rabies, with hyperactivity, hallucinations, hydrophobia (fear of water), and sometimes aerophobia (fear of drafts or fresh air); and (ii) paralytic rabies, characterized by growing paralysis. The paralytic form of rabies is often misdiagnosed, contributing to the under-reporting of the disease in humans and the underestimation of the disease in dogs. In each form, once the symptoms of the disease develop, rabies is incurable and fatal to both animals and humans. However, rabies is 100% preventable in humans through thorough wound washing with water and soap,

*Corresponding author: nadal.deborah@gmail.com

DOI: 10.1079/9781800622975.0001

as well as post-exposure prophylaxis (PEP), which includes the use of rabies vaccine and immunoglobulins (WHO, 2018b).

Rabies is widely distributed across the world. It is estimated that 59,000 people die of rabies each year, mainly in Asia (59.6%) and Africa (36.4%) (Hampson *et al.*, 2015). The number of dogs and other mammals who die of it is unknown. Most human deaths, 99%, are caused by exposure to an infected dog. About 40% of human victims are children under the age of 15. Most human victims are males and belong to geographically, economically and socially marginalized areas (WHO, 2018b).

Rabies costs US$8.6 billion globally each year (Hampson *et al.*, 2015). About half of this cost is for medical care and this mainly includes the direct cost of care, where the average cost of the whole vaccination schedule (not including immunoglobulin) is estimated at US$108. When this cost is borne out-of-pocket by patients, it can be catastrophic in low-income settings. When PEP is provided for free by government healthcare facilities, it represents a substantial expenditure for health systems, especially the under-resourced ones. Medical care-related cost also includes lost income while seeking care, and the cost of travelling to the healthcare facility. Fifty-four per cent of the total cost is incurred through productivity losses due to premature death. When the victim is the main income earner in the family, their death can easily push the relatives into deeper poverty. Most of the remaining cost is caused by the loss of infected livestock, which can severely affect the economic stability of rural families. In addition to this, psychological trauma for individuals and their communities must be taken into account, even though it remains uncalculated.

1.2 History of Rabies and Rabies Control in Ancient Times

As one of the oldest infectious diseases and zoonoses known to humans, rabies is referenced in a wide range of texts spanning most of early human civilization (King *et al.*, 2004; Wasik and Murphy, 2012; Tarantola, 2017), with possible close links to the history of the domestication of dogs (Rupprecht *et al.*, 2020). The disease is documented in texts from diverse locations, such as the laws of Eshunna from ancient Sumerian and Akkadian civilizations (1930–1770 BCE) in present-day Iraq (Tarantola, 2017), the *Susruta Samhita*, an ancient Indian treatise on Ayurveda (1000 BCE – first or second century CE) and a medical treatise of the Greek philosopher Aristotle (384–322 BCE) (King *et al.*, 2004), among others (Pankhurst, 1970; Liu, 2013). All these texts recognize the links between the bite of a rabid animal and the development of rabies symptoms. Indeed, an Arabic translation of the Greek physician Dioscorides' (40–90 CE) *De Materia Medica* includes a drawing of a rabid dog biting a man (Tarantola, 2017). These texts also recognize the existence of a latent or incubation period between bite exposure and the development of symptoms, and that once symptoms of rabies, such as hydrophobia, develop in human patients, death is inevitable. Furthermore, the disease was often considered a punishment from the gods, a belief that persists in some communities today (Nadal *et al.*, 2022b). Perhaps as a consequence, treatments often comprised religious chants as well as herbal remedies or procedures such as applying parts from the brain of the biting dog to bite wounds (Wasik and Murphy, 2012) or sucking out the 'poison' injected through the bite. Notably, the *Susruta Samhita* and Dioscorides proposed cauterizing bite wounds as a treatment for rabies. Avoiding dog bites was recognized as an essential preventive measure in the Persian Avesta (200–400 CE) (Tarantola, 2017).

In subsequent centuries, there was little progress in advancing the understanding of the prevention or treatment of rabies, although the importance of washing wounds after animal bites was increasingly recognized from the Renaissance period in Europe. This latter period also saw an increase in the scientific understanding of the pathophysiology of the disease (Tarantola, 2017).

Rabies was thus probably enzootic throughout most parts of ancient Asia, the Middle East and Europe. The history of rabies in the African continent is less clear. Dog rabies appears to have been known in the Mediterranean Basin, North Africa (King *et al.*, 2004), such as in ancient Egypt nearly 5000 years ago (Tarantola, 2017; Rupprecht *et al.*, 2020), Kenya (Kuwert *et al.*, 1985) and in 18th-century Amharic texts in

Ethiopian communities (Pankhurst, 1970). In contrast, British travellers visiting southern Africa in the early 19th century commented on the absence or relative rarity of canine rabies in this part of the world. This lack of historical knowledge about rabies in Africa may partly have been due to the inability of extant Europeans to engage with the rich oral traditions of African communities at the time (Brown, 2011). Later on, canine rabies became established throughout Africa after the European colonization. This is also true of North, Central and South America, the Caribbean and Australia, where canine rabies was rare, if at all present, and the disease appears to have been maintained only by bats and other wildlife hosts (Rupprecht *et al.*, 2020).

1.3 History of Rabies Control in the Victorian Period in the UK and the Colonization Period in Asia and Africa

With the advent of European sea travel and eventual colonialism, there are increasing records of canine rabies caused by RABV in sub-Saharan Africa and the Americas, as well as detailed descriptions of canine rabies in several regions in Asia (Kuwert *et al.*, 1985; Ward, 2014; Rupprecht *et al.*, 2020; Dande, 2021). At the same time, rabies was also reported in various wildlife hosts (Rupprecht *et al.*, 2020), resulting in sustained attempts to cull wildlife (Kuwert *et al.*, 1985; Brown, 2011; Radhakrishnan *et al.*, 2020). British military personnel and doctors frequently recorded details of cases of human and animal rabies, such as in India, and various forms of treatment were vigorously debated in medical circles (Radhakrishnan *et al.*, 2020). As a result, rabies control efforts generally consisted of removing or destroying dogs or wildlife, implementing laws restricting dog movement and requiring owned dogs to be muzzled in public (Pemberton and Worboys, 2007; Radhakrishnan *et al.*, 2020; Rupprecht *et al.*, 2020). Such preventive measures were especially successful in eliminating dog-mediated human rabies in a number of Scandinavian countries in the early 1800s (King *et al.*, 2004) and the UK by 1902 (Pemberton and Worboys, 2007), well before the development of preventive animal rabies vaccines.

Before its elimination, canine rabies was common in the Victorian period of the UK, and several people died of dog-mediated rabies every year (King *et al.*, 2004). The disease was frequently attributed to the allegedly irresponsible way poor and working-class persons reared 'curs' – mongrels with no particular breed characteristics. In many ways, rabies was thus often conflated with poverty, low class and criminality (Pemberton and Worboys, 2007). In the 1830s, legislation requiring dogs to be muzzled was viewed as symbolic of political or gender-based oppression and was often opposed. Later, such measures were also objected to on the grounds of animal welfare. While public concern over rabies in the UK waxed and waned in the intervening decades, by the late 1890s, there were concerted efforts to eliminate dog rabies from the country. Nevertheless, there continued to be vehement opposition to the implementation of muzzling laws. However, in combination with laws requiring the registration and confinement of dogs and quarantine of imported animals, such legislation caused a gradual decline in the number of dog and human rabies cases. In 1900, no rabies cases were reported in England and Wales. The last indigenous cases of animal rabies were reported in 1902 in South Wales (Pemberton and Worboys, 2007).

While similar control measures were implemented in colonies in Asia and Africa, there appears to have been limited success in controlling the spread of rabies (Brown, 2011; Radhakrishnan *et al.*, 2020). In Africa, measures such as 'tie up' orders for free-roaming dogs, most of which were owned, were unpopular (Brown, 2011; Rupprecht *et al.*, 2020; Dande, 2021). In the early 1890s, an outbreak of dog rabies in Port Elizabeth in present-day South Africa prompted authorities to enforce the muzzling of all dogs, movement restrictions, licensing of owned dogs and destruction of 'stray' or unmuzzled dogs. As observed in the UK, such measures disproportionately targeted poorer African neighbourhoods and their indigenous breed dogs, which were viewed as inferior or diseased and often considered vermin to be exterminated (Brown, 2011; Dande, 2021). In British India, where canine rabies was enzootic, legislation permitting the destruction of ownerless dogs was present as early as 1813. Other measures, such as levying a dog tax and issuing badges for owned dogs, were also attempted with limited success (Radhakrishnan *et al.*, 2020).

The pioneering work of Louis Pasteur, Emile Roux and other colleagues in developing the first human rabies vaccine in 1885 led to the establishment of Pasteur Institutes worldwide (Tarantola, 2017). These institutes enabled the PEP of bitten individuals and prevented countless human rabies deaths. The first patients outside Europe to receive PEP were two children in Saigon (present-day Ho Chi Minh City, Vietnam) in 1891 (Tarantola, 2017). Pasteur's vaccine was improved upon by various scientists (Tarantola, 2017), most notably David Semple at the Pasteur Institute in Kasauli, British India, established in 1900. At one point, the Kasauli institute treated more rabies patients than any other Pasteur Institute globally. The Semple vaccine itself was used for decades worldwide for human PEP. Its use was discouraged by the World Health Organization (WHO) in the 1980s because of its severe adverse side effects and its lower efficacy compared to modern cell culture vaccines. Many countries, such as India, officially discontinued its use only in the new millennium, so the memory of the 14 painful injections into the abdomen required by the nerve tissue vaccine is often still vivid nowadays. This may negatively affect access to PEP by at-risk individuals (Nadal, 2018), so awareness campaigns need to address this issue.

The development and use of human rabies vaccines were followed by experimental efforts in British India, Japan, the USA and Italy to develop animal rabies vaccines to prevent disease in primary animal reservoirs, particularly dogs (Radhakrishnan *et al.*, 2020). The first dog rabies vaccine was developed in Japan and later refined for use in preventive mass dog vaccination in this country from the early 1920s (Kurosawa *et al.*, 2017) and 1930 in Taiwan (then a Japanese colony) (Liu, 2013). However, while animal rabies vaccines were used to treat valuable pets or livestock (Radhakrishnan *et al.*, 2020), there is little evidence from the early 1900s for similar attempts at dog mass vaccination in colonies in Asia and Africa. In these regions, rabies control continued to rely on the culling of dogs and wildlife, restriction of dog movement and attempts to regulate dog ownership (Brown, 2011; Radhakrishnan *et al.*, 2020).

1.4 History of Rabies Control After Independence in Asia and Africa

Following the development of animal rabies vaccines, mass dog vaccination gradually became an invaluable tool for canine rabies control. The successful elimination in 1957 of animal rabies in Japan through mass vaccination (in combination with other measures described above) (Kurosawa *et al.*, 2017) and similar efforts in the USA (Rupprecht *et al.*, 2020) provided proof of concept for the possibility of rabies elimination. In the 1940s, red fox (*Vulpes vulpes*) rabies emerged as a significant concern in North America and Europe (King *et al.*, 2004), prompting the development of oral rabies vaccines (ORV) during the 1970s. Following the first successful ORV field trial in 1978 in Switzerland, ORV campaigns during the 1980s and 1990s steadily eliminated wildlife rabies in Western Europe (King *et al.*, 2004). In South American and Caribbean countries, the implementation of mass dog vaccination and enhanced disease surveillance efforts from the 1980s, coordinated by the Pan American Health Organization (PAHO), has resulted in significant reductions in human and canine rabies deaths in the region (Del Rio Vilas *et al.*, 2017; Rupprecht *et al.*, 2020). In 2016, Mexico reported its last case of canine rabies caused by RABV (Rupprecht *et al.*, 2020). With increasing evidence of the futility and inhumanity of culling for rabies control (Morters *et al.*, 2013), attention has shifted globally to using humane population management strategies such as surgical or chemical animal birth control to support mass dog vaccination campaigns (WHO, 2018b). Other recent developments have included the adoption of dose-sparing intradermal rabies vaccines for human PEP (WHO, 2018a), the development of rabies monoclonal antibodies (Dias de Melo *et al.*, 2022), and research on the control of bat rabies in South America using vaccines and reproductive suppressants (Benavides *et al.*, 2020).

Despite these developments, rabies control efforts in large parts of Asia and Africa have been haphazard and uncoordinated (Kuwert *et al.*, 1985). In the decades after the Second World War (1939–1945), several Asian and African regions gained independence from colonial rule. In these newly independent countries, nearly all of which were poor and economically underdeveloped, rabies was among a host of competing health priorities. As a result, the disease remained under-prioritized by various national and regional public health authorities. Despite the availability of effective canine rabies vaccines, rabies control has continued to rely on population reduction methods such as culling dogs or wildlife and vaccination of owned dogs (Kuwert *et al.*, 1985; Rupprecht *et al.*, 2020). While some countries have never managed to control rabies, some nations eliminated it at local or national levels, while others have seen outbreaks of dog or wildlife rabies occur in areas previously free of the disease (Yang *et al.*, 2018; Rupprecht *et al.*, 2020).

Rabies spread throughout southern Africa in the 1940s and 1950s, moving through present-day Angola, Zambia, Namibia, Botswana, Zimbabwe, Mozambique and South Africa (Brown, 2011). In the face of new rabies outbreaks, canine vaccines were first imported from the USA by Southern Rhodesia (present-day Zimbabwe) for mass dog vaccination campaigns, followed by Bechuanaland (present-day Botswana) and South Africa (1952) (Brown, 2011). At the same time, authorities also taxed dog owners and killed free-roaming dogs. As before, such measures affected native African dog owners and their dogs more severely, so local Africans opposed these measures or actively circumvented them (Dande, 2021). Such actions significantly reduced the effectiveness of rabies control efforts in these areas. Similar vaccination campaigns, combined with the destruction of free-roaming dogs, have also been conducted in Kenya, Tanzania and various other African nations from the 1950s onwards. Still, canine rabies remains enzootic in these regions (Kuwert *et al.*, 1985). Field trials in some African countries have demonstrated the value of ORV for dog rabies control on the continent, as a complementary measure to parenteral vaccination useful to target hard-to-catch free-roaming dogs (Cliquet *et al.*, 2018). In 2015, the Pan African Rabies Control Network (PARACON) was established as a network of rabies experts in Africa to strengthen and streamline rabies control efforts on the continent.

In Asia, the picture has been more heterogeneous. Only Japan and Singapore are rabies free. In India, which gained independence in 1947 and nowadays accounts for one-third of the global human rabies burden (Hampson *et al.*, 2015), rabies control was first discussed in the national 5 year plans for national development only in 2002. Although dog culling was outlawed in 2001, reactive culling continues to occur throughout the country, and animal vaccination has consistently focused on owned dogs, but with very low vaccination coverage (Radhakrishnan *et al.*, 2020). India's first formal policy for rabies control was released only in 2021 when human rabies was also made notifiable (Benavides *et al.*, 2020). The Indian approach to rabies control, and overall dog population management, is currently characterized by a strong focus on dog population control (Nadal, 2020). In 2008, the previously rabies-free island of Bali in Indonesia reported its first dog rabies outbreak, resulting in mass dog vaccination campaigns and widespread dog culls that have so far failed to control the disease (Ward, 2014). In contrast, Taiwan was free of canine rabies from 1961 until the emergence of the disease in ferret badgers (*Melogale moschata*) in 2013 (Liu, 2013). Similarly, Malaysia lost its rabies-free status after the declaration of a rabies epidemic in 2015. Canine rabies continues to be enzootic in China (Liu, 2013) and several other Asian countries (Yang *et al.*, 2018). Like in Africa, field trials in Sri Lanka, the Philippines and Thailand have demonstrated the potential benefits of ORV for canine rabies control (Cliquet *et al.*, 2018). In 2018, some Asian countries followed the example of PARACON and formed the Asian Rabies Control Network (ARACON).

1.5 The Zero by 30 Goal

In 2015, the world called for action by setting the global goal of achieving zero human dog-mediated rabies deaths by 2030 (WHO and OIE, 2015). In 2018, the Tripartite (the Food and Agriculture Organization of the United Nations (FAO); the World Organization for Animal Health (WOAH), previously OIE; and the WHO) joined forces with the Global Alliance for Rabies Control (GARC) under the United Against Rabies

Forum (Tidman *et al.*, 2022) and launched the *Zero by 30: the Global Strategic Plan to End Human Deaths from Dog-mediated Rabies by 2030* (WHO *et al.*, 2019). Tools and expertise are provided to rabies-endemic countries to empower, engage and enable them, according to their budget, capacity and local context, to control rabies in dogs and eliminate the human burden of this disease.

As we read in the document, the rationale for rabies elimination is that rabies is 100% preventable with the current tools and knowledge, but still, it takes many human lives, especially among the world's most vulnerable populations. Eliminating rabies also strengthens health systems because the same infrastructure built to provide human vaccination, dog vaccination and community awareness in marginalized settings can be used for responding to other human and animal health needs, including emerging zoonoses, at the local level. Moreover, rabies elimination is considered a model for One Health collaboration and it is also aligned with the United Nations Sustainable Development Goal 3 – to 'ensure healthy lives and promote well-being for all at all ages' – and 1 – to 'end poverty in all its forms' without leaving anyone behind.

The plan has three objectives. The first is to use vaccines, medicines, tools and technologies effectively. Human rabies risk will be reduced by improving awareness, increasing access to healthcare and mass vaccinating dogs – in a One Health fashion. The second objective is to generate, innovate and measure impact, by ensuring reliable data to enable effective decision making. The third objective is to sustain commitment and resources, by harnessing multi-stakeholder engagement. Countries are expected to mobilize domestic and international resources to sustainably finance their rabies control activities.

The path towards the 2030 goal is divided into three phases. In 2018–2020, the Start Up phase involves the building of a strong foundation for rabies elimination by preparing and improving normative tools and structures to catalyse action, such as robust, budgeted, effective and sustainable national rabies elimination plans following a One Health approach. In 2021–2025 in the Scale Up phase, the plan reaches its maturity, thanks to the learning and experience gained along the way, and goes global. In 2026–2030, the Mop Up phase is the last mile, where remaining countries will be engaged and supported in the achievement of 'Zero by 30'.

1.6 The One Health Approach

As a concept – human, animal and environmental health are deeply interlinked – One Health goes as far back as ancient Greece, if not pre-modern times. As a phrase, it was formalized in 2004, when the Wildlife Conservation Society held the 'One World, One Health' meeting in New York City and issued the 'Manhattan Principles', 12 recommendations that included the need to look at human, animal and environmental health as a unit, and to adopt interdisciplinary approaches to disease prevention, health awareness and policy development (Bresalier *et al.*, 2021).

In 2022, the One Health High-Level Expert Panel, an interdisciplinary body created to advise the Quadripartite (which consists of the Tripartite and the newly added United Nations Environment Programme, UNEP), proposed a working definition for One Health. This definition is as follows:

> One Health is an integrated, unifying approach that aims to sustainably balance and optimize the health of people, animals, and ecosystems. It recognizes the health of humans, domestic and wild animals, plants, and the wider environment (including ecosystems) are closely linked and interdependent. The approach mobilizes multiple sectors, disciplines, and communities at varying levels of society to work together to foster well-being and tackle threats to health and ecosystems, while addressing the collective need for healthy food, water, energy, and air, taking action on climate change and contributing to sustainable development.
>
> (OHHLEP *et al.*, 2022)

Many examples are available across the different areas of One Health application, including rabies elimination, to demonstrate the efficacy, sustainability and cost-saving of the interventions that adopt this approach. Strengthening and expanding the operationalization of One Health and giving programmes the political, financial and organizational stability that they need to achieve their target is now the priority. The 4Cs of Communication, Coordination, Collaboration and Capacity building are instrumental to this next step (OHHLEP *et al.*, 2022).

In late 2022, the Quadripartite launched the *One Health Joint Plan of Action (2022–2026): Working Together for the Health of Humans, Animals, Plants and the Environment* to create a framework to integrate systems and capacity for better preventing, predicting, detecting and responding to health threats, while contributing to sustainable development (WHO *et al.*, 2022). Rabies elimination is part of Action Track 2, 'Reducing the risks from emerging and re-emerging zoonotic epidemics and pandemics'.

1.7 Impact of the COVID-19 Pandemic on Rabies Elimination Efforts

The pandemic has taken the lives not only of the 6.5 million people who died of COVID-19 but also of those who couldn't access the healthcare services they needed, or couldn't receive the necessary healthcare due to the overwhelming pressure of the crisis on health systems. Low- and middle-income countries experienced this double effect of the pandemic more than rich ones. Further, as rabies is a notorious disease of poverty, rabies-prone communities are likely to have been severely affected by the pandemic, even though the real impact, in terms of human and animal rabies deaths, of this 3-year-long health crisis may remain unknown. This is because rabies surveillance, already weak before the pandemic, collapsed during it, mainly due to restrictions on the movement of the field surveillance staff that led to cases being missed or investigated late (Raynor *et al.*, 2021; Nadal *et al.*, 2022a). Moreover, a vicious cycle was observed: access to and delivery of PEP declined, so fewer dog-bite cases were reported, bite reports were not sent to investigators and incidents remained unaddressed.

Access to PEP was impacted by the pandemic in most countries because people were worried about COVID-19 infections at hospitals and were unable to reach them due to reduced public transportation and reluctance to share private transportation. PEP delivery was disrupted too, due to a shortage of human vaccines (because of supply issues and financial constraints) and staff shortages (because of quarantine, illness or redeployment) (Gongal *et al.*, 2022; Nadal *et al.*, 2022a). In a war-torn, endemic country, where human vaccines have usually been available in private clinics, none could be found during the first year of the pandemic. The postponement of the vaccine investment strategy for rabies by Gavi, the Vaccine Alliance, has been a major concern for under-resourced countries.

Awareness activities for children, usually carried out at schools, survived the beginning of the pandemic only in very few countries. Online events, both for at-home children and the larger community, were able to reach only those who could afford a computer at home and a good Internet connection (Nadal *et al.*, 2022a).

Yet the most severely disrupted element of the rabies elimination strategy was mass dog vaccination (Raynor *et al.*, 2021; Nadal *et al.*, 2022a). Disruptions included delays of at least 6 months, prolonged duration of vaccination campaigns (when performed), increased costs and failure to reach targets. The main hindrances were the restrictions on the movement of dog vaccinators and the struggle to organize vaccination campaigns that adhered to COVID-19 safety guidelines (especially where the campaigns are usually carried out by non-governmental organizations rather than the government).

Overall, in the endemic countries that were just beginning their journey to the Zero by 30 goal, 'the momentum that was gaining was lost', while the countries that were progressing towards it well experienced an unfortunate step back. Additionally, the pandemic has shown once more the global inequity in vaccine access. This is a well-known problem in the case of rabies, where effective vaccines, tools and strategies are there, but they fail to reach those who most need them. That said, the past 3 years could have had a potentially positive effect in the long term, in terms of: (i) an improved cold chain; (ii) strengthened diagnostic capacity; (iii) increased regional coordination; and (iv) augmented awareness about the importance of safeguarding animal health and the key role of the veterinary sector.

1.8 2030 and Beyond

Looking at the future, the main lessons learnt during the pandemic can be summarized as follows (Nadal *et al.*, 2022a):

- It is crucial to mobilize long-term political commitment and sufficient and sustainable financial, infrastructural and workforce-related resources to catch up on the delay accumulated during the pandemic and advance fast in phases 1 and 2 of the Zero by 30 plan.
- Efforts should be directed to the support of the animal health sector, in particular, to make sure that mass dog vaccination becomes a consolidated rabies control strategy in all rabies-endemic countries.
- A rabies-dedicated budget should be created to ensure the stable procurement of human and animal vaccines, which should be considered essential biologicals even in times of crisis.
- It is vital to identify the most cost-effective and sustainable methods of meeting the needs of local communities, to facilitate the access and delivery of PEP and dog vaccination.
- A simple but effective and sustainable participatory disease surveillance mechanism should be developed, for human and animal healthcare professionals to quickly receive information about dog bites and rabies cases directly from local communities.
- A well-rounded rabies communication strategy should be designed to target both children and adults, using the channels and tools that people can easily access in normal times and especially during crises.

Authors' Declaration

All authors declare that they have no conflict of interest.

All authors have approved this manuscript, agree with its submission, and share collective responsibility and accountability.

This manuscript has not been published or is not under review elsewhere.

References

Benavides, J.A., Valderrama, W., Recuenco, S., Uieda, W., Suzán, G. *et al.* (2020) Defining new pathways to manage the ongoing emergence of bat rabies in Latin America. *Viruses* 12(9), 1002. DOI: 10.3390/v12091002.

Bresalier, M., Cassidy, A. and Woods, A. (2021) One Health in history. In: Zinsstag, J., Schelling, E., Crump, L., Whittaker, M., Tanner, M. *et al.* (eds) *One Health: The Theory and Practice of Integrated Health Approaches*, 2nd edn. CAB International, Wallingford, UK, pp. 1–14.

Brown, K. (2011) *Mad Dogs and Meerkats*. Ohio University Press, Athens, Ohio.

Cliquet, F., Guiot, A.L., Aubert, M., Robardet, E., Rupprecht, C.E. *et al.* (2018) Oral vaccination of dogs: a well-studied and undervalued tool for achieving human and dog rabies elimination. *Veterinary Research* 49(1), 61. DOI: 10.1186/s13567-018-0554-6.

Dande, I. (2021) The colonial state, African dog-owners, and the political economy of rabies vaccination campaigns in Southern Rhodesia in the 1950s and 1960s. *Journal of the History of Biology* 54(4), 689–717. DOI: 10.1007/s10739-021-09661-6.

Del Rio Vilas, V.J., Freire de Carvalho, M.J., Vigilato, M.A.N., Rocha, F., Vokaty, A. *et al.* (2017) Tribulations of the last mile: sides from a regional program. *Frontiers in Veterinary Science* 4, 4. DOI: 10.3389/fvets.2017.00004.

Dias de Melo, G., Hellert, J., Gupta, R., Corti, D. and Bourhy, H. (2022) Monoclonal antibodies against rabies: current uses in prophylaxis and in therapy. *Current Opinion in Virology* 53, 101204. DOI: 10.1016/j.coviro.2022.101204.

Fooks, A. and Jackson, A. (2020) *Rabies. Scientific Basis of the Disease and its Management*, 4th edn. Elsevier, Amsterdam.

Gongal, G., Sampath, G., Kishore, J., Bastola, A., Punrin, S. *et al.* (2022) The impact of COVID-19 pandemic on rabies post-exposure prophylaxis services in Asia. *Human Vaccines & Immunotherapeutics* 18(5), 2064174. DOI: 10.1080/21645515.2022.2064174.

Hampson, K., Coudeville, L., Lembo, T., Sambo, M., Kieffer, A. *et al.* (2015) Estimating the global burden of endemic canine rabies. *PLoS Neglected Tropical Diseases* 9(4), e0003709. DOI: 10.1371/journal.pntd.0003709.

King, A.A., Fooks, A.R., Aubert, M. and Wandeler, A.I. (2004) *Historical Perspective of Rabies in Europe and the Mediterranean Basin*. World Organisation for Animal Health, Paris.

Kurosawa, A., Tojinbara, K., Kadowaki, H., Hampson, K., Yamada, A. *et al.* (2017) The rise and fall of rabies in Japan: a quantitative history of rabies epidemics in Osaka Prefecture, 1914-1933. *PLoS Neglected Tropical Diseases* 11(3), e0005435. DOI: 10.1371/journal.pntd.0005435.

Kuwert, E., Mérieux, C., Koprowski, H. and Bogel, K. (1985) *Rabies in the Tropics*. Springer, Berlin. DOI: 10.1007/978-3-642-70060-6.

Liu, C.H. (2013) History of rabies control in Taiwan and China. Taiwan Epidemiology Bulletin 29(133), S44–52.

Morters, M.K., Restif, O., Hampson, K., Cleaveland, S., Wood, J.L.N. *et al.* (2013) Evidence-based control of canine rabies: a critical review of population density reduction. *Journal of Animal Ecology* 82(1), 6–14. DOI: 10.1111/j.1365-2656.2012.02033.x.

Nadal, D. (2018) Pregnant with puppies: the fear of rabies in the slums of New Delhi, India. *Medicine Anthropology Theory* 5(3), 130–156.

Nadal, D. (2020) *Rabies in the Streets: Interspecies Camaraderie in Urban India*. Pennsylvania State University Press, University Park, Pennsylvania. DOI: 10.1515/9780271086866.

Nadal, D., Abela-Ridder, B., Beeching, S., Cleaveland, S., Cronin, K. *et al.* (2022a) The Impact of the first year of the COVID-19 pandemic on canine rabies control efforts: a mixed-methods study of observations about the present and lessons for the future. *Frontiers in Tropical Diseases* 3. DOI: 10.3389/fitd.2022.866811.

Nadal, D., Hampson, K., Lembo, T., Rodrigues, R., Vanak, A.T. *et al.* (2022b) Where rabies is not a disease. Bridging healthworlds to improve mutual understanding and prevention of rabies. *Frontiers in Veterinary Science* 9, 867266. DOI: 10.3389/fvets.2022.867266.

OHHLEP (One Health High-Level Expert Panel), Adisasmito, W.B., Almuhairi, S., Behravesh, C.B., Bilivogui, P. *et al.* (2022) One Health: a new definition for a sustainable and healthy future. *PLoS Pathogens* 18(6), e1010537. DOI: 10.1371/journal.ppat.1010537.

Pankhurst, R. (1970) The history and traditional treatment of rabies in Ethiopia. *Medical History* 14(4), 378–389. DOI: 10.1017/s0025727300015829.

Pemberton, N. and Worboys, M. (2007) *Mad Dogs and Englishmen*. Palgrave Macmillan, London. DOI: 10.1057/9780230589544.

Radhakrishnan, S., Vanak, A.T., Nouvellet, P. and Donnelly, C.A. (2020) Rabies as a public health concern in India. A historical perspective. *Tropical Medicine and Infectious Disease* 5(4), 162. DOI: 10.3390/tropicalmed5040162.

Raynor, B., Díaz, E.W., Shinnick, J., Zegarra, E., Monroy, Y. *et al.* (2021) The impact of the COVID-19 pandemic on rabies reemergence in Latin America: the case of Arequipa, Peru. *PLoS Neglected Tropical Diseases* 15(5), e0009414. DOI: 10.1371/journal.pntd.0009414.

Rupprecht, C.E., Freuling, C.M., Mani, R.S., Palacios, C., Sabeta, C.T. and Ward, M. (2020) A history of rabies. The foundation for global canine rabies elimination. In: Fooks, A.R. and Jackson, A.C. (eds) *Rabies*, 4th edn. Elsevier, Amsterdam, pp. 1–42.

Tarantola, A. (2017) Four thousand years of concepts relating to rabies in animals and humans, its prevention and its cure. *Tropical Medicine and Infectious Disease* 2(2), 5. DOI: 10.3390/tropicalmed2020005.

Tidman, R., Thumbi, S.M., Wallace, R., de Balogh, K., Iwar, V. *et al.* (2022) United Against Rabies Forum: the One Health concept at work. *Frontiers in Public Health* 13(10), 854419. DOI: 10.3389/fpubh.2022.854419.

Ward, M.P. (2014) Rabies in the Dutch East Indies a century ago. A spatio-temporal case study in disease emergence. *Preventive Veterinary Medicine* 114(1), 11–20. DOI: 10.1016/j.prevetmed.2014.01.009.

Wasik, B. and Murphy, M. (2012) *Rabid. A Cultural History of the World's Most Diabolical Virus.* Penguin Books, London.

WHO (World Health Organization) (2018a) Rabies vaccines: WHO position paper. *Weekly Epidemiological Record* 16(93), 201–220.

WHO (World Health Organization) (2018b) *WHO Expert Consultation on Rabies: Third Report*. WHO, Geneva, Switzerland.

WHO and OIE (WorldHealth Organization and Office International des Epizooties) (2015) *Global Elimination of Dog-Mediated Human Rabies*. WHO, Geneva, Switzerland.

WHO, FAO and OIE (World Health Organization, Food and Agriculture Organization of the United Nations and Office International des Epizooties) (2019) *Zero by 30: The Global Strategic Plan to End Human Deaths from Dog-Mediated Rabies by 2030*. WHO, Geneva, Switzerland.

WHO, FAO, WOAH and UNEP (World Health Organization, Food and Agriculture Organization of the United Nations, World Organization for Animal Health and United Nations Environment Programme) (2022) *One Health Joint Plan of Action (2022–2026): Working Together for the Health of Humans, Animals, Plants and the Environment*. WHO, Geneva, Switzerland.

Yang, D.K., Cho, I.S. and Kim, H.H. (2018) Strategies for controlling dog-mediated human rabies in Asia: using "One Health" principles to assess control programmes for rabies. *Revue Scientifique et Technique* 37(2), 473–481. DOI: 10.20506/rst.37.2.2816.

2 FAO, WOAH and WHO Working Together in the Asia Pacific Region to Eliminate Dog-mediated Human Rabies by 2030

Katinka de Balogh[1*], Yooni Oh[1], Mary Joy Gordoncillo[1], Yoenten Phuentshok[1], Eric Brum[1], Luuk Schoonman[1], Scott Newman[1], Hirofumi Kugita[2], Kinzang Dukpa[2], Lesa Thompson[2], Pasang Tshering[2], Ronello Abila[3], Tikiri Wijayathilaka[3], Supalak Prabsriphum[3], Gyanendra Gongal[4], David Sutherland[4] and Aya Yajima[5]

[1]Food and Agriculture Organization of the United Nations, Regional Office for Asia and the Pacific (FAORAP), Bangkok, Thailand; [2]World Organisation for Animal Health (WOAH), Regional Representation for Asia and the Pacific, Tokyo, Japan; [3]World Organisation for Animal Health (WOAH), Sub-Regional Representation for South-east Asia, Bangkok, Thailand; [4]World Health Organization, Regional Office for South-east Asia (WHO-SEARO), New Delhi, India; [5]World Health Organization, Regional Office for the Western Pacific (WHO-WPRO), Manila, Philippines

Abstract

Rabies is endemic in humans and animals in the Asia Pacific region. Dog-mediated rabies accounts for more than 95% of the cases in humans. Although tools, strategies and mechanisms are available, rabies continues to be neglected resulting in preventable human and animal deaths. The Regional Tripartite for Asia and the Pacific, comprising the Food and Agriculture Organization of the United Nations (FAO), the World Organisation for Animal Health (WOAH), and the World Health Organization (WHO), has been advocating policy changes, building capacity in diagnosis and control, enhancing multisectoral collaboration, sharing tools and practices, and rapidly responding to the needs of countries. Controlling the disease in animals would be more sustainable in the long run. The use of oral rabies vaccines in dogs needs to be promoted to complement parenteral vaccination of dogs. Furthermore, a better understanding of the dog–human bond would contribute to the design of more efficient rabies control programmes.

2.1 Introduction

The Asia Pacific region (AP region) is home to about 4.3 billion people (60% of the world's population) and includes the world's most populous countries, China and India (UNFPA, 2022). The region has an estimated 400 million people living in extreme poverty (i.e. below

*Corresponding author: Katinka027@gmail.com

DOI: 10.1079/9781800622975.0002

US$1.90 income/day). It remains vulnerable to the impacts of climate change and pandemics (ESCAP, 2019), and an estimated 75–80 million people in Asia and the Pacific were pushed into extreme poverty because of disruptions in economic activity due to coronavirus 2019 (COVID-19) (Devex, 2021). Most people in the region live in rural areas and depend on agriculture and livestock farming.

Rabies is endemic to much of Asia and almost all the reported rabies cases in humans are transmitted by dog bites. Dog bites are common due to the abundance of dogs in the region; dogs that are confined, free-roaming or feral. Free-roaming dog populations vary between countries in relation to their habitat, socio-cultural factors and human population density (Matter and Daniels, 2000). For example, the mean density of dogs is 14 dogs/km^2 in Bangladesh (Hossain *et al.*, 2013); 42 dogs/km^2 in Bhutan (Rinzin *et al.*, 2016); 468 dogs/km^2 in the Philippines (Childs *et al.*, 1998); and 719 dogs/km^2 in rural Maharashtra, India (Belsare and Gompper, 2013). The high density of free-roaming dogs in Asia plays a major role in rabies transmission and endemicity. It is estimated that the number of dog bites in India alone is 17.4 million, every year. In Vietnam, there are 350,000 cat and dog bites each year (Lee *et al.*, 2018); and 20,000–40,000 animal bites in Nepal of which 90% were dog bites (Pantha *et al.*, 2020).

The importance of rabies in the region means that a regional approach to rabies prevention and control is supported by the Regional Tripartite (comprising the Food and Agriculture Organization of the United Nations (FAO), the World Organization for Animal Health (WOAH, founded as OIE) and the World Health Organization (WHO)) through its regional and sub-regional offices and partner institutions. This chapter outlines the key Tripartite activities to tackle rabies in the AP region (in countries from Afghanistan in the west to New Zealand in the east).

2.2 Rabies Prevalence and Burden

Rabies is still endemic in the AP region except for Australia, Brunei, Japan, Maldives, New Zealand, Papua New Guinea, Singapore, Timor Leste and the Pacific Island Countries. Though, even these rabies-free countries face a constant threat of rabies incursion owing to travel and trade. Bhutan, Malaysia and Mongolia were able to maintain zero human deaths for many years; however, in recent years they have reported human rabies deaths. The Association of Southeast Asian Nations (ASEAN) launched a rabies elimination initiative in 2008 that led to the endorsement of the ASEAN Rabies Elimination Strategy (ARES) for the region in 2014 (WOAH, 2015).

A recent outbreak of dog rabies in previously rabies-free islands of Indonesia, at the Indonesia-Malaysia border into Sarawak State of Malaysia, clearly demonstrated that inter-island and cross-border collaboration is critical to contain rabies outbreaks, and prevent incursions to maintain the rabies-free status of countries or regions.

Rabies is sustained through two interrelated cycles: (i) urban; and (ii) sylvatic. An urban cycle involves maintenance of infection in dog populations and a sylvatic cycle involves wildlife. The rabies virus can spill over from dogs to wildlife and vice-versa. For example, mongoose (*Herpestes* spp.), jackals (*Canis aureus*), foxes (*Vulpes bengalensis*) and wolves (*Canis lupus*) have been incriminated as wildlife reservoirs of rabies in rural areas of Bangladesh, India and Nepal (Gongal, 2006). Nevertheless, dog bites remain the primary source of rabies in all rabies endemic countries, accounting for 95% of human rabies cases.

The disease causes an estimated 31,000 human deaths in Asia every year, and an annual expenditure of over US$563 million (Knobel *et al.*, 2005) is incurred for direct and indirect costs of post-exposure prophylaxis (PEP) in humans, and in dog rabies control efforts. Comprehensive surveillance data is not available but the economic impact of rabies on livestock production is potentially high.

Progress in preventing human rabies through the control of the disease in the animal reservoir has been slow as attempts to control rabies through dog culling have not been sustained, or are not socially acceptable due to the public, religious and animal welfare concerns. There are successful programmes of dog rabies vaccination alongside animal birth control in limited urban areas coordinated by leading non-governmental organizations (NGOs). However, they are location specific and have generally not been replicated in rural areas with community participation (Gongal and Wright, 2011).

During the COVID-19 pandemic, 67% of countries reported severe impacts on health-care services: (i) reductions in outpatient care attendance during the lockdown; (ii) staff redeployment; (iii) unavailability of services due to closures of health facilities; and (iv) supply-chain disruptions (WHO, 2020). Some countries have been facing difficulties in procuring human rabies PEP and human rabies immunoglobulins (hRIG) during the pandemic partly due to limitations in production systems, compounded by a lack of demand forecast and disruption of international air transportation. In addition, in various cases persons after dog bites were reluctant to visit clinics due to fear of contracting SARS-CoV-2 (severe acute respiratory syndrome coronavirus 2) at health facilities. It is difficult to assess the impact of limited access to PEP on human rabies incidence. Similarly, many mass dog vaccination and animal birth control programmes were frequently suspended.

2.3 Tripartite Actions Against Rabies at the Global Level

The FAO, WOAH and WHO have institutionalized a tripartite coordination mechanism since 2010 to support countries with surveillance, prevention and control of emerging and re-emerging infectious diseases including zoonoses at the animal–human–ecosystem interface. In September 2011, the WOAH organized the Global Conference on Rabies Control in Incheon, Republic of Korea. The conference recommended that the Tripartite considers rabies a priority and supports countries in raising funds to initiate and sustain rabies control programmes, update their legislation and adopt a 'One Health' approach to disease control (WOAH, 2011).

In September 2015, WHO, WOAH, FAO and the Global Alliance for Rabies Control (GARC) called for investment to defeat human rabies transmitted by dogs and published a document which demonstrated the feasibility of global elimination (WHO, 2015). The global rabies conference held in December 2015 in Geneva, Switzerland called for the global elimination of dog-mediated human rabies (WHO *et al.*, 2016) and in 2018, the global strategic plan for the elimination of dog-mediated rabies by 2030 based on the STOP-R (Socio-cultural, Technical, Organization, Political and Resources) framework was launched (WHO *et al.*, 2018).

The search for an approach that would enable countries to improve their rabies control programmes led to the development of a 'Stepwise approach towards rabies elimination' (SARE). SARE (Anonymous, 2016) is a template that countries can use to develop activities, measure progress towards a national strategy, and programme for sustainable rabies prevention, control and eventual elimination.

Rabies has also been included in Tripartite publications promoting wider One Health multisectoral approaches to zoonoses. In particular, in the FAO, WOAH and WHO-South East Asia Regional Office (SEARO) and WHO-Western Pacific Regional Office (WPRO) *Zoonotic Diseases: a Guide to Establishing Collaboration Between Animal and Human Health Sectors at the Country Level* in Asia and the Pacific (WHO, 2009) and the global level *Taking a Multisectoral, One Health Approach: a Tripartite Guide to Addressing Zoonotic Diseases in Countries* also denominated the 'Tripartite Zoonoses Guide' (WHO *et al.*, 2019a). This work aligns with the United Against Rabies (UAR) Forum established in 2020 to create a more inclusive network of state and non-state actors who share a common vision and wish to work together for rabies elimination. The purpose of the UAR Forum is to provide a mechanism for implementing the objectives of the Global Strategic Plan (WHO *et al.*, 2020a, b), while enabling participants to benefit from sharing knowledge, experience, ideas and information under the leadership of the global Tripartite. Working groups of the UAR Forum have been formed and are open to all institutions that wish to actively engage. The combined efforts, political will and commitment of UAR Forum members, together with the resources they provide, will help rabies elimination become a global reality by 2030.

The Tripartite faces various challenges during its efforts to implement actions against rabies at the global and regional levels. Challenges specific to working within the Tripartite include the complexities of working across organizations, and difficulties experienced when there is a need for agreeing through consensus, or to obtain high-level clearances from all partners. Other challenges for the Tripartite centred around multisectoral collaboration. More generally, there is often an imbalance in resources and level of progress between sectors, which can make a uniform or consistent approach to disease prevention, control and response challenging. In the context of rabies, there is often weak collaboration between key sectors at the country level. The animal health sectors of endemic countries are often mainly concerned with the prevention and control of economically important transboundary animal diseases of livestock. With competing priorities, rabies may not be classified as a priority disease for the animal health sector, resulting in insufficient resource allocation and political support towards the prevention and control of rabies at the source, which is known to be the more cost-effective and sustainable control measure in the long term. Although human health sectors are often not willing to share resources to support rabies control in animal populations and instead focus on pre-exposure prophylaxis and PEP provision, the Ministry of Health of Bangladesh and the Phillippines have set an example by successfully funding dog rabies vaccination campaigns. These countries may serve as models for other lower-middle-income countries (Republic of the Philippines Department of Health, 2020; WOAH, 2021c).

2.4 Tripartite Actions Against Rabies at the Regional Level

2.4.1 Introduction to the Asia Pacific Regional Tripartite

In the same year that the FAO/WOAH/WHO Tripartite Concept Note was presented (WHO *et al.*, 2010), the first Asia Pacific multisectoral workshop on zoonoses was held in Sapporo, Japan. Although the initial focus was a response to the avian influenza epidemic, rabies was quickly added to the list of zoonoses of regional concern. By 2020, eight Asia Pacific multisectoral collaboration workshops had been held, demonstrating an ever-closer working relationship of the Regional Tripartite and country partners as these workshops sought to deepen understanding of each other's sectors and strengthen systems and functions.

On 23 October 2020, the regional representatives in Asia of FAO, WOAH and WHO (SEARO and WPRO) signed a *Joint Statement of Intent to Coordinate* in which they committed to establish and support the Tripartite One Health Coordination Group for Asia and the Pacific (WHO *et al.*, 2020a). This Coordination Group consolidates the multisectoral work carried out over many years in the region, including the eight regional workshops.

The Regional Tripartite coordination seeks to provide full-time attention to One Health coordination, and to collect and disseminate information (on focal points, activities, case studies and videos). The Regional Tripartite further monitors progress and arranges technical/administrative support and communicates between global, regional and country levels, and between technical, developmental and financing institutions.

The Asia Pacific Tripartite works with countries and partners through the Tripartite multisectoral workshops, along with various disease-specific regional and sub-regional events, capacity-building opportunities, emergency missions that are undertaken in countries, and the organization of World Rabies Day events including webinars.

2.4.2 Activities of the Asia Pacific Regional Tripartite

2.4.2.1 Enhancing multisectoral collaboration mechanisms to address rabies

The eight regional workshops on 'Asia-Pacific multisectoral collaboration at the animal–human–ecosystems interface' mentioned above were initially organized on an annual basis, but since 2015 they have been biennial to allow more time to report meaningful progress. As the multidisciplinary and intersectoral One Health approach is increasingly adopted at the global, regional and country levels to address common health threats, the Regional Tripartite has institutionalized this regional One Health mechanism by providing a platform for countries to share their achievements and challenges in addressing health threats (WHO *et al.*, 2019b). Rabies has always featured as an important topic to illustrate how countries can use the One Health approach to address endemic as well as emerging health threats.

The Tripartite regularly collects up-to-date information from participating countries on their disease situation and multisectoral coordination mechanisms through questionnaires and/or poster presentations. Rabies remains the most frequently detected zoonosis and is ranked as a top priority zoonotic disease in the region. Comparing inter-annual survey results has shown that more countries have gradually established national One Health coordination mechanisms, and more government funding has become available, often including rabies as the focus of One Health collaboration. However, obtaining high-level commitment remains challenging.

The Regional Tripartite also actively contributes to the UAR Forum by sharing expertise, knowledge, experience, ideas and information.

2.4.2.2 Regional/sub-regional rabies workshops

The Regional Tripartite regularly brings countries together in partnership with the regional economic communities in Asia – ASEAN and the South Asian Association for Regional Cooperation (SAARC).

In January 2012, together with the ASEAN Secretariat, the Regional Tripartite organized a rabies workshop in Chiang Mai, Thailand, which focused on developing a regional strategy and road map for rabies control and elimination (see Fig. 2.1). Country discussions were used to develop a step-wise elimination of rabies, which included progressive steps or stages (WOAH, 2015). The importance of social support or acceptance (S), technical support (T) and political support (P) across each stage was recognized as critical elements for rabies control and elimination (Fig. 2.2). This framework later became STOP-R by including R for 'resources' and became part of the 'The Global Strategic Plan to end human deaths from dog-mediated rabies by 2030'.

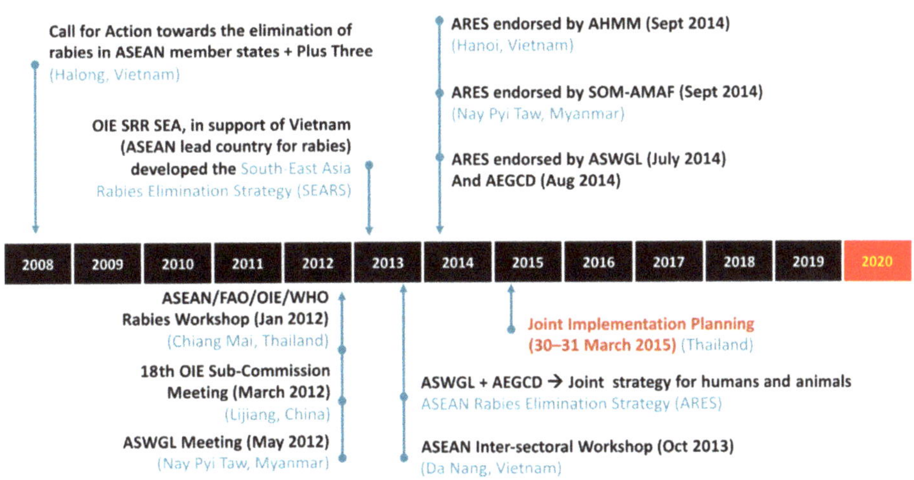

Fig. 2.1. Development of the ASEAN Rabies Elimination Strategy (ARES) (2008–2015). Initially the ARES (ASEAN rabies elimination strategy) had set its goal to reach rabies elimination by 2020. This goal, however, has been reviewed to 2030 and is now in line with the global goal of elimination of dog-mediated rabies by 2030. AEGCD, ASEAN Experts Group on Communicable Diseases; AHMM, ASEAN Health Ministers Meeting; ASWGL, ASEAN Sectoral Working Group for Livestock; SOM-AMAF, Senior Officials of ASEAN Ministers of Agriculture and Forestry; SRR SEA, Sub-Regional Representation for South-east Asia.

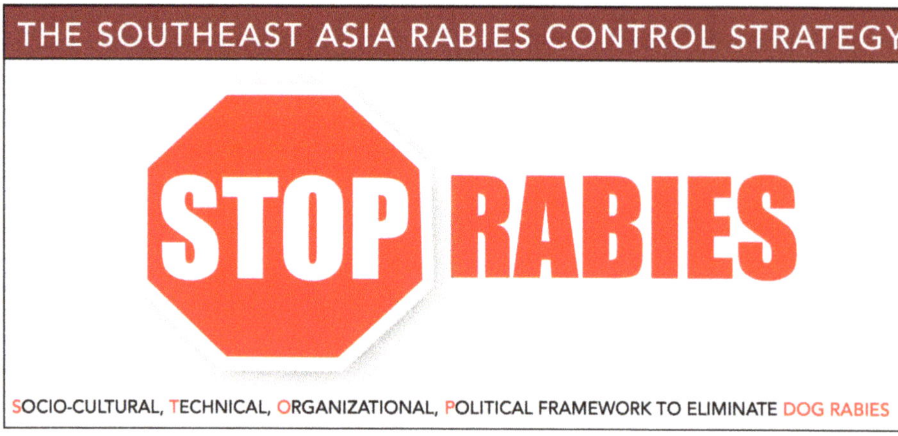

Fig. 2.2. STOP-Rabies framework developed in 2012.

In December 2018, the Tripartite along with the ASEAN Secretariat, partners and Vietnam (the ASEAN lead country for rabies) organized the ASEAN Tripartite Rabies Meeting 'Towards Rabies Elimination in ASEAN Region' in Hanoi, Vietnam. The ASEAN workshop was attended by the ten ASEAN member states and the meeting resolved to align ARES with the global plan.

As with ASEAN, a workshop on rabies prevention and control in SAARC countries was organized by the Regional Tripartite in August 2015 in Colombo, Sri Lanka. Together with the SAARC Secretariat, GARC, Humane Society International (HSI), World Animal Protection (WAP) and Vets Beyond Borders, the Tripartite reviewed ongoing initiatives on rabies elimination, best practices and lessons learnt on rabies control/elimination in SAARC countries. Based on the identified priority gaps, the Tripartite provided a road-map template for countries to consider necessary actions, a timeline and responsible bodies to conduct those actions.

In 2019, together with the SAARC Secretariat, GARC and WAP, the Regional Tripartite organized the SAARC Tripartite Rabies Workshop on 'Enhancing Progress Towards Rabies Elimination 'Zero by 30' in the SAARC Region' in Kathmandu, Nepal. The meeting was preceded by in-country SARE workshops in Bangladesh, Bhutan, India, Nepal and Sri Lanka. The SARE tool uses the concept of progressive stages which highlights completed and pending activities together with an automatically associated score and identifies activities that need to be included in rabies national action plans (NAPs). Based on the information entered, SARE categorizes countries against one of six stages, ranging from 0 indicating the absence of basic information to 5 indicating that a country has self-declared freedom from dog-mediated rabies (Fig. 2.3). The SARE supports operationalization by identifying the strengths and weaknesses of rabies control programmes at country level, thereby providing a basis for the development of country NAPs.

The SAARC country SARE scores (in Fig. 2.4) indicate the prevailing situation in SAARC. The score should not be used to compare countries but rather as an indicator for countries to monitor their progress over time when reassessed in the future. At the SAARC rabies meeting in 2019, the outputs of the SARE exercises were used to develop country-specific work plans to outline how countries can implement their NAP. Following the SARE in-country assessment workshops in 2019: (i) in 2021 India finalized their first joint NAP for dog-mediated rabies elimination by 2030; (ii) Sri Lanka is reviewing and updating its national strategic plan for the elimination of dog-mediated human rabies by 2025; and (iii) Bhutan has updated their strategic plan for the elimination of dog-mediated human rabies by 2023 and rabies freedom by 2030. Indonesia has also finalized their 'One Health national road map for rabies

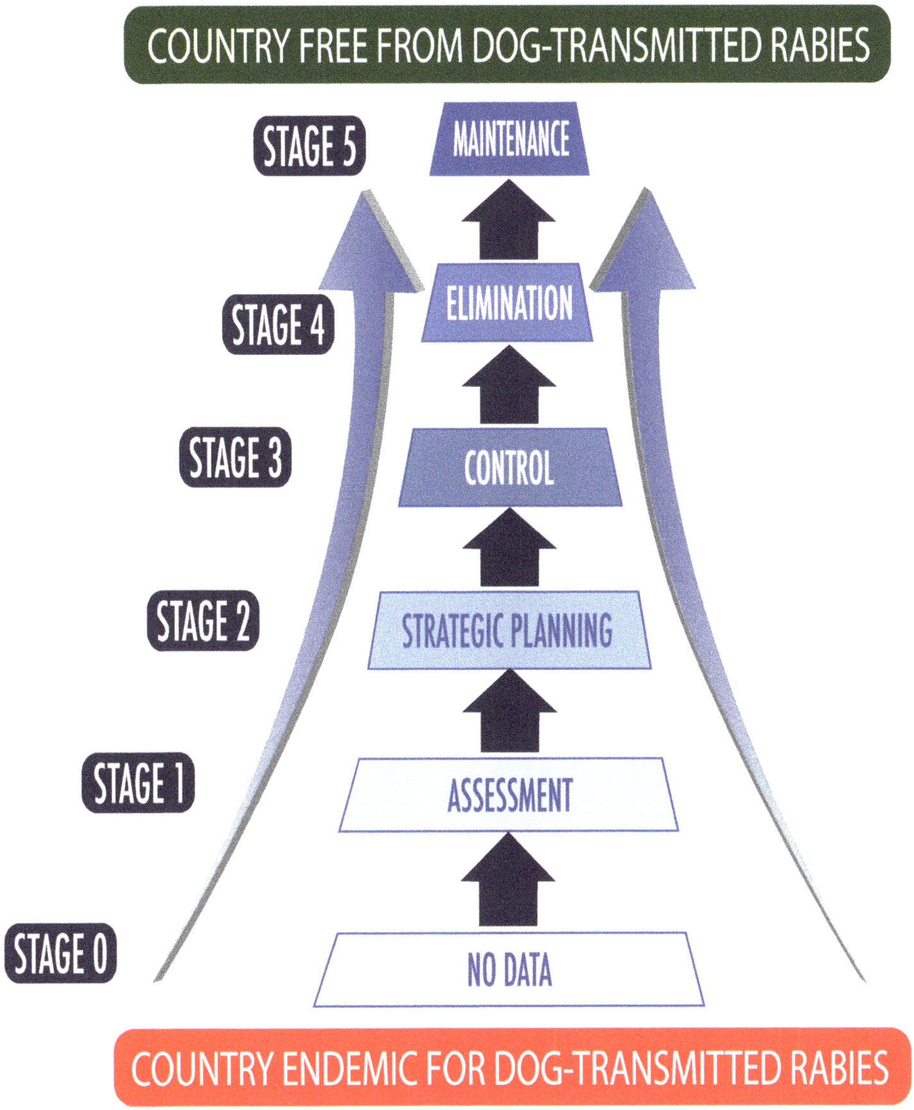

Fig. 2.3. Stages of the Stepwise approach towards rabies elimination (SARE) to move from endemic to free from dog-transmitted human rabies.

elimination by 2030' in 2019 with the involvement of the Ministry of Health and the Ministry of Agriculture, supported by WHO and FAO. While all country NAPs are in line with reaching our goal of zero dog-mediated human rabies by 2030, much needs to be done to coordinate and implement them.

The ASEAN (2018) and SAARC (2019) meetings aimed to keep rabies high on the agenda for countries and to catalyse country actions towards achieving the global target of 'Zero by 30' in the AP region. Participants at these meetings included representatives from human health, animal health, wildlife/

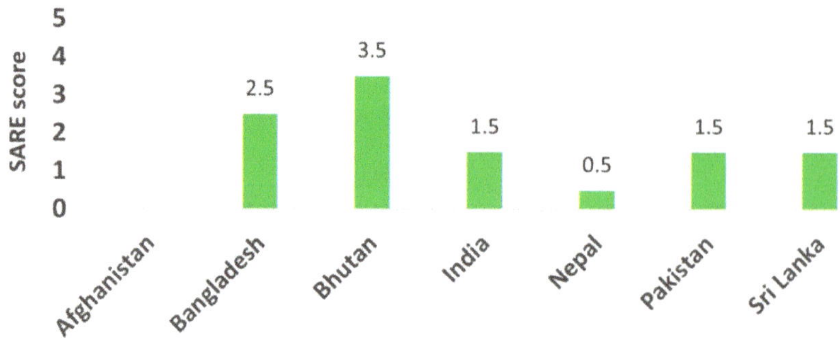

Fig. 2.4. SARE scores in the South Asian Association for Regional Cooperation (SAARC) region in 2019. Scores range from 0 indicating absence of basic information to 5 indicating self-declared freedom from dog-mediated rabies (see Fig. 2.3). Note, Pakistan's score is from its SARE exercise in 2017 and there has been no SARE done in Afghanistan.

environmental sectors, local governments and other stakeholders working on rabies control. Priority actions identified in these meetings included: (i) maintaining political engagement; (ii) focusing on dog vaccination; (iii) making human PEP accessible and affordable; (iv) enhancing intersectoral coordination; (v) considering dog population management to control numbers of free-roaming dogs; (vi) enhancing awareness and community engagement; and (vii) mobilizing resources.

Following these workshops, countries took actions such as developing and/or revising their NAPs (Bangladesh, Bhutan, India, Indonesia, Laos PDR, Myanmar, Nepal, Philippines), making rabies notifiable (Nepal, India, Indonesia) and/or expanded their mass dog vaccination programmes (Myanmar, Sri Lanka, Bangladesh).

2.4.2.3 Regional reference centres in Asia

WOAH REFERENCE LABORATORIES. There are three WOAH reference laboratories for rabies in the AP region: (i) Animal and Plant Quarantine Agency (APQA, Republic of Korea); (ii) Changchun Veterinary Research Institute (CVRI, People's Republic of China); and (iii) the most recent addition, KVAFSU-CVA Rabies Diagnostic Laboratory, Karnataka Veterinary, Animal and Fisheries Sciences University (KVAFSU, India).

Experts in these reference laboratories provide capacity building for rabies diagnosis and control to countries in the region through hands-on laboratory training, such as those organized in 2014 in Japan, in 2017 at CVRI China, or through country-level training sessions, such as those in Myanmar in 2018, and Malaysia and the Philippines in 2019. Due to the COVID-19 pandemic, laboratory training had to be adapted to a virtual format. The WOAH, in collaboration with the WOAH reference laboratory in India, organized a virtual training session on rabies diagnosis for SAARC in November 2020 which focused on the occipital foramen approach for brain sampling, lateral flow immunoassays (LFA) and direct fluorescent antibody testing (DFA) to enhance dog rabies surveillance in South Asia. The WOAH and KVAFSU also organized virtual training on rabies serology, focusing on the rapid fluorescent focus inhibition test for SAARC in October 2021. WOAH with CVRI also conducted two training sessions on molecular epidemiological techniques in 2021 and 2022.

WHO COLLABORATING CENTRES (WHO CCS) AND REFERENCE LABORATORIES. There are four WHO CCs for research and training on rabies in Asia. The first WHO CC created in the region was the Queen Saovabha Memorial Institute in Bangkok, Thailand. It pioneered a cost-effective intradermal rabies vaccination, now widely used by most Asian countries, which has been adapted and promoted by WHO as a 1-week schedule for PEP. The institute has also organized training on human rabies prophylaxis for the member states.

Chulalongkorn University, also in Bangkok, is a WHO CC on viral zoonoses and hosted a global expert consultation on rabies in 2017, the first such meeting outside Geneva. The National Institute of Mental Health and Neuro-Sciences (NIMHANS) in Bangalore, India, is a WHO CC actively involved in rabies diagnostics, and conducts immunological studies to generate evidence-based information for policy decisions. Hands-on training on rabies can be organized at the request of member states and referral serological services can be provided to the member states. The National Centre for Disease Control is another WHO CC based in Delhi, India, which is a nodal agency for the national rabies control programme. All WHO CCs participate in the global network of WHO CCs for rabies to contribute to achieving zero human deaths due to dog-mediated rabies by 2030.

2.4.2.4 Tripartite support to countries' rabies control activities in Asia

As introduced above, over the last 15 years, the Regional Tripartite has coordinated support to several countries in the AP region through emergency missions, projects and training sessions as well as multisectoral consultations to develop national multisectoral strategic plans to eliminate rabies. The FAO/WOAH Crisis Management Centre (CMC) – Animal Health, established following the avian influenza crisis in 2006, fielded emergency missions to respond to the introduction of rabies in Bali, Indonesia, in December 2008 and to address the worrying rabies situation in the Northern Provinces of Vietnam in 2013. Subsequent FAO rabies projects in Bali supported the government in its efforts to control rabies through large-scale dog rabies vaccinations, the development of Integrated Bite Case Management (IBCM) to enhance a concerted One Health response (FAO, 2016a) and established 'A-teams' consisting of highly skilled dog capture and vaccination teams (FAO, 2016b). In Vietnam, FAO coordinated studies on traditional practices following dog bites and identified the most appropriate vaccination strategies to address the re-current outbreaks in affected areas.

The Tripartite has been supporting the development of important national multisectoral action plans for rabies control. In 2014, WHO SEARO organized the first Tripartite review mission to Sri Lanka to evaluate their national rabies elimination programme. Subsequently, their National Strategic Plan for Rabies was reviewed.

Through the Australian-funded STANDZ project, WOAH provided support to Myanmar and the Philippines in the development of their action plans for rabies elimination. Myanmar's National Action Plan for Rabies Elimination (2018–2030) was endorsed by the Ministry of Agriculture, Livestock and Irrigation. The Philippines National Action Plan for Rabies Elimination (2021–2025) was endorsed by the Department of Agriculture and the Department of Health and subsequently also officially endorsed by the WOAH in 2021.

WHO helped Lao PDR's Ministry of Health hold a national stakeholders' meeting, attended by other relevant ministries, including the Ministry of Agriculture and Forestry and the Ministry of Education, as well as non-governmental partners, to develop the Strategic Plan on Rabies Prevention and Control in Lao PDR 2020–2024. This was endorsed by the Ministers of Health and of Agriculture and Forestry in 2020.

Sarawak State in Malaysia had been rabies free until three suspected human cases were notified on 29 June 2017. By October 2018, 13 individuals between the ages of 3 and 59 years, and 268 animals (253 dogs and 15 cats) from 46 areas in 11 (of 12) divisions were reported as infected with rabies and on 28 October of the same year the Government of Sarawak requested assistance from the Tripartite via the WHO country office in Malaysia. In response, the Tripartite organized an expert mission, comprising human and animal rabies experts, to provide a rapid assessment of the rabies situation and to recommend a control strategy. The strategy included: (i) logistics of animal and human rabies vaccination strategies; (ii) surveillance including tracing back and forward the (suspected) rabid animals; (iii) diagnosis; (iv) awareness raising; (v) identification of high-risk areas and management of suspected animals; and (vi) an IBCM programme. The Tripartite is also committed to facilitating the procurement of human rabies vaccine and rabies immunoglobulin and to exploring hands-on training for Malaysian animal health laboratory staff on up-to-date diagnostics and alternative sampling methods. Subsequently, in the WOAH hands-on laboratory diagnosis training conducted in Myanmar in November 2018, a veterinarian from the Sarawak State of Malaysia also participated.

Brunei Darussalam remained rabies free but was worried by the risk of introducing rabies from neighbouring Sarawak. In early 2019, following a request from the Ministry of Health, the Tripartite organized an expert mission team of human and animal rabies experts to assess the risks of rabies from endemic countries coming into Brunei Darussalam and the state of preparedness for prevention, detection and response measures against possible introduction of rabies into the country. The mission recommended mass dog vaccination and enhanced public awareness in the Sungai Belait area bordering with Miri, Sarawak, and was offered to facilitate the procurement of veterinary vaccines through the WOAH vaccine bank and resolve shortcomings in laboratory diagnostic capacity through training from international reference facilities. Assistance was also given to revise the national clinical guidelines for the management of rabies and suspected rabies cases, and facilitation offered for bilateral or trilateral cross-border meetings with Malaysia and Indonesia. No cases of suspect animal or human rabies have been reported to date.

Since 2019, FAO has supported Bhutan, Bangladesh, Nepal and Sri Lanka through a Regional Technical Cooperation project to strengthen the countries' capacities to control rabies and the establishment of a Rabies Centre of Excellence (RACE) for the region.

The WOAH also provided technical support for the development of India's National Action Plan for Dog Mediated Rabies Elimination from India by 2030, which was launched on 28 September 2021. Currently, India is now working to develop State Action Plans for Rabies Elimination through regional workshops.

2.4.2.5 *Webinars on rabies*

Since the onset of the global pandemic in early 2020, the Regional Tripartite has not been able to conduct physical events (workshops, meetings, training sessions, etc.) on rabies and instead had to move activities online. As a follow up to the SAARC rabies workshop in 2019, the Regional Tripartite, with support from rabies experts in the region, organized the 'SAARC Rabies Webinar' (WOAH, 2020b) for the SAARC countries in May 2020. The webinar focused on mass dog vaccination including use of oral rabies vaccines, multisectoral coordination mechanisms in responding to rabies elimination, and developing a National Strategic Plan for the elimination of dog-mediated human rabies. The virtual format enabled the participation of a much greater number of participants than would have been possible in a face-to-face event. The Goa, India, example showed that we can bring human deaths down to zero by focusing on mass dog vaccination. The webinar generated a lot of interest in the use of oral rabies vaccines as complementary to parenteral dog vaccination, greatly enhancing vaccination coverage, especially in free-roaming dogs.

In response to a request from Sri Lanka, another webinar was organized in November 2020 with expert inputs from Friedrich-Loeffler-Institut, Germany, on the use of oral rabies vaccines as a complementary tool (WOAH, 2020c). Following outbreaks of rabies in both domestic and wild animals in the Kalutara district of Sri Lanka, the webinar was organized to provide information about rabies in wildlife and share European experiences of using oral rabies vaccines to eliminate wildlife rabies in Europe. In March 2021, a webinar was organized on mass dog vaccination methods and tools for rabies elimination in the SAARC region (WOAH, 2021b). This was a follow up to the SAARC rabies webinar in May 2020. Besides the concepts and tools for upscaling dog vaccination, the webinar covered dog population estimation methods which were well received by the participants.

2.4.2.6 *World Rabies Day (WRD) activities*

World Rabies Day, or WRD, is commemorated every year on the 28 September to raise awareness and encourage the prevention and control of rabies. International communities come together to promote the fight against the disease. The first WRD was in 2007 with 28 September selected as it is the anniversary of the death of Louis Pasteur who, with his colleagues, developed the first effective vaccines against rabies.

Fig. 2.5. Students are made aware that mass dog vaccination is key to rabies elimination in Bhutan. Photo courtesy of Y. Phuentshok.

The Regional Tripartite in Asia and the Pacific and its partners celebrate WRD with innovative ideas. The Regional Tripartite collaborated with the University of Sydney in Australia, the European Union and AusAid to produce a rabies board game, 'Dogsville', which was trialled in the Philippines before distribution to schools in South-east Asia for WRD in 2013.

Since 2018, the Regional Tripartite has jointly organized a webinar annually for WRD. In 2018, the Tripartite collaborated with GARC with the webinar theme 'Rabies: Share the message. Save a life'. For WRD 2019, the Tripartite and GARC again collaborated, with the theme 'Rabies: Vaccinate to Eliminate' (WOAH, 2019). The theme for WRD 2020 (WOAH, 2020a) was 'End Rabies: Collaborate, Vaccinate'. The theme for the Tripartite webinar on WRD 2021 (WOAH, 2021a) was 'Rabies: Facts, not Fear'. This webinar doubled as a side event for the 9th Asia-Pacific Workshop on Multisectoral Collaboration at the Animal–Human–Ecosystems Interface.

2.5 Conclusion

The key to successfully eliminating dog-mediated human rabies cases by 2030 is to address rabies at its source, in dog populations. The tools are available for this focused effort, consisting largely of concerted and sustained dog vaccination campaigns, and continuing to maintain a high level of awareness about rabies prevention and control (Fig. 2.5). Perhaps a less obvious but equally necessary element of achieving rabies freedom is maintaining high-level commitment from national authorities, regional bodies and global partners. Some fatigue has been observed in this long-term campaign against rabies, and it is essential for the global community to maintain enthusiasm for rabies elimination programmes, and for both human and financial commitments to remain in place until we reach our goal.

When looking at the practical elements of a sustained and concerted mass vaccination programme, we must take into consideration several important elements. It is essential that quality dog rabies vaccines are readily available at regional, national and sub-national levels. Unfortunately, in the AP region, we have seen several examples of poor-quality vaccines being administered which do not result in proper protection from rabies, and can seriously jeopardize national programmes. Next, it is essential to have a well-organized and funded community-based approach. This is particularly important where governments lack the capacity and effective governance mechanisms to mobilize community efforts. This includes a community-based approach for awareness, vaccination of owned/cared-for dogs as well as capture and vaccination of roaming street dogs, as well as ensuring that community leaders provide backing and sustained support. Finally, as is recognized in the region and globally, regardless of whether a government programme or community-based approach is used, budgetary constraints often limit the number of dog rabies vaccines available. There is a need to deliver vaccines over a longer duration in order to maintain progress against rabies and not let gains become losses due to an unsustainable situation. While human vaccine production capacities of Asian countries have improved greatly, and this has saved human lives, the long-term success and reaching our goal of zero dog-mediated human rabies by 2030 will depend on the determination, and capacity to ensure at least 70% coverage during dog rabies vaccination campaigns.

Over the last several years, the Regional Tripartite has demonstrated its commitment to the prevention, control and elimination of rabies from the AP region. From developing tools, strategies and guidelines, to coordinating meetings and also providing technical and fund support to countries in the region, the Regional Tripartite has shown by example how a One Health approach can be used to deal with a zoonotic disease such as rabies.

Acknowledgements

We would like to acknowledge the support and collaboration of countries in the region, partners and other collaborators.

Authors' Declaration

All authors declare that they have no conflict of interest.

All authors have approved this manuscript, agree with its submission, and share collective responsibility and accountability.

This manuscript has not been published or is not under review elsewhere.

References

Anonymous (2016) Canine Rabies Blueprint – the Stepwise Approach towards Planning and Evaluation (SARE). Available at: https://caninerabiesblueprint.org/Canine-Rabies-Blueprint-PDF-of (accessed 2 November 2022).

Belsare, A.V. and Gompper, M.E. (2013) Assessing demographic and epidemiologic parameters of rural dog populations in India during mass vaccination campaigns. *Preventive Veterinary Medicine* 111(1–2), 139–146. DOI: 10.1016/j.prevetmed.2013.04.003.

Childs, J.E., Robinson, L.E., Sadek, R., Madden, A., Miranda, M.E. *et al.* (1998) Density estimates of rural dog populations and an assessment of marking methods during a rabies vaccination campaign in the Philippines. *Preventive Veterinary Medicine* 33(1–4), 207–218. DOI: 10.1016/S0167-5877(97)00039-1.

Devex (2021) Extreme poverty rose by 80 million in Asia and Pacific due to COVID-19. Available at: www.devex.com/news/extreme-poverty-rose-by-80-million-in-asia-and-pacific-due-to-covid-19-100672 (accessed 1 November 2022).

ESCAP (United Nations Economic and Social Commission for Asia and the Pacific) (2019) Why can't dynamic Asia-Pacific beat poverty? Available at: www.unescap.org/blog/why-cant-dynamic-asia-pacific-beat-poverty (accessed 1 November 2022).

FAO (Food and Agriculture Organization of the United Nations) (2016a) Bali agrees on SOP for Integrated Bite Case Management under the One Health Umbrella. Available at: www.fao.org/indonesia/news/detail-events/en/c/431438/ (accessed 1 November 2022).

FAO (Food and Agriculture Organization of the United Nations) (2016b) Bali fights against rabies with mass dog vaccination. Available at: www.fao.org/indonesia/news/detail-events/ru/c/411271/ (accessed 1 November 2022).

Gongal, G.N. (2006) The epidemiological trend of animal rabies in Nepal and future control strategy. *Association for Prevention and Control of Rabies in India Journal* 8(1), 8–12. Available at: https://www.apcrijournal.com/JournalImage/CoverImage/APCRI%20Journal%20Vol-%20VIII,%20Issue%201%20July%202006.pdf (accessed 1 November 2022).

Gongal, G. and Wright, A.E. (2011) Human rabies in the WHO Southeast Asia region: forward steps for elimination. *Advances in Preventive Medicine* 2011, 383870. DOI: 10.4061/2011/383870.

Hossain, M., Ahmed, K., Marma, A.S.P., Hossain, S., Ali, M.A. *et al.* (2013) A survey of the dog population in rural Bangladesh. *Preventive Veterinary Medicine* 111(1–2), 134–138. DOI: 10.1016/j.prevetmed.2013.03.008.

Knobel, D.L., Cleaveland, S., Coleman, P.G., Fèvre, E.M., Meltzer, M.I. *et al.* (2005) Re-evaluating the burden of rabies in Africa and Asia. *Bulletin of the World Health Organization* 83(5), 360–368.

Lee, H.S., Vu, D.T., Dang, D.A., Tran, N.D., Mihye, L. *et al.* (2018) Geographical and temporal patterns of rabies post exposure prophylaxis (PEP) incidence in humans in the Mekong River Delta and Southeast Central Coast regions in Vietnam from 2005 to 2015. *PLoS ONE* 13(4), e0194943.

Matter, H.C. and Daniels, T.J. (2000) Dog ecology and population biology. *Dogs, Zoonoses and Public Health* 2, 17–62.

Pantha, S., Subedi, D., Poudel, U., Subedi, S., Kaphle, K. *et al.* (2020) Review of rabies in Nepal. *One Health* 10, 100155. DOI: 10.1016/j.onehlt.2020.100155.

Republic of the Philippines Department of Health (2020) Rabies prevention and control program. Available at: https://doh.gov.ph/national-rabies-prevention-and-control-program (accessed 14 February 2022).

Rinzin, K., Tenzin, T. and Robertson, I. (2016) Size and demography pattern of the domestic dog population in Bhutan: implications for dog population management and disease control. *Preventive Veterinary Medicine* 126, 39–47. DOI: 10.1016/j.prevetmed.2016.01.030.

UNFPA (United Nations Population Fund) (2022) Population trends. Available at: https://asiapacific.unfpa.org/en/node/15207 (accessed 26 September 2021).

WHO (World Health Organization) (2009) *Zoonotic Diseases: A Guide to Establishing Collaboration Between Animal and Human Health Sectors at the Country Level*. Available at: https://apps.who.int/iris/handle/10665/207731 (accessed 2 November 2022).

WHO (World Health Organization) (2015) Rabies: rationale for investing in the global elimination of dog-mediated human rabies. Available at: www.who.int/publications/i/item/9789241509558 (accessed 14 December 2022).

WHO (World Health Organization) (2020) Pulse survey on continuity of essential health services during the COVID-19 pandemic: interim report. Available at: www.who.int/publications-detail-redirect/WHO-2019-nCoV-EHS_continuity-survey-2020.1 (accessed 2 November 2022).

WHO (World Health Organization), FAO (Food and Agriculture Organization of the United Nations) and WOAH (World Organization for Animal Health) (2010) The FAO-OI-WHO Collaboration: a Tripartite Concept Note. Available at: www.who.int/publications/m/item/the-fao-oie-who-collaboration (accessed 2 November 2022).

WHO (World Health Organization), FAO (Food and Agriculture Organization of the United Nations), WOAH (World Organization for Animal Health) and GARC (Global Alliance for Rabies Control) (2016) *Global Elimination of Dog-Mediated Human Rabies: Report of the Rabies Global Conference*. WHO, Geneva, Switzerland.

WHO (World Health Organization), FAO (Food and Agriculture Organization of the United Nations), WOAH (World Organization for Animal Health) and GARC (Global Alliance for Rabies Control) (2018) *Zero by 30: The Global Strategic Plan to End Human Deaths from Dog-Mediated Rabies by 2030*. WHO, Geneva, Switzerland.

WOAH (World Organization for Animal Health) (2015) *ASEAN Rabies Elimination Strategy*. WOAH, Bangkok.

WHO (World Health Organization), FAO (Food and Agriculture Organization of the United Nations) and WOAH (World Organization for Animal Health) (2019a) *Taking A Multisectoral One Health Approach: A Tripartite Guide to Addressing Zoonotic Diseases on Countries*. Available at: https://apps.who.int/iris/handle/10665/325620 (accessed 2 November 2022).

WHO (World Health Organization), FAO (Food and Agriculture Organization of the United Nations) and WOAH (World Organization for Animal Health) (2019b) Covered by longstanding agreement between FAO, OIE and WHO. YouTube video. Available at: www.youtube.com/watch?v=fm9sQDfWAQQ (accessed 2 November 2022).

WHO (World Health Organization), FAO (Food and Agriculture Organization of the United Nations) and WOAH (World Organisation for Animal Health) (2020a) *Joint Statement of Intent to Coordinate*. Available at: https://rr-asia.woah.org/wp-content/uploads/2020/11/soi-on-tripartite-coordination-final-20201023.pdf (accessed 2 November 2022).

WHO (World Health Organization), FAO (Food and Agriculture Organization of the United Nations) and WOAH (World Organisation for Animal Health) (2020b) United against rabies forum: zero by 30: One Health in action. Available at: www.who.int/publications/i/item/WHO-UCN-NTD-VVE-2021.1 (accessed 2 November 2022).

WOAH (World Organization for Animal Health) (2019) World Rabies Day – Regional activities in 2019. Available at: https://rr-asia.oie.int/en/projects/rabies/world-rabies-day-regional-activities/ (accessed 23 April 2020).

WOAH (World Organisation for Animal Health) (2011) *Global Conference on Rabies Control – Recommendations.* Incheon-Seoul (Republic of Korea), 7–9 September. Available at: https://former.woah.org/eng/A_RABIES/presentations.htm (accessed 2 November 2022).

WOAH (World Organisation for Animal Health) (2020a) Activities for World Rabies Day in our region in 2020. Available at: https://rr-asia.oie.int/en/projects/rabies/activities-for-world-rabies-day-in-our-region-in-2020/ (accessed 29 September 2020).

WOAH (World Organisation for Animal Health) (2020b) SAARC rabies webinar. Available at: https://rr-asia.woah.org/en/events/saarc-rabies-webinar/ (accessed 2 November 2022).

WOAH (World Organisation for Animal Health) (2020c) Technical consultation on oral rabies vaccines – a complementary tool for vaccination in animals. Available at: https://rr-asia.oie.int/en/events/technical-consultation-on-oral-rabies-vaccines-a-complementary-tool-for-vaccination-in-animals/ (accessed 10 November 2020).

WOAH (World Organisation for Animal Health) (2021a) Activities for World Rabies Day in our region in 2021. Available at: https://rr-asia.oie.int/en/projects/rabies/activities-for-world-rabies-day-in-our-region-in-2021/ (accessed 2 November 2022).

WOAH (World Organisation for Animal Health) (2021b) Webinar on mass dog vaccination methods and tools for rabies elimination in the SAARC region. Available at: https://rr-asia.oie.int/en/events/oie-webinar-on-mass-dog-vaccination-methods-and-tools-for-rabies-elimination-in-the-saarc-region/ (accessed 30 March 2021).

WOAH (World Organisation for Animal Health) (2021c) Agenda: 29–30 March 2021. Webinar on mass dog vaccination methods and tools for rabies elimination in the SAARC region. Available at: https://rr-asia.woah.org/wp-content/uploads/2021/04/final-agenda_website-1.pdf (accessed 2 November 2022).

3 Emerging Opportunities for Enhanced Regional One Health Approach in the Prevention, Control and Elimination of Rabies and Other Zoonoses in South-east Asia

Noel Lee Jose Miranda[1] and Mary Elizabeth Gutierrez Miranda[2]*
[1]Independent Regional Consultant for Asia; [2]Field Epidemiology Training Program Alumni Foundation Inc, Manila, Philippines

Abstract
The Association of South-east Asian Nations (ASEAN) launched the ASEAN Rabies Elimination Strategy officially endorsed by the ASEAN Ministers of Health and Agriculture and Forestry in 2014. Because of coronavirus 2019 (COVID-19), global and regional stakeholders realized the urgency of strengthening One Health collaborations. In November 2020, the ASEAN endorsed its ASEAN Strategic Framework for Public Health Emergencies, with coordination being done through the ASEAN Centre for Public Health Emergencies and Emerging Diseases. ASEAN aims to synergize regional preparedness and response capacities with existing international frameworks to promote multisectoral collaboration, including the One Health approach. This is cognizant of the ASEAN Coordinating Centre for Animal Health and Zoonoses aimed at providing comprehensive regional coordination. We call on stakeholders to seek the opportunity for how best to establish a functional entity in ASEAN to deal with zoonoses prevention and control and achieve One Health objectives and the global goal of Zero by 30.

3.1 Introduction

In South-east Asia, more than 600 million people are at risk of rabies exposure. Dog-mediated rabies is endemic in eight out of ten member states of the Association of South-east Asian Nations (ASEAN) (i.e. Cambodia, Indonesia, Lao PDR, Malaysia, Myanmar, Philippines, Thailand and Vietnam). In Malaysia, the disease re-emerged and has remained in Sarawak since 2017. Only Singapore and Brunei are rabies free. The ASEAN Rabies Elimination Strategy (ARES) is a regional initiative officially endorsed by the ASEAN Ministers of Health and Agriculture and Forestry in 2014, with Vietnam as the lead coordinating country (ASEAN, 2016a).

The ARES provides a strategic framework for rabies control activities in ASEAN countries. The ARES notes the importance of periodic reviews to highlight progress and the critical areas for improvement as a basis to maintain political engagement and momentum towards regional rabies elimination in the ASEAN.

*Corresponding author: megmiranda@fetpafi.org

DOI: 10.1079/9781800622975.0003

The World Health Organization (WHO), the World Organization for Animal Health (OIE/WOAH), the Food and Agriculture Organization of the United Nations (FAO), and the Global Alliance for Rabies Control formed the United Against Rabies collaboration to achieve the goal of zero human dog-mediated rabies deaths by 2030, worldwide (WHO *et al.*, 2018).

The ASEAN member countries recommended implementing the following priority activities to expedite the achievement of the goals of ARES:

1. Increase dog vaccination coverage in the region through advocacy for mass dog vaccination and dog population management.
2. Develop a regional preparedness plan on rabies for humans and animals, focusing on capacity-building activities and human rabies vaccine advocacy and stockpiling.
3. Promote Integrated Bite Case Management.
4. Develop a regional platform for regular information sharing and monitoring and evaluation for rabies in ASEAN.

The two key objectives of ARES are: (i) to establish and continuously strengthen the coordinating and supporting mechanisms among stakeholders; and (ii) to obtain and sustain high-level governmental engagement. ARES also advocates the implementation of formal One Health multisectoral engagements driven by a high-level coordination mechanism.

3.2 ASEAN Regional Health Mechanisms Relevant to One Health and Rabies/Zoonoses Risk Management

Within ASEAN, several regional mechanisms at an operational level and additional high-level meeting arrangements and platforms are important for decision making, coordination and management of regional health-related issues, including public health emergencies.

ASEAN health sector oversight is provided through four health clusters, which are functional bodies as extensions of the Senior Officials Meeting on Health Development (SOMHD). Of the four clusters, the ASEAN Health Cluster 2 (AHC-2) currently implements activities under the theme: Responding to All Hazards and Emerging Threats. The ASEAN Plus Three (with China, Japan and Korea) SOMHD serves as a significant venue for collective decisions among plus three member states' senior health officials, which the AHC-2 also works through. This is according to the Governance and Implementation Mechanism of the *ASEAN Post-2015 Health Development Agenda* (ASEAN, 2018).

The Health Division under the ASEAN Socio-Cultural Community Department of the ASEAN Secretariat (ASEC) provides secretariat and facilitation support to the various relevant meetings and committees. Because of the multidimensional and expansive impacts of COVID-19, the ASEAN Coordinating Council Working Group on Public Health Emergencies (ACCWG-PHE) was established, with senior official representation from all three Community Pillars of ASEAN, to support the ASEAN Coordinating Council (ACC) in coordinating regional response efforts (Philippine Department of Foreign Affairs News, 2020).

Because of COVID-19, global and regional stakeholders realized the urgency of strengthening One Health collaborations and public health emergency risk management. In November 2020, the ASEAN endorsed its ASEAN Strategic Framework for Public Health Emergencies (ASF-PHE) (ASEAN, 2020). This document elaborates the current arrangements and procedures for dealing with emerging infectious diseases (EIDs) utilizing existing ASEAN and WHO mechanisms like the International Health Regulations (IHR) 2005 (WHO, 2016), with coordination, command and action expected to be done through the newly established ASEAN Centre for Public Health Emergencies and Emerging Diseases (ACPHEED), and the ASEAN Emergency Operations Centre (EOC) Network (ASEAN, 2021b). These new arrangements aim to support the fulfilment of national obligations to the IHR 2005-Asia Pacific Strategy for Emerging Diseases III (APSED III) to strengthen core capacity and enable efficient regional preparedness and response (WHO, 2005, 2017).

The ASF-PHE mentions that regional coordination of health sector agencies shall be under the ACPHEED and the EOC Network, while coordination of agencies outside the health

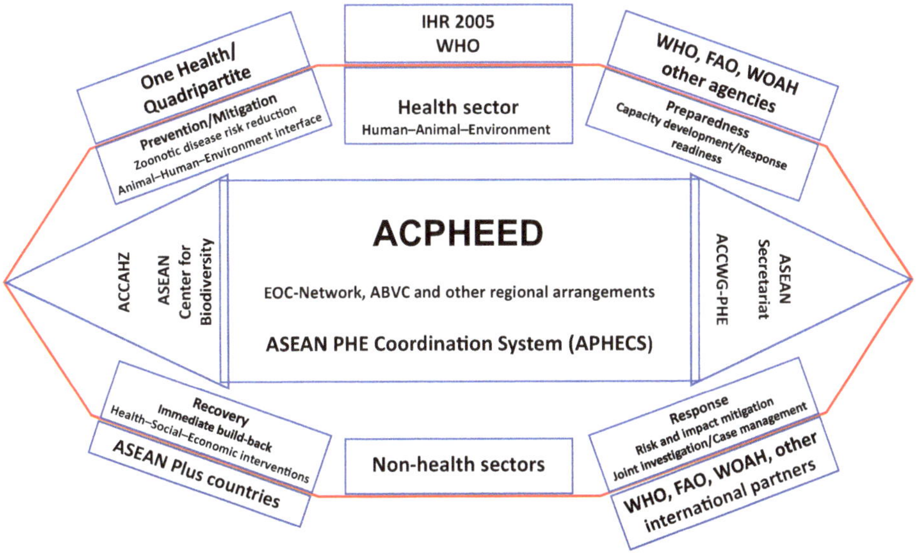

Fig. 3.1. Foreseen ASEAN regional health security coordination and management structure. ABVC, ASEAN Biodiaspora Virtual Center; ACCAHZ, ASEAN Coordinating Centre for Animal Health and Zoonoses; ACPHEED, ASEAN Centre for Public Health Emergencies and Emerging Diseases; ACCWG-PHE, ASEAN Coordinating Council Working Group on Public Health Emergencies; ASEAN, Association of South-east Asian Nations; EOC, Emergency Operations Centre; FAO, Food and Agriculture Organization of the United Nations; IHR, International Health Regulations; WHO, World Health Organization; WOAH, World Organization for Animal Health. Created by N. Miranda.

sector shall be under the ACCWG-PHE. The ASF-PHE intends to enhance ASEAN capacities for preparedness, detection, response and resilience. Of importance is the clear request for ASEAN leaders to further develop the ASF-PHE, with a strategic map to define outputs, outcome indicators and deliverables for the regional guidance for the ASEAN member states (AMS) in strengthening preparedness, detection and response and, more generally, health security and to efficiently prepare for and mitigate public health emergency and biosafety risks. Fig. 3.1 depicts the foreseen ASEAN regional multisectoral health security coordination and management structure.

The ACPHEED is planned to be a physical facility and coordinating centre that will provide the ASEAN technical oversight. ACPHEED will serve as a centre of excellence and regional hub to strengthen capabilities to prepare for, prevent, detect and respond to public health emergencies and EIDs. It is also expected to be a key mechanism for rabies initiatives. It was established with the financial assistance of the governments of Japan and Australia and is projected to be operational in 2023. The existing ASEAN EOC Network and other mechanisms will probably be integrated with ACPHEED. The initial scope of work includes:

- prevention and preparedness;
- detection and risk assessment (laboratory network, surveillance/field epidemiology);
- response including risk communication;
- information sharing and analytics;
- capacity building; and
- innovation and coordination/support (including R&D).

ASEAN is expected to define what is feasible regional assistance to the AMS and ensure these are appropriately covered under ACPHEED, including funding, supplies assistance and regional vaccine acquisition, capacity building, outbreak response coordination, and closing the gap in regional information sharing, communication and collaboration. It is envisioned that the ACPHEED would be positioned to accelerate countries' progress in achieving IHR (WHO, 2016) core capacities. ASEAN's dedicated support to countries is key to collective regional progress.

ASEAN aims to synergize all its capacities with existing international frameworks through the ongoing development of an ASEAN Public Health Emergencies Coordination System (APHECS). This is expected to include the One Health approach and should be cognizant of the coordination that needs to happen through the ASEAN Coordinating Centre for Animal Health and Zoonoses (ACCAHZ) established in 2016. The ACCAHZ will provide a comprehensive, integrated and concerted regional approach to coordinate national approaches in animal health and zoonoses measures, including disease surveillance, diagnosis and control, and quick response (ASEAN, 2016b).

There are several key mechanisms under the ASEAN health sector, and several other mechanisms related to public health and EID concerns coordinated through the animal health and non-health sectors. These are operated independently under lead/host countries, with the ASEC providing coordinative support. These mechanisms synergize and are complementary in strengthening regional collaborations in the following focus areas: (i) information/data sharing and management; (ii) risk assessment (surveillance and epidemiology); (iii) laboratory networking; and (iv) risk communication.

Several ASEAN declarations and decisions impacting disease risk management have been published recently. On 21 June 2006, the ASEAN Leaders' Declaration of the 8th ASEAN Health Ministers' Meeting, highlighted the importance of regional cooperation in information sharing and assistance during times of crisis and of strengthening capabilities, close cooperation, and leading national and regional responses to health emergencies posed by disasters and disease outbreaks in the region (ASEAN, 2006). They proposed preparing for potential public health emergencies, including natural disasters, bioterrorism, outbreaks of communicable diseases and potential pandemics. They committed further to rapid and transparent disease notification, epidemiological data and sample sharing, and essential information. This would require a regional agreement to: (i) institutionalize regional monitoring, reporting and response; (ii) facilitate the deployment of multinational ASEAN outbreak response teams; and (iii) standardize respective procedures, protocols and institutional arrangements (ASEAN, 2006).

As State Parties in the WHO, countries are obligated to conform to the provisions of the IHR 2005 related to core capacities, specifically, to the APSED III. ASEAN fully recognizes and supports this WHO mechanism and member states' obligations. The APSED III, updated in 2017, is an important international common framework to address shared threats as required by the IHR 2005. It points out the importance of close coordination with other international frameworks and initiatives such as the Sustainable Development Goals, universal health coverage, and the Global Health Security Agenda (GHSA), as well as enhanced One Health collaboration on zoonoses. While APSED III focuses on enhancing and sustaining the core public health functions required to sustain and strengthen the entire health system, it underlines the importance of linking other sectors and health security initiatives to prevent, respond to and mitigate the impact of public health emergencies using an all-hazards approach. The strategy aims to enhance further the core public health systems and regional collaboration and connectedness (WHO, 2017). This will continue to play an important role in guiding the AMS and ASEAN and the functionalities within the APHECS, including the ACPHEED. In the context of One Health development, APSED III performance targets should be augmented with the technical targets described in the OIE's Performance of Veterinary Services (PVS) frameworks (OIE, 2021a).

The ASEAN mechanisms/entities are realized through cooperation with Development and Dialogue Partners. At the operational level, coordinating regional efforts in the technical areas potentially relevant for rabies risk management are under the following mechanisms.

3.2.1 Human health sector

3.2.1.1 ASEAN emergency operations centre (EOC) network for public health emergencies

This is made up of the EOCs of each AMS, and serves as ASEAN's key centre for regional surveillance and response towards EID and public health emergencies and ensures information sharing. It supports AMS in capacity-building efforts for their EOCs and the development of common procedures and protocols. In addition, the EOC Network also managed workshops and training activities for the Risk Assessment and Risk Communication Centre (RARC) and the regional laboratory network. As part of the regional COVID-19 response, the ASEAN EOC Network provides ASEAN Plus Three countries with daily situation updates and shares virological and epidemiological information (ASEAN, 2021a).

3.2.1.2 ASEAN plus 3 field epidemiology training network

This promotes capacity building of field epidemiology training in ASEAN Plus Three countries, and enhances joint efforts to prevent and control public health events through collaboration in surveillance, investigation, study, research, etc. It also ensures active and effective network management for sustainable development – promoting sharing of best practices and experiences (ASEAN, 2019a).

3.2.1.3 Mekong Basin disease surveillance

This has supported countries in the Mekong Basin for ~20 years. The focus is to strengthen national and sub-regional capabilities in infectious disease surveillance and outbreak response, harmonize protocols and support local cross-border communication and collaboration for outbreak investigation and response management (Phommasack *et al.*, 2013).

3.2.1.4 The public health laboratories network and regional public health laboratories network (RPHL)

The Public Health Laboratories Network is under the purview of AHC-2 on Responding to All Hazards and Emerging Threats, while the RPHL is led by Thailand through the Global Health Security Agenda platform. The RPHL facilitates exchanges on laboratory readiness and response and disease surveillance actions (Fernando, 2020).

3.2.1.5 ASEAN BioDiaspora regional virtual centre

The objective of this activity is to build regional capacity in big data predictive analytics that strengthens ASEAN's epidemic and pandemic preparedness and response capabilities. It supports regional data and information sharing, for example risk reports on the potential spread of COVID-19 (ASEAN, 2021c). The ASEAN Portal for Public Health Emergencies, which was recently established, is also important for information sharing (ASEAN, 2021d).

3.2.1.6 Regional strategic and action plan for ASEAN vaccine security and Self-reliance 2021–2025

Endorsed by the ASEAN Health Ministers Meeting in May 2021, this has become urgent because of COVID-19 (ASEAN, 2021e). This is relevant to all vaccine-preventable diseases.

3.2.2 Animal health sector-led mechanisms relevant to One Health and EIDs

3.2.2.1 ASEAN sectoral working group on livestock (ASWGL)

This is an important stakeholder and partner for coordination between animal health and human health sectors. It is supported by the Food, Agriculture and Forestry Division of ASEC, and is in charge of animal health and veterinary public health, including as a lead body for ARES implementation (ASEAN, 2016c).

3.2.2.2 Regional strategic framework for veterinary epidemiology capacity development and networking in ASEAN (Epi Framework)

This was endorsed in 2013 and it led to the establishment of the ASEAN Veterinary Epidemiology Group (ASEAN, 2019b).

3.2.2.3 ASEAN Coordinating Centre for Animal Health and Zoonoses

This is envisaged to protect the health of people and livestock to ensure a quick response against animal health emergencies and zoonotic disease emergencies and to provide a framework of cooperation and coordination for the prevention, control and eradication of transboundary animal diseases and zoonoses in ASEAN (ASEAN, 2016b).

3.2.2.4 ASEAN regional animal health information system

This was developed by OIE in cooperation with the ASWGL to share timely information on livestock diseases and improve regional disease control (OIE, 2018). This links to the OIE World Animal Health Information System (OIE, 2021b).

Generally, all the ASEAN health mechanisms are focused on the prevention and control of communicable diseases, emerging infectious diseases and neglected tropical diseases, including zoonotic diseases and the elimination of rabies by 2030.

3.3 Regional One Health Initiatives

In 2019, the WHO, FAO and OIE developed the Tripartite zoonoses guide (WHO *et al.*, 2019). Three operational tools were rolled out to support national staff in developing technical capacities: (i) the multisectoral coordination mechanism; (ii) the joint risk assessment; and (iii) the surveillance and information sharing. These tools can be used independently or in coordinated efforts to support national and intercountry capacity for preparedness and response. The original tripartite collaboration now includes the United Nations Environment Programme (UNEP) (OIE, 2021c).

Rabies prevention and control being a key One Health concern should be among the priority infectious disease and zoonoses that the emerging regional public health risk management system should address. We call on stakeholders to seek the opportunity to plan how best to establish a functional entity within the ACPHEED to deal with rabies prevention, control and eventual elimination.

The One Health approach is particularly relevant to the conduct of joint emerging zoonosis surveillance and continuing surveillance during epidemics, where there could be cycles of spillback and spillover for certain pathogens.

While ASEAN human and veterinary health sectors have collaborated to a certain extent, the desired level of One Health collaboration involving other relevant sectors, such as environment/biodiversity, education and social services sectors, has yet to be achieved. In this regard, ASEAN needs to reform its traditional operational structure by separating sectoral bodies. This is beginning to be realized with the integration of non-health sectors in public health emergency strategies directed at the underlying social, economic, environmental and political determinants of health.

There is currently no high-level ASEAN agreement or declaration promoting cross-sectoral One Health collaborations related to zoonotic disease threats. ASEAN must prioritize this to address risks from zoonoses and other public health threats existing and emerging at the human–animal–ecosystems interface and provide guidance on reducing these risks. For instance, the existing health mechanisms could be effectively utilized to promote: (i) One Health coordination; (ii) capacity building for joint surveillance and outbreak response; (iii) the sharing of data and best practices (e.g. vaccination approaches); (iv) animal and human vaccines security and stockpiling; (v) the operationalization of vaccination drives; (vi) risk communication and public awareness; (vii) the alleviation of socio-economic issues related to increased zoonoses risks; (viii) the harmonization of relevant policies and legal frameworks; (ix) joint strategic planning; and (x) performance monitoring and evaluation. The ASEAN Leaders' Declaration on Antimicrobial Resistance (AMR): Combating AMR through One Health Approach (ASEAN, 2017)can be referred to. The ARES devotes a section to One Health, which reflects broad statements encouraging formal multisectoral engagements driven by a multi-stakeholder committee or high-level coordination mechanism relevant to rabies and other zoonotic diseases. The ASEAN senior officials on forestry have agreed to develop an ASEAN strategy for preventing transmission of zoonotic diseases from the wildlife trade, with the aim of forming a regional One Health group for implementation.

Under the GHSA, Indonesia and Vietnam are contributing countries to the action package on zoonotic disease, which means they commit to building relevant capacity at national, regional and/or global levels.

3.4 Regional One Health Core Elements

One Health collaborations, through cross-sectoral meetings, mechanisms and plans, will need to be organized involving relevant mechanisms under the concerned sectors – principally through the ACCAHZ and ACPHEED. External mechanisms are to be linked to these meetings to include the following One Health bodies and initiatives: (i) the Tripartite/Quadripartite; (ii) South-east Asia One Health University Network; and (iii) proponents of the Preventing the Next Pandemic Initiative, and others. The One Health Stakeholders' Consultations need to deliberate on human-to-animal transmission risk factors and challenges, agree on risk reduction and mitigation objectives, strategies, activities and expectations, and formulate appropriate plans. Countries should present their relevant approaches and capacities: laboratory and field-based capacity, biosafety and biosecurity procedures, pathogen genomics data, etc. Correspondingly, countries are expected to strengthen their animal–human–ecosystem/wildlife interface risk management efforts to deal with the threat of zoonosis spillover and spillback at the source.

The health sector should assume its lead stakeholder role because of the focus on immediate zoonotic threats to public health. Animal health services should support field investigations and laboratory capacities. Further, the agriculture/livestock sector should remain cognizant of the urgent need to manage a zoonosis's impact on food security and livelihoods, including misinformation and potential disruptions in food supply chains.

As specified in the IHR, WHO must support assessments and guide the national action planning on health security and implementation of capacity-strengthening plans. Most countries are expected to have sufficiently fulfilled their obligations. However, based on the recent Joint External Evaluations (JEEs) conducted for each country, many AMS are yet to fully reach scores of 'demonstrated and sustained capacities' in the majority of the IHR core capacities except for Singapore (WHO, 2021). Based on the JEE results, most AMS identified the largest performance gaps in AMR, biosafety and biosecurity, regional preparedness, surveillance, alert/response, risk communication, and capability building. The JEE results signify the urgent need to improve capacities relevant to enhancing multisectoral collaborations to address outbreak risks at the human–animal–environment interface.

3.5 Conclusion

Implementing rabies and other zoonosis control measures must be well prepared and known to the various stakeholders to maximize multisectoral and community participation. Regional mechanisms should develop actionable work plans to ensure readiness for zoonotic disease outbreaks. There should be defined roles in support of the national and regional response. Depending on the ASEAN decisions, they should include support for corresponding national capacity building, which includes workforce development on technical and procedural areas of responsibility, and support for harmonizing SOPs (standard operating procedures), protocols and expertise complementarity between countries. In this regard, the various stakeholders and organizations should be predesignated and assigned their respective roles in the rabies/zoonosis risk management system. In addition, the One Health governance and oversight structure for decision making must be laid out. Sustained mechanisms for regional coordination should allow the member states to support each other.

To ensure an effective response after confirmation of a rabies outbreak, early mitigation will require well-prepared sectoral and multisectoral plans. These would describe joint One Health command and control arrangements, coordination and collaboration, guiding risk communication and public information campaigns, infection prevention and control and other measures to be implemented quickly at all levels, with the involvement of health and non-health sectors, such as law enforcement, border control, etc. Plans need to be regularly tested in different types of simulation exercises and updated regularly. Once an actual outbreak occurs, the information received during the rapid assessment phase would be used to adapt plans for specific situations.

Considering ASEAN's recent adoption of additional policies and arrangements favourable to EID risk management, there is an enhanced opportunity to achieve One Health objectives and the Zero by 30 goal. Stakeholders should seek to consolidate the regional mechanisms for zoonoses prevention and control effectively.

Authors' Declaration

All authors declare that they have no conflict of interest.

All authors have approved this manuscript, agree with its submission, and share collective responsibility and accountability.

This manuscript has not been published or is not under review elsewhere.

References

ASEAN (Association of Southeast Asian Nations) (2006) ASEAN Declaration of the 8th Health Ministers Meeting, ASEAN Unity in Health Emergencies. Yangon, Myanmar. Available at: https://asean.org/wp-content/uploads/images/2012/news/documents/Declaration%20of%20the%208th%20ASEAN%20Health%20Ministers%20Meeting.pdf (accessed 5 December 2022).

ASEAN (Association of Southeast Asian Nations) (2016a) *ASEAN Rabies Elimination Strategy*. OIE World Organisation for Animal Health, OIE Sub-Regional Representation for South-East Asia (OIE SRR-SEA), Bangkok. Available at: https://asean.org/wp-content/uploads/2017/02/ASEAN-Rabies-Elimination-Strategy.pdf (accessed 20 January 2023).

ASEAN (Association of Southeast Asian Nations) (2016b) Agreement on the Establishment of ASEAN Coordinating Centre for Animal Health and Zoonoses. Available at: http://agreement.asean.org/media/download/20161108071810.pdf (accessed 5 December 2022).

ASEAN (Association of Southeast Asian Nations) (2016c) ASEAN Strategic Plan of Action for Cooperation on Livestock (2016–2020). Available at: https://asean.org/wp-content/uploads/2016/10/Strategic-Plan-of-Action-for-the-ASEAN-Cooperation-in-Livestock-2016-2020.pdf (accessed 5 December 2023).

ASEAN (Association of Southeast Asian Nations) (2017) ASEAN Leaders' Declaration on Antimicrobial Resistance (AMR): combating AMR through One Health approach. 31st ASEAN Summit in Manila, Philippines on 13 November 2017. Available at: https://asean.org/asean-leaders-declaration-on-antimicrobial-resistance-amr-combating-amr-through-one-health-approach/ (accessed 20 January 2023).

ASEAN (Association of Southeast Asian Nations) (2018) *ASEAN Post-2015 Health Development Agenda (2016–2020)*. ASEAN Secretariat, Jakarta.

ASEAN (Association of Southeast Asian Nations) (2019a) ASEAN Plus 3 FETN [Field Epidemiology Training Network]. ASEAN Portal for Public Health Emergencies. Available at: www.aseanplus3fetn.net/whoweare.php#strategic (accessed 29 January 2023).

ASEAN (Association of Southeast Asian Nations) (2019b) ASEAN regional strategic framework for veterinary epidemiology capacity development and networking. Available at: http://164.115.45.89/th/file/ghsa2019/05Regional_Strategic_Framework_for_Veterinary.pdf (accessed 5 December 2022).

ASEAN (Association of Southeast Asian Nations) (2020) ASEAN Strategic Framework for Public Health Emergencies. Available at: https://asean.org/asean-strategic-framework-for-public-health-emergencies/ (accessed 5 December 2022).

ASEAN (Association of Southeast Asian Nations) (2021a) ASEAN health sector response to COVID-19. ASEAN Portal for Public Health Emergencies. Available at: ASEAN Emergency Operation Centre (A SEAN EOC) Network - PHE - Public Health Emergencies (aseanphe.com) (accessed January 2023).

ASEAN (Association of Southeast Asian Nations) (2021b) ASEAN Portal for Public Health Emergency Launch. Available at: https://asean.org/asean-portal-for-public-health-emergency-launch/ (accessed 20 January 2023).

ASEAN (Association of Southeast Asian Nations) (2021c) ASEAN BioDiaspora Virtual Centre (ABVC). ASEAN Portal for Public Health Emergencies. Available at: https://aseanphe.com/phe-mechanism/asean-biodiaspora-virtual-centre-abvc/ (accessed 29 January 2023).

ASEAN (Association of Southeast Asian Nations) (2021d) Data & Publications. ASEAN Portal for Public Health Emergencies. Available at: https://aseanphe.com/data-publications/ (accessed 29 January 2023).

ASEAN (Association of Southeast Asian Nations) (2021e) ASEAN Regional Strategic and Action Plan for Vaccine Security and Self-Reliance (AVSSR) 2021–2025. Available at: https://asean.org/wp-content/uploads/2021/10/AHMM-ADOPTED_AVSSR-Strategic-and-Action-Plans-2021-2025-cleaned-version_FINAL.pdf (accessed 20 January 2023).

Fernando, F.M. (2020) ASEAN health sector collective response to address COVID-19 and future public health emergencies & pandemics. Available at: https://aric.adb.org/pdf/rcipod/episode_14/01%20FMFernando_ASEAN.pdf (accessed 29 January 2023).

OIE (World Organization for Animal Health) (2021a) PVS Pathway. Available at: www.oie.int/en/what-we-offer/improving-veterinary-services/pvs-pathway/ (accessed 5 December 2022).

OIE (World Organization for Animal Health) (2021b) World Animal Health Information System. Available at: https://wahis.oie.int/#/home (accessed 5 December 2022).

OIE (World Organisation for Animal Health) (2021c) One Health: the Tripartite Alliance works with UNEP to better integrate the health of ecosystems into Tripartite work. *OIE Bulletin*. Available at: https://oiebulletin.fr/?officiel=08-3-03-2021-1_unep (accessed 5 December 2022).

OIE (World Organization for Animal Health) (2018) Regional animal health information systems. In: *Manual 6: Animal Health Information Systems*. OIE, Paris, p.3. Available at: https://rr-asia.woah.org/wp-content/uploads/2020/02/seacfmd-manual-6.pdf (accessed 29 January 2023).

Philippine Department of Foreign Affairs News (2020) ASEAN Coordinating Council Working Group on Public Health Emergencies holds its Inaugural Meeting. Available at: https://dfa.gov.ph/dfa-news/dfa-releasesupdate/26540-asean-coordinating-council-working-group-on-public-health-emergencies-holds-its-inaugural-meeting?fbclid=IwAR2EgVXfLQdJeuifWwyLW0xeJmsa1dMBUthKac03gDn3kwevjWCMtI8xKfw (accessed 29 January 2023).

Phommasack, B., Jiraphongsa, C., Oo, M.K., Bond, K.C., Phaholyothin, N. *et al.* (2013) Mekong basin disease surveillance (MBDS): a trust-based network. *Emerging Health Threats Journal* 2013(6). DOI: 10.3402/ehtj.v6i0.19944.

WHO (World Health Organization) (2005) *World Health Organization International Health Regulations*, 2nd edn. WHO, Geneva, Switzerland.

WHO (World Health Organization) (2016) *World Health Organization International Health Regulations (2005)*, 3rd edn. WHO, Geneva, Switzerland. Available at: https://www.who.int/publications/i/item/9789241580496 (accessed 29 January 2023).

WHO (World Health Organization) (2017) Asia Pacific Strategy for Emerging Diseases and Public Health Emergencies (APSED III): advancing implementation of the International Health Regulations (2005). License: CC BY-NC-SA 3.0 IGO. World Health Organization Regional Office for the Western Pacific, Manila.

WHO (World Health Organization) (2021) World Health Organization Joint External Evaluation (JEE). Available at: https://extranet.who.int/sph/jee (accessed 5 December 2022).

WHO, FAO and OIE (World Health Organization, Food and Agriculture Organization of the United Nations and Office International des Epizooties) (2018) *Zero by 30: The Global Strategic Plan to End Human Deaths from Dog-Mediated Rabies by 2030*. WHO, Geneva, Switzerland.

WHO, FAO and WOAH (World Health Organization, Food and Agriculture Organization of the United Nations and World Organisation for Animal Health) (2019) Taking a multisectoral one health approach: a tripartite guide to addressing zoonotic diseases in countries. Available at: https://apps.who.int/iris/handle/10665/325620 (accessed 2 November 2022).

4 One Health Approach to Control Canine Rabies in Thailand: The Chiang Mai Model

Wilaiwan Petsophonsakul[1,2]* and Satoshi Inoue[3,4]
[1]Lanna Dog Welfare, Chiang Mai, Thailand; [2]AMS Clinical Service Centre, Chiang Mai University, Chiang Mai, Thailand; [3]National Institute of Infectious Diseases, Tokyo, Japan; [4]The Institute of Social Communication, Meiji University, Tokyo, Japan

Abstract

The Chiang Mai province is a top tourist spot in Thailand so local governments have adopted an intensive policy to eliminate human rabies deaths. The Chiang Mai Model strengthens rabies control efforts by integrating a Buddhist approach to encourage humane compassion towards animals, rather than killing for population management. A major strength of this model is that it uses a scientific approach with a high level of cooperation among government and non-government sectors, academics and temples. This chapter covers the three components of the Chiang Mai Model (animal birth control and rabies vaccination, animal welfare education, and effective enforcement of animal welfare laws), as well as some challenges. The One Health approach combined with religious and scientific applications are essential components of the Chiang Mai Model for humane rabies control. This model, which strongly focuses on the role of community, could guide other Asian countries to overcome challenges in rabies control.

4.1 Introduction

Thailand is an endemic area for rabies, and dogs and cats are at risk of rabies infection. In 2012, the World Health Organization (WHO), World Organization for Animal Health (OIE) and Food and Agriculture Organization of the United Nations (FAO) collectively urged all nations to fight against rabies caused by dogs. The Association of South-east Asian Nations (ASEAN) implemented a regional elimination strategy intending to eliminate human rabies in the region by 2020, before the global strategic plan of Zero by 2030 was set. Accordingly, the Thai government has adopted a guideline and announced a national policy for Thailand's mission to eliminate human rabies by 2030. The Ministry of Health has persuaded local communities in both urban and rural areas to create rabies-free zones by vaccinating animals. They have categorized a rabies control area into three levels:

- **Level A:** No report of human or animal rabies infection from active control or surveillance by laboratory tests for at least 2 years. A rabies-free zone will be announced.

*Corresponding author: wilaiwan_pet@yahoo.com

 One Health for Dog-mediated Rabies Elimination in Asia (eds V. Slack *et al.*)
DOI: 10.1079/9781800622975.0004

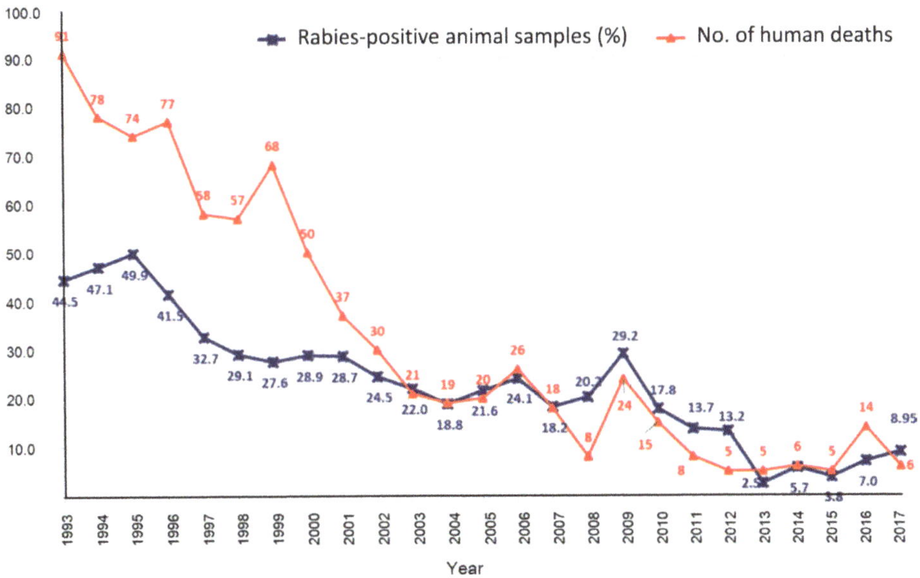

Fig. 4.1. Rabies statistics in Thailand between 1993 and 2017. The red line is the number of human deaths from rabies each year. The blue line is the percentage of rabies-positive animal samples. Image modified and translated into English from the original report: Bureau of General Communicable Disease, 2017, p.3.

- **Level B:** No report of human rabies infection, but rabies has still been found in animals for at least 2 years with active control and surveillance.
- **Level C:** Rabies infection is still found in both humans and animals in the area.

In 1992, the Rabies Prevention Act was announced and used nationwide. Since then, rabies cases have gradually declined each year. According to the policy issued by the Ministry of Public Health, starting in 2002, all provinces should have had no rabies-related deaths. The Bureau of Health and Strategy, Ministry of Public Health, stated that although the number of human deaths from rabies has declined dramatically throughout Thailand, the number of post-exposure prophylaxis cases had actually increased annually between 1995 and 2002 (from 153,483 to 401,181, respectively). Therefore, considering only the decrease in the number of rabies cases might not be sufficient (Fig. 4.1). Nevertheless, several factors such as the cost of post-exposure treatment, which is related to the incidence of rabies cases, must be considered.

Rabies re-emergence in Thailand derived from the budget control's bureaucratic problem in 2015. Thus, rabies vaccination was interrupted nationwide. Because rabies can be controlled by mass vaccination of dogs and cats, reducing it caused the re-emergence of the disease in many provinces. The rabies outbreak reached its peak with 18 human deaths in 2018, and a large percentage of 1,475 animal samples were rabies positive, mainly from dogs but also from some cats and livestock such as cows and buffaloes (Thai Rabies Net, 2018; Thanapongtharm *et al.*, 2021). Accordingly, Professor Dr Her Royal Highness Princess Chulabhorn Mahidol of Thailand has passionately promoted the mass vaccination and management of dogs and enhanced awareness of rabies with the project 'Animals Free of Rabies: Humans Are Safe from the Disease'. Her Royal Highness' involvement has helped to engage and motivate the public to prevent rabies, from the village to the province to the national level.

The Princess is a champion of efforts to make Thailand rabies free by 2020, in line with the worldwide initiative to end human rabies deaths by 2030 (WHO, 2017).

Chiang Mai, in northern Thailand, is considered the second most important area in the country, after the capital, Bangkok. Chiang Mai has an area of 20,107 km^2 and a total population of 1.78 million, 0.96 million of which live in the urban centre (http://chiangmai.nso.go.th/ (accessed 29 September 2021)). The city is surrounded by high mountain ranges, with various ethnic groups living in the highland area; the ethnic groups are linked to a wide system of mountain ranges in neighbouring Myanmar and Laos. The highland geography and ethnic diversity make it challenging to ensure rabies control via mass vaccination, undertake dog population control, monitor dog migration along the border, and introduce rabies education programmes.

Chiang Mai province is one of the top tourist spots in Thailand, and tourism is one of the most significant sources of income in Chiang Mai. The Thai government has focused on building the country's wealth via the tourism industry. A tourist satisfaction survey revealed that many visitors cited sick, free-roaming dogs as a negative factor in Chiang Mai. As such, improvements should be seriously considered to protect the health and welfare of people and dogs. In the previous decade, roaming canines were considered a nuisance and were often poisoned, kept in cages or sold off to the meat market; this occurred in many of the towns and villages of Thailand, creating the wrong impression for locals and tourists alike. The idea of turning Chiang Mai into a dog management model – a compassionate city, where humans and dogs can live together in harmony – has slowly been introduced. The community and local authorities have been informed that periodically culling free-roaming dogs does not solve the rabies problem, nor does it create a healthy environment for Chiang Mai. Retrospective reports from several counties have shown that culling dogs does not reduce the number of dog bites or eliminate rabies (Taylor *et al.*, 2017). Killing without neutering means that the dog population continues to increase; new groups of free-roaming dogs always migrate into areas where local dog populations have been killed and rapidly begin to repopulate the area. This increase in the movement of dogs also increases the possibility of disease transmissions, such as rabies or other zoonotic diseases. Lanna Dog Welfare has been working in conjunction with the Office of Health Development, and the Provincial Livestock Office has been able to identify the problems during 2003–2008 as follows:

1. Although Chiang Mai is one of the world's top tourist destinations, the city lacks governmental unity regarding animal health and welfare.
2. The city did not have a government-run veterinary clinic to provide pet care for lower-income residents.
3. Communities solved the free-roaming dog problem by killing, poisoning or abandoning dogs at temples and public places, but this approach only moves the problem from one location to another.
4. The government sectors and communities believed that if they could physically eliminate dogs, all other problems would disappear.
5. The cost of spaying/neutering was US$30–100 per animal, more than pet owners could afford.
6. When dogs have puppies, often the owners could not afford to take care of the animals and then abandoned them in the community or temples, thus creating a burden for all concerned.
7. Free-roaming dogs were difficult to catch in order to provide rabies vaccination and spay/neuter.
8. Free-roaming dogs were often in bad health, and unspayed/unneutered dogs fought with each other when in heat. This led to abuse of the dogs by people and problems such as dog bites.

The Chiang Mai Model was initiated when Lanna Dog Welfare began a project in 2009 with the National Institute of Infectious Diseases, Tokyo, Japan. Since Dr Satoshi Inoue's first visit to Chiang Mai regarding rabies control, there has been progressive cooperation with local government sectors such as the Chiang Mai Provincial Livestock Office, sub-district municipalities and Chiang Mai University Hospital. The One Health approach of the Chiang Mai Model has employed monitoring rabies through scientific measurements (assays) and has involved more stakeholders. Their goals have been to:

1. Encourage every sector to prevent and control rabies.
2. Encourage rabies vaccination in all dogs.
3. Control ownerless/free-roaming dog populations.
4. Be active and continuously monitor and control rabies in humans and animals.
5. Educate and bring awareness about rabies prevention to the community.
6. Enforce laws involving rabies.

Based on information released by the Bureau of Epidemiology in Chiang Mai, one person was infected with rabies in 2005 and two people in 2006. In 2006, Chiang Mai was one of the ten leading provinces in terms of the morbidity rate for rabies. Since the implementation of the rabies control policy, there have been no human rabies cases; however, rabies was still found in dogs in 2009 and 2011. For nearly 10 years, the city had neither human nor dog rabies cases until recently. However, in 2020 and 2021, there were three cases and one case of canine rabies, respectively (Thai Rabies Net, 2021). Therefore, the rabies virus appears to exist in the area, and its re-emergence is dependent on rabies control management.

In this study, we summarize the rabies management practices implemented in Chiang Mai from 2003 to 2021, specifically how the One Health approach has been used to improve the efficiency of rabies control towards the goal of zero rabies cases by 2030. Initially, most animal organizations and non-governmental organizations (NGOs) focused on the sterilization of free-roaming dogs at temples or within local communities, whereas local government sectors mainly concentrated on rabies vaccination. Vaccination alone cannot control rabies, so there has been intense campaigning for humane dog population control; however, culling did not reduce the dog population or prevent rabies infection. The Chiang Mai Model has followed WHO and the International Companion Animal Management Coalition (2019) guidelines for controlling rabies and dog population management by using the Animal Birth Control (ABC) and the Anti-rabies Vaccination (ARV) programmes. The programme in which free-roaming dog populations are managed via sterilization, and then returned to their home area, is considered the most effective way to control rabies and population growth in the Chiang Mai community. In addition, education and animal welfare laws must be implemented into the guidelines to establish a sustainable solution. We consider the following challenges of eliminating dog-mediated human rabies in Chiang Mai.

4.2 Challenge I: Enhancing Local Community Measures via a Competitive Award for Rabies Education

Rabies control awards have been given to sub-districts after field evaluations by rabies experts from Thailand and Japan. The competitive award for rabies education was initiated in 2005, when a local animal organization, Lanna Dog Welfare, collaborated with provincial livestock organizations to initiate a Rabies Zero Day when a rabies management competition was held between sub-districts. This event continued to grow to the provincial level, and local communities were slowly made aware of rabies control. In accordance with the national policy issued by the Ministry of Public Health, local governments adopted intensive policies to achieve a rabies-free zone; at the time, the desired timeline was by 2020, but it is currently set for 2030 (Lanna Dog Welfare, 2012). Rabies control was used to approach local communities to introduce animal welfare education. A self-made 'Chiang Mai Model', including the humane management of the dog population and rabies control programmes, was initiated. The success of this model was demonstrated by two sub-districts winning first (Mae Hia) and second (San Klang) places in the national competition (the MoPH MoAC Sanofi Pasteur Rabies Award, currently named the Thailand Rabies Awards) in 2010 and 2013, respectively (Office of Regional Livestock 5, 2011; *The Medical News*, 2014).

Competitions run by the government have been restricted and the number of prizes awarded is limited, such that several local authorities that had performed outstanding work have not received acknowledgement.

Additionally, different conceptions of management and issues have made judging the best performance difficult. Therefore, this contest has been improved by assessing the projects based on quality to allow for improvement within the local authority, rather than as a competition with other local authorities. In cooperation with the World Society for the Protection of Animals (WSPA) (UK), Lanna Dog Welfare (Thailand), the Japan Society of Clinical Study for Rabies (Japan) and the National Institute of Infectious Diseases (Japan), the Tokyo-Chiang Mai conference was held to celebrate World Rabies Day in 2014 (Lanna Dog Welfare, 2014a). Tokyo-Chiang Mai World Rabies Day targeted policy makers and the heads of government sectors in the communities. Rabies experts from Thailand and Japan, which is a rabies-free model country, assessed their management practices and awarded prizes after visiting each community area, providing consultations based on social or scientific points of view and assessing each community's work from three different perspectives: (i) project management; (ii) animal welfare; and (iii) project sustainability. Rabies evaluation feedback was provided in written form; therefore, it encouraged the districts to improve – for example, by allowing them to analyse their weaknesses and strengths. The pilot conference was a great success, with six sub-districts participating. By using this approach, local authorities improved and maintained the quality of their work; they were delighted and hoped to hold this conference annually. Moreover, the awards given by the Chiang Mai governors provided excellent publicity, as 250 sub-districts became aware of the Chiang Mai Model for rabies control, allowing it to achieve an international reputation.

4.3 Challenge II: Enhancing Animal Birth Control/Anti-Rabies Vaccination via a Joint Project Between Academia, Non-Profit Organizations and the Communities

Dog population management is an integral part of rabies control programmes. Contraceptives via hormone injection in female dogs have been widely used for dog population control because this approach is inexpensive and no veterinary personnel are needed, especially in suburban and rural areas. However, this raised welfare concerns regarding the side effect of uterine infection or pyometra after withdrawal. Spaying/neutering has slowly been introduced where livestock or veterinary personnel were available. Nevertheless, spaying/neutering has to be improved to ensure it is safe for animals. Mobile sterilization clinics and rabies vaccination have been performed by following the standard international protocol. The team includes personnel from public health, the veterinary sector (e.g. veterinary students and livestock veterinarians) as well as volunteers from various fields. Between 2006 and 2007, the WSPA (currently World Animal Protection) supported Lanna Dog Welfare as one of its member societies, particularly regarding humane dog population management, by promoting capacity building and by providing technical advice. The activities supported by WSPA included two workshops on dog sterilization in 2006 and 2007: a clinical technique update and a review of international standards for efficiency.

The workshops aimed to train government department veterinarians in humane dog population control principles and to improve sterilization techniques. Support from the local mayor's office, the Buddhist community and government connections enabled the workshops to proceed, in anticipation of 'Anti-rabies Month'. Later, a mobile sterilization clinic was developed, instrumental in humane dog population control and animal welfare education. The mobile sterilization programme involves several organizations, including the veterinary faculty's staff and students, livestock and animal husbandry veterinarians, village livestock volunteers and local community staff.

In 2014, paperless, real-time mobile sterilization clinic registration software was developed, including pre-registration (GARC, 2018). The pre-registration programme was designed to facilitate recording patient history (dogs or cats) in the same manner as humans and to pre-calculate anaesthesia drug doses; this is safer for both the animals and the management personnel. This pre-registration programme has been improved and optimized based on trials in different communities. The registration software architecture includes a hypertext preprocessor (PHP) web-based

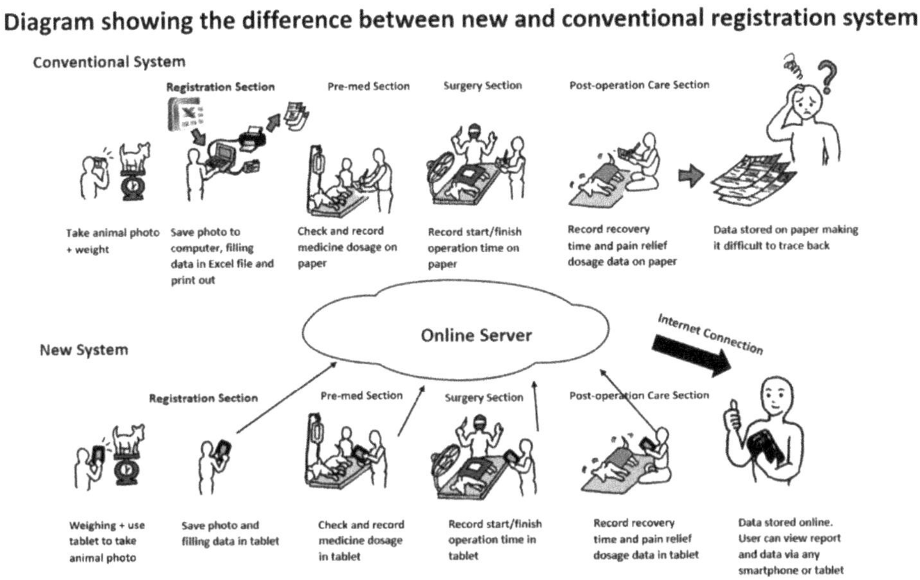

Fig. 4.2. Diagram of a mobile sterilization clinic showing the new real-time computerized pre-registration system compared with the conventional paper-based system. Image created by W. Petsophonsakul and S. Inoue.

application that runs on a remote server; users can access this application via an Internet connection. The necessary equipment comprises a 17.8 cm (7 inch) Android tablet for recording data at each station and a wireless router with a 3G Internet package for an Internet connection. A remote database server and host are required to store data.

During sterilization, patient history records must be linked to each animal at each station. In the original implementation, each station produced paper records, which included a photograph, weight and health check, pre-medical anaesthesia, surgery, aftercare, recovery and discharge. This approach ensured all records would be transferred from station to station. Finally, all paper records would be saved as a computer file. In 2015, pre-registration was computerized and extended to all field sterilization stations, as shown in Fig. 4.2. There were no paper records because everything was computerized in real time on the tablets at each station. From their station, each user could track data from another station because patient status and data were saved on a server via an Internet connection. All data were saved and analysed to determine the most economical strategy (e.g. the cost of each drug, the prevalence of common diseases and rabies vaccination). This paperless and real-time pre-registration system was the first of its kind established for animals in Thailand. The system uses more advanced technology than that utilized by humans, for which no mobile clinics use real-time registration.

4.4 Challenge III: Education Through Workshops and Training

Workshops on animal welfare and rabies vaccination techniques have been presented to various stakeholder groups, including monks, military members and village and livestock volunteers. Without education, dog population control cannot be implemented successfully. In 2010, the 'Fight for Five Freedoms for All Animals' was introduced during the first World Animal Day in Chiang Mai, a collaboration between NGOs and local government sectors (Lanna Dog Welfare, 2010). The event increased public awareness of cruelty to animals and focused on the benefits of having animals in one's daily life. Through the event, the public understood that animals are not only pets but also an essential part of the ecology of our natural environment. The event emphasized that animals should be treated with respect and that they are not here for us, but with us. A new campaign was introduced for free-roaming dogs, to promote them as community or university members. This campaign later became a well-known Chiang Mai University project: 'Ma-CMU' or 'Chiang Mai University's dogs'. Because dogs transmit over 90% of human rabies cases, efforts should be focused on dog management. Different populations of free-roaming dogs (i.e. temple, community and forest dogs) require different approaches.

In Thailand, temples act as traditional shelters for abandoned dogs, with hundreds of dogs finding shelter in Buddhist temples each year. Dedicated monks feed and care for these dogs every day. The construction of new shelters or pounds can be avoided by helping temples to sterilize and treat sick and neglected animals. Therefore, the project aimed to integrate free-roaming animals into society. This, in turn, would allow the government to save money by implementing sustainable and cost-effective methods to solve free-roaming dog-related problems. Animal organizations (i.e. Lanna Dog Welfare, Thailand; World Animal Protection, UK; and Blue Tail Animal Aid International, France) collaborated to improve the welfare of free-roaming dogs by providing knowledge and animal welfare and handling training to Buddhist monks in several temples in Chiang Mai province between 2014 and 2016 (Lanna Dog Welfare, 2015). The training programme was developed in accordance with Lanna Dog Welfare's experience working with various temples in Chiang Mai and was conducted as follows: (i) practising approaching, handling and managing wounds on dummies; (ii) administering medication and record keeping; (iii) animal health and diet; and (iv) rabies vaccination.

In addition to helping temples by training monks, training for rabies vaccination and animal welfare handling were also expanded to local sub-districts. Participants included livestock and health volunteers from the villages of Sobtia, Suthep and Muanglen sub-districts; mass rabies vaccinations were performed by appointed village volunteers (Blue Tail Animal Aid, 2014; Lanna Dog Welfare, 2014b). The programme included basic information and training in animal welfare, including how to approach animals safely, the management and storage of rabies vaccines and practising injection techniques on artificial skin. This training was followed by in-house visits to practise efficient vaccinations.

Several military camps in the territory are located in forests, where animals are frequently abandoned. This leads to issues related to environmental sanitation, dog bites and the spread of zoonotic diseases, particularly rabies. Therefore, a free-roaming dog management programme was initiated at the 33rd Military Circle, which included instruction on humane handling and catching methods. The training and instructions for soldiers included dog behaviour, dog catching methods (including the pros and cons of each method), anaesthesia and monitoring of dogs, wound management, nursing care and rabies vaccination. This dog management training programme has been expanded to other military camps.

Nationally conserved forests are attractive to ecological tourists. Pet dogs are often abandoned in these areas. This situation raises several problems, including increased dog populations, tourists being bitten and the possible spread of several zoonoses, especially rabies, between other wildlife and humans. A management model has been developed to capture forest dogs in national conservation areas. A humane catching method (e.g. wire net pen) has been implemented at the edge of the Mae Kuang Dam forest. By providing food in the pen, forest dogs become familiar with it. Then, when the dogs enter the pen, they are captured. Several dog packs were captured, and 104 dogs were spayed or neutered, vaccinated against rabies and then returned to their home area. A real-time animal database has been developed and used to locate the forest dogs in each area. Forest dogs are scattered around the Mae Kuang Dam in various regions, as shown in Fig. 4.3 (a and b).

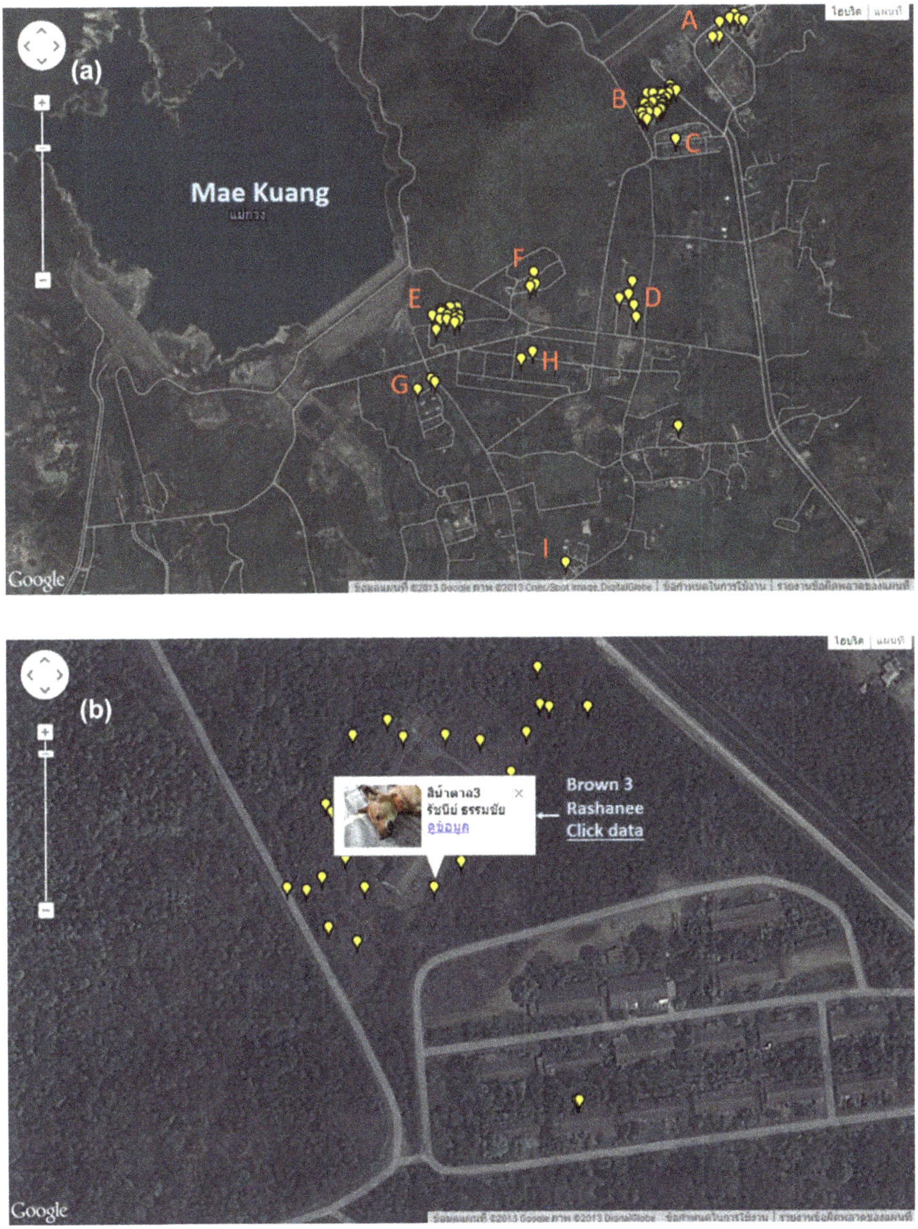

Fig. 4.3. Geographic mapping of forest dog packs in different areas of the Mae Kuang Dam. (a) Letters A–I refer to each pack of dogs. (b) Each yellow spot represents one dog linked to a patient's history. After sterilization, each dog was photographed and wore a collar indicating it had been sterilized. Images created by W. Petsophonsakul and S. Inoue.

4.5 Challenge IV: Evaluating Rabies Vaccination in High-Risk Exposure Groups and Community Dogs

Regarding the rabies-free zone, the local communities and government understand that rabies vaccination covers 70% of animals according to WHO rabies criteria. Many local reports demonstrated that 100% of animals were vaccinated. In 2008, Lanna Dog Welfare collected data on 1000 animals that had been vaccinated against rabies by its mobile sterilization clinic in Chiang Mai city. An average of 39% (0–79%) of dogs and cats had been vaccinated, and three sub-districts showed nearly 80% of animals had been vaccinated. These three sub-districts underscore the potential of working with both active surveillance and healthcare. Several annual reports indicated that 70–80% effective rabies vaccination from central Thailand (i.e. Bangkok and the surrounding areas) should be sufficient for controlling rabies. However, human rabies cases were still found, so this effectiveness should be reconsidered.

In 2011, a case of rabies in a puppy imported from Bangkok to Chiang Mai led to the initiation of the One Health approach by the Chiang Mai Provincial Livestock Office, which called for an urgent meeting with the local governments, public health office, municipality, academia and animal NGOs (Petsophonsakul *et al.*, 2013). According to the rabies prevention guidelines, the puppy rescuer and shelter were examined for their provision of rabies vaccination. Rapid rabies tests help monitor risky areas and are crucial for making medical/veterinary decisions. During surveillance, many shelter dogs did not have protective antibodies, even though they had been vaccinated for rabies after their rescue. Because of the imported rabies case in Chiang Mai in 2011, all sectors learned about the benefits of a rabies vaccination programme and monitoring protective antibody levels to enable action against an outbreak. Follow-up rapid tests for protective antibodies make it easier to manage an outbreak while improving animal care and welfare. Moreover, determining the levels of protective antibodies allows stakeholders to assess the quality of vaccines and vaccination skills. Deficiencies in these areas could be addressed to improve protective rabies immunity in dogs. Therefore, the remaining issues in the rabies control zone in Chiang Mai are to examine the vaccination status to ensure dogs carry protective antibodies.

Regarding mass rabies vaccination campaigns covering ≥70% of the dog population, the vaccination programme in rabies control areas was evaluated to ensure that dogs had adequate immunity. Canine sera were collected from four sub-districts (Sobtia, Nongjom, Nongpakrang and Nonghoi) and were examined for protective anti-rabies antibodies by the rapid fluorescent foci inhibition test (RFFIT). This test was kindly performed by the National Institute of Infectious Diseases, Japan, 3 and 12 months after rabies vaccination. After 3 months, neutralizing antibody levels in the four areas were 4.90 ± 6.68 IU/ml, and 70.8% of the samples had adequate antibody titres (> 0.5 IU/ml). The average antibody levels in Sobtia, Nongpakrang, Nongjom and Nonghoi were 3.38, 5.71, 0.76 and 9.70 IU/ml, respectively. After 12 months, only 48.9% of the sera collected had adequate antibody titres, with a mean neutralizing antibody level of 2.25 ± 3.27 IU/ml; the average antibody levels in Sobtia, Nongpakrang, Nongjom and Nonghoi were 1.40, 1.58, 4.09 and 3.38 IU/ml, respectively.

As mentioned previously, pet dogs are often abandoned around the Mae Kuang Dam, where they become de-domesticated and form packs. A total of 75 forest dogs were caught from different packs, of which 65 were monitored for anti-rabies antibodies; 47.7% of those observed had protective antibodies (> 0.5 IU/ml). The average antibody level was 8.34 ± 26.11 IU/ml (< 0.02–191.18 IU/ml). Less than half of the forest dogs had protective immunity against rabies; this failed to achieve the targeted vaccination coverage of ≥70% (Petsophonsakul *et al.*, 2019). The immunity of forest dogs can act as a buffer zone for rabies control along the borders between the forest and the city. As such, in addition to domestic pets and free-roaming dogs in the city, forest dogs should not be neglected. The four rabid dog cases that re-emerged in 2020 and 2021 were in Chiang Dao and Wiang Haeng, two rural sub-districts that share a border with Myanmar.

In a notable case study of the One Health approach, Lanna Dog Welfare in collaboration with Chiang Mai University and several other stakeholders helped investigate a 14-month-old child who had been bitten by a pack of free-roaming dogs in Doi Saket, Chiang Mai. The case was managed in two ways: (i) laboratory tests, including collecting sera from two free-roaming dogs and determining the sero-titre by using a cost-effective easy competitive enzyme-linked immunosorbent assay (CEE-cELISA) (Aronthippaitoon *et al.*, 2019, p. 99); and (ii) a One Health seminar held at the AMS (Associated Medical Sciences) Clinical Service Centre, with participants from the government sector, including the chief of the district, livestock officers, a community leader, a hospital director, expert clinicians, an NGO and the press. The One Health seminar had 35 participants from different sectors involved in rabies control. Over 40 free-roaming dogs were captured, sterilized and moved to a local NGO shelter. Although newspaper reports were inaccurate, they monitored and followed the case effectively. The patient received adequate treatment in the form of post-exposure prophylaxis for rabies. Antibodies to the rabies vaccine were monitored in two dogs: one had a level of 2.7 EU/ml (where EU is ELISA unit), whereas the other had undetectable levels. All participants learned more about the One Health approach. In addition to the involvement of livestock officials in the management of free-roaming dogs and clinicians for patient treatment, this case engaged the community, an NGO and mass media, all working together to solve the problem. This case study offers a nationwide model of the One Health approach applied to managing dog bites and rabies control (AMS CMU, 2018).

4.6 Challenge V: Developing an ELISA Suitable for Mass Monitoring of Protective Immunity Against the Rabies Virus

Quantifying anti-rabies antibodies is necessary to determine immune status and ensure that sufficient protective antibodies have been produced to prevent and treat after rabies virus exposure. RFFIT is the gold standard for measuring anti-rabies antibodies in serum samples; however, it is expensive, time-consuming and cumbersome to run samples. As such, only a few institutes have the RFFIT test. The CEE-cELISA (Cost Effective Easy competitive ELISA) is a cost-effective and easy-to-use test to monitor protective immunity, thus ensuring the efficacy of rabies immunization and seroprevalence in vaccinated dogs (Aronthippaitoon *et al.*, 2019). Sero-titres measured by CEE-cELISA and RFFIT show a very strong correlation, with *R*-values of 0.958 and 0.931 in humans and dogs, respectively. CEE-cELISA can detect rabies antibodies in human sera and other animal species within the same test.

CEE-cELISA has been used to evaluate mass rabies vaccination in veterinary students who followed an annual rabies pre-exposure prophylaxis (PrEP) vaccination scheme at Chiang Mai University (Aronthippaitoon, Thananchai, Thongkorn, Samer, Surajinda, Inoue, Noguchi, Park, Kawai, Petsophonsakul, 2020, unpublished data). It has also been used to evaluate the immune response to the rabies vaccine in village health volunteers in two sub-districts, Maeka and Doi Lor. These individuals are at a high risk of exposure because they provide rabies vaccinations to dogs and cats in their villages during the rabies campaign period from March to May each year. The antibody levels were determined in village volunteers who had received the current rabies PrEP recommended by WHO in 2018 (two intradermal doses provided 7 days apart). With this economical strategy, 92% of the vaccinated individuals had an adequate antibody titre (> 0.5 EU/ml) with a median of 12.4 EU/ml (Khamduang *et al.*, 2022). Thus, the new WHO recommendation is effective for pre-exposure vaccination. In addition to vaccinating village health volunteers, future studies will evaluate a single shot or a smaller vaccine dose, which may produce a sufficient immune response. A one-shot protocol for rabies vaccination may be implemented in children in high-endemic areas; it is a simple, short protocol that would save money while being convenient for children, especially those who are underprivileged or live in rural areas.

4.7 Conclusion

Based on what has been accomplished over the past 18 years (2003–2021), the Chiang Mai Model for rabies control has provided numerous long-term welfare benefits to the local communities. The model uses the One Health approach, which includes engagement with all stakeholders and incorporates their beliefs and traditional religious way of life. The importance of implementing this model in other communities/countries involves comprehensive identification of problems in the area, the use of humane methods of control and maintaining long-term sustainable strategies. Many factors facilitate successful rabies management when implementing the Chiang Mai Model in other communities. These include: (i) providing knowledge and information to local authorities that rabies control is not merely rabies vaccination in dogs, but combines several measures; (ii) organizing and running training or workshops by inviting international experts with interpreters; (iii) operating more scientific data and research to support the management; (iv) collaborating with global institutions to exchange information and to gain experience; (v) integrating new activities/ideas within the routine or local activities to reduce costs; and (vi) promoting academic or research institutes to engage with the community or to perform grass-roots research and, vice versa, to encourage the community to participate in scientific activities. Of note, it is necessary to regularly reinforce appropriate long-term rabies control practices and humane animal treatment.

We will continue to follow the lead of the Global Alliance for Rabies Control (GARC) and encourage groups to work together to help eliminate rabies. The World Rabies Day Awards encourages us to continue our work on the Chiang Mai Model. This action involves three key projects. First, we will use gamification to educate underprivileged and ethnic children – who have restricted access to information and language barriers – about rabies. A survey of the media environment for school children to develop a media education system for rabies prevention has been performed (Oda *et al.*, 2018). Second, we will continue to develop our real-time animal database application for dog registration and rabies control. Dogs can be surveyed and registered on smartphones. This helps monitor the dog population and enables the administration of rabies vaccinations in the community. Third, we will develop an easy-to-use immunochromatography test kit, which will provide semi-quantitative anti-rabies antibody detection to evaluate mass rabies vaccination in humans and animals. The test kit will help assess the immune response in the field without sending samples to a laboratory.

Our model using the One Health approach, which includes all stakeholders, has demonstrated success in two main areas: (i) animal welfare; and (ii) the tourism industry. First, the model consists of an education and information service for animal owners across a wide variety of age groups regarding the routine care and welfare of domestic animals. We have witnessed remote communities increase their understanding of the needs of animals, from simple feeding programmes to methods of carriage and learning that animals have feelings. These small lessons are understood and passed down over the years. Second, the tourism industry suffers from free-roaming dogs; on many occasions, visitors feel that our communities are uncaring about these animals. It is refreshing to hear tourists mention that they have noticed fewer free-roaming dogs on subsequent visits to Chiang Mai. We truly believe our free-roaming dog problem is being reduced via our sterilization efforts and the corresponding reduction in numbers, preventing exponential growth.

The Chiang Mai Model has been very successful. Indeed, in 2018 Lanna Dog Welfare won the World Rabies Day Awards in the regional category for the Asian Rabies Control Network (ARACON) Global Alliance for Rabies 2018 (GARC, 2018). This award was a great honour for Lanna Dog Welfare, a non-profit animal organization based in Chiang Mai, and for the city of Chiang Mai and the country of Thailand. The award encourages all sectors to continue working on the Chiang Mai Model for Rabies Control.

Finally, we believe that 'Animal Welfare is Human Welfare', and this will lead to a sustainable solution in accordance with 'Animals Free of Rabies: Humans Are Safe from the Disease' project, which was initiated in 2016 under the wish of Her Royal Highness Princess Chulabhorn Mahidol, who is a champion of using the One Health approach to eliminate rabies in both Thailand and South-east Asia (WHO, 2017).

Acknowledgements

This work has been possible thanks to the collaboration among: (i) non-profit organizations (Lanna Dog Welfare, Thailand; World Animal Protection, UK; Royal Society for the Prevention of Cruelty to Animals, UK; Blue Tail Animal Aid International, France; Thai Society for the Prevention of Cruelty to Animals, Thailand); (ii) Thai government bodies (Chiang Mai Provincial Livestock office; Veterinary Research and Development Centre, Upper Northern Region); (iii) Thai academia (Faculty of Medicine, Chiang Mai University; Faculty of Veterinary Medicine, Chiang Mai University; Faculty of Associated Medical Sciences, Chiang Mai University; Faculty of Associated Medical Sciences, Khon Kaen University; Queen Saovabha Memorial Institute, Thai Red Cross Society); (iv) international government bodies (Department of Veterinary Sciences, National Institute of Infectious Diseases, Japan); (v) international academia (Institute of Social Communication, Meiji University, Japan; School of Information and Communication, Meiji University, Japan); and (vi) local government bodies and monastery institutions (the sub-districts of Sobtia, Nong Pa Krang, Nong Hoi, Nongjom, Mae Hia, San Klang, Suthep, Luang Nuer and Muanglen; the Venerable Abbot of Chiang Mai province; the Venerable Abbot of Nonghoi sub-district).

This study was partially supported by the Research Programme on Emerging and Re-emerging Infectious Diseases of the Japan, Agency for Medical Research and Development (grant numbers 16fk0108311j1903 and 17fk0108122j2201) and Grants-in-Aid for Scientific Research (Pioneering/Exploratory) (grant number 19K21787).

Authors' Declaration

All authors declare they have no conflict of interest.

All authors have approved this manuscript, agree with its submission and share collective responsibility and accountability.

This manuscript has not been published or is not under review elsewhere.

References

AMS CMU (Faculty of Associated Medical Sciences, Chiang Mai University) (2018) One health: a case study of a child bitten by stray dogs in Doi Saket, Chiang Mai. Available at: www.ams.cmu.ac.th/main/web/index.php/site/article?id=1546 (accessed 29 September 2021).

Aronthippaitoon, Y., Samer, W., Atuntee, T., Thananchai, K., Thongkorn, K. *et al.* (2019) A cost-effective easy competitive enzyme-linked immunosorbent assay suitable for monitoring protective immunity against the rabies virus in the serum of humans and dogs. *Japanese Journal of Infectious Diseases* 72(2), 99–105. DOI: 10.7883/yoken.JJID.2018.248.

Blue Tail Animal Aid (2014) Rabies vaccination for stray animals. Lanna Dog Welfare and Blue Tail Association YouTube video. Available at: www.youtube.com/watch?v=cJ9LyjC5EsY (accessed 29 September 2021).

Bureau of General Communicable Disease (2017) Guideline for rabies control. Available at: https://ddc.moph.go.th/dcd/journal_detail.php?publish=13755 (accessed 23 January 2023).

GARC (Global Alliance for Rabies Control) (2018) Animal data management for field sterilization by Lanna Dog Welfare. Available at: https://rabiesalliance.org/world-rabies-day/nominee/lanna-dog-welfare (accessed 26 September 2021).

International Companion Animal Management Coalition (2019) Humane dog population management guidance. Available at: www.icam-coalition.org/wp-content/uploads/2019/09/2019-ICAM-DPM-guidance-Interactive-updated-15-Oct-2019.pdf (accessed 26 September 2021).

Khamduang, W., Hongjaisee, S., Pothita, T., Kongyai, N., Surajinda, S. *et al.* (2022) Antibody response to rabies pre-exposure vaccination among village health volunteers in a northern region of Thailand. *Preventive Veterinary Medicine* 199, 1–5. DOI: 10.1016/j.prevetmed.2021.105555.

Lanna Dog Welfare (2010) Chiang Mai World Animal Day 2010. Available at: https://cmwad.wordpress.com/ (accessed 26 September 2021).

Lanna Dog Welfare (2012) Chiang Mai model towards a rabies free zone. Available at: www.lannadog.net/blog/?p=1424 (accessed 26 September 2021).

Lanna Dog Welfare (2014a) Tokyo-Chiang Mai World Rabies Day announcement. Available at: www.lannadog.net/blog/?p=1852 (accessed 26 September 2021).

Lanna Dog Welfare (2014b) Training on rabies vaccination and animal welfare handling. Available at: www.lannadog.net/blog/?p=1839 (accessed 29 September 2021).

Lanna Dog Welfare (2015) Completion of the 3rd monk training program on animal welfare. Available at: www.lannadog.net/blog/?p=2094 (accessed 29 September 2021).

Oda, M., Hirokawa, M., Na Bang Chang, P., Petsophonsakul, W. and Aronthippaitoon, Y. (2018) The questionnaire survey of media environment and information behavior of the students at primary and secondary schools in Chiang Mai Province, Thailand. In: *Presentation at the International Association for Media and Communication Research (IAMCR)*, Eugene, Oregon, June 20–24.

Office of Regional Livestock 5 (2011) MoPH MoAC Sanofi Pasteru Rabies Awards. Available at: http://region5.dld.go.th/webnew/index.php/en/news-menu/2019-08-07-09-22-48/160-moph-moac-sanofi-pasteur-rabies-awards (accessed 26 September 2021).

Petsophonsakul, W., Sripanya, T., Khawplod, P., Kheuanrat, W., Phornwisedsirikun, S. *et al.* (2013) Learning from a case of imported rabies to establish a rabies control zone. In: *Presentation at the 4th International Meeting on Emerging Diseases and Surveillance*, Vienna, Austria, February 15–18.

Petsophonsakul, W., Aronthippaitoon, Y., Atuntee, T., Jaisuda, S., Yodgomleo, A. *et al.* (2019) Monitoring of anti-rabies antibody in sera of stray dogs living at borders between forest and city. In: *Presentation at the 4th International Conference on One Medicine One Science*, Chiang Mai, Thailand, February 11–14.

Taylor, L.H., Wallace, R.M., Balaram, D., Lindenmayer, J.M., Eckery, D.C. *et al.* (2017) The role of dog population management in rabies elimination – A review of current approaches and future opportunities. *Frontiers in Veterinary Science* 4, 109. DOI: 10.3389/fvets.2017.00109.

Thai Rabies Net (2018). Available at: www.thairabies.net/trn/report/rpt040301?Year=2018 (accessed 26 September 2021).

Thai Rabies Net (2021). Available at: http://www.thairabies.net/trn/report/rpt040401?Year=2020&AllRegionIds=5&AllProvinceIds=38 and http://www.thairabies.net/trn/report/rpt040201 (accessed 26 September 2021).

Thanapongtharm, W., Suwanpakdee, S., Chumkaeo, A., Gilbert, M. and Wiratsudakul, A. (2021) Current characteristics of animal rabies cases in Thailand and relevant risk factors identified by a spatial modeling approach. *PLoS Neglected Tropical Diseases* 15(12), e0009980. DOI: 10.1371/journal.pntd.0009980.

The Medical News (2014) MoPH-MoAC. Rabies Awards 2013. *The Medical News* 16(414), 17–19. Available at: www.wongkarnpat.com/upfilepat/Medicalnews%20414.pdf (accessed 1 February 2023).

WHO (World Health Organization) (2017) Thailand committed to defeating Human Rabies by 2020. Available at: www.who.int/news/item/04-09-2017-thailand-committed-to-defeating-human-rabies-by-2020 (accessed 29 September 2021).

5 The 'World Café': Strengthening Rabies Prevention with the Government–Academia Collaboration in Japan

Chiho Kaneko[1], Kentaro Yoshii[2,3], Yumi Kirino[1], Shintaro Kobayashi[2], Genki Arikawa[4], Akitoyo Hotta[5], Naoaki Misawa[1,6] and Satoshi Inoue[1,5]*

[1]Center for Animal Disease Control, University of Miyazaki, Miyazaki, Japan; [2]Hokkaido University, Sapporo, Japan; [3]National Research Center for the Control and Prevention of Infectious Diseases, Nagasaki University, Nagasaki, Japan; [4]Public Welfare and Health Department, Miyazaki Prefecture, Miyazaki, Japan; [5]National Institute of Infectious Diseases, Tokyo, Japan; [6]University of Miyazaki, Miyazaki, Japan

Abstract

The Japanese national government strengthened its rabies action plan in accordance with three guidelines: (i) for risk management and rabies detection through a reporting system; (ii) for the road map after an outbreak; and (iii) for animal surveillance. The government has held a national rabies training course, including a hands-on nationwide programme of laboratory diagnosis and workshops, since 2015, targeting local government veterinary public health officers. This chapter discusses a participatory interactive workshop, the 'World Café', which was introduced to the training course. The key points of the World Café are its participatory approach, hence small group conversations and discussion. The workshops changed participants' perceptions effectively and increased their awareness for sustained preparedness for a rabies outbreak. The World Café targets idea exchange to generate breakthroughs in sustainable rabies control. Participatory, practical rabies training courses in Japan, a rabies-free country, could serve as a model for rabies control and prevention through public education and awareness for Zero by 30 worldwide.

5.1 Introduction

No rabies cases in humans or any other animals have been recorded in Japan since 1957, except for four human cases imported from endemic countries (in 1970, 2006 and 2020) (Nosaki *et al.*, 2021). Japan achieved the dog-rabies-free status by enforcing the Rabies Prevention Law enacted in 1950 by the Ministry of Health, Labour and Welfare (MHLW) (Anonymous, 1968). This law specifically stipulates, among other things, owners' registration of dogs, vaccination by veterinarians, detention of stray dogs, the administrative response at the scene of rabies outbreaks, provisions for rabies prevention officers who are veterinary public health officers and the import and export quarantine of dogs at the Animal Quarantine Station of the Ministry of Agriculture, Forestry and Fisheries.

*Corresponding author: sinoue@niid.go.jp

 One Health for Dog-mediated Rabies Elimination in Asia (eds V. Slack *et al.*)
DOI: 10.1079/9781800622975.0005

However, a large number of animals, including wild animals other than dogs, had been imported into Japan from rabies-endemic countries without quarantine. In particular, many raccoons imported from the USA as pets multiplied in the wild after escaping from breeding facilities that had been abandoned since the 1970s. In 1999, the Rabies Prevention Law was amended to include raccoons, cats, skunks and foxes in addition to dogs (Takahashi-Omoe *et al.*, 2008; Inoue *et al.*, 2018).

Around that time, emerging and re-emerging infectious diseases, many of which are zoonotic, occurred as a result of globalization in the latter half of the 20th century (Smith *et al.*, 2007; Aguirre, 2017; Allen *et al.*, 2017). Human infectious disease control in Japan was also reviewed and revised in April 1999 as a part of the MHLW's new Infectious Diseases Control Law (IDCL) (NIID, 1999; Nomura *et al.*, 2003). Under the new law, national epidemiological surveillance was established to reinforce and expand the existing surveillance system, and the law also introduced a new infectious disease classification system that divided diseases into new categories from I to IV. Rabies was classified as Type IV, and infectious diseases in this category must be reported within 7 days to the nearest health centre, along with other information from the physician who diagnosed the patient. Following a review of the IDCL, a revised law came into force in 2003, the main amendment being strengthened measures against zoonoses. In particular, the law established a system for notifying animal importation and strengthening measures against zoonoses, including rabies (Nakajima, 2005).

Along with changes in the Rabies Prevention Law, the MHLW developed Rabies Guidelines 2001 – a guide for dealing with a suspected rabies outbreak – (MHLW of Japan, 2001; Takahashi-Omoe *et al.*, 2008) to ensure that medical and veterinary facilities report rabies cases to local and national public health departments. Laboratory confirmation was also reinforced through the network between the local prefectural institutes of health and the National Institute of Infectious Diseases (NIID). In addition, each municipality prepared an administrative response manual based on the city's guidelines for suspected rabies cases. The initial response to two cases in humans in 2006 was to follow the guidelines and confirm rabies with the government. This led to a surge in the demand for post-exposure prophylaxis (PEP) for returnees from rabies-endemic areas, even in domestic dog-bite cases, and the depletion of shareable rabies vaccines in the rabies-free country became a major problem. The first two imported cases of rabies in 36 years prompted local authorities to strengthen their systems following the rabies guidelines, and in conjunction with the development of the rabies response manual, each local authority began to carry out desktop drills to simulate an outbreak. These activities had a significant ripple effect.

In 2013, MHLW developed the new Rabies Guidelines 2013 – a supplement to the Rabies Guidelines 2001 on Rabies Countermeasures – (MHLW of Japan, 2013), to enable local authorities across Japan to prepare for and respond to a possible outbreak in case an animal is suspected of having rabies and is detected as positive. For the preparation of these new guidelines, lessons were learnt from the experience of countries such as France with dealing with the (illegal) entry of rabid animals (Mailles *et al.*, 2011). In Taiwan, which had been considered a rabies-free area for half a century, the discovery of rabies in wild animals (ferret-badgers) (Huang *et al.*, 2015; Shih *et al.*, 2018) and the isolation of a new species of *Lyssavirus* from bats (Hu *et al.*, 2018), which had not been reported in Asia, demonstrated the need for wildlife surveillance even in rabies-free countries. In Japan, this led to the development of the Guidelines for Animal Rabies Survey (MHLW of Japan, 2014).

Since 2015, the MHLW has held the national rabies training course targeting veterinary public health officers, and the workshop is based on three guidelines to promote rabies testing and rabies control systems in local governments (Fig. 5.1). Training courses have been held in the Hokkaido, Tohoku, Hokuriku, Kanto, Chubu, Kinki, Chugoku, Shikoku and Kyushu-Okinawa regions. Advanced municipalities have developed their own manuals

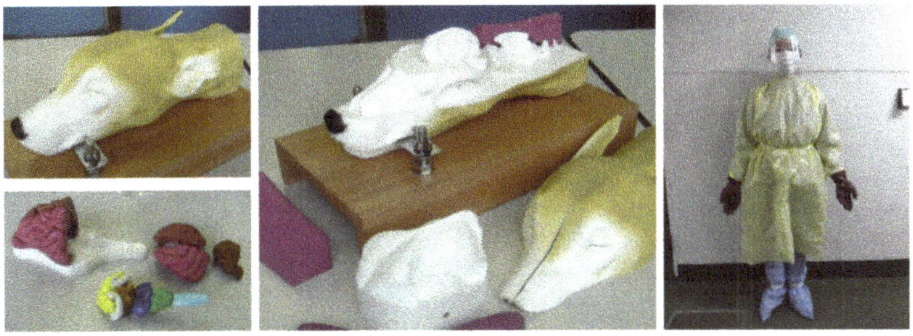

- Outline of program

I. LECTURE
   a. Overview and topic of rabies in the world
   b. Current status and issues in the promotion of rabies control in Japan
   c. Dissection facilities and biosafety, testing equipment and methods required for rabies diagnosis in animals

II. PRACTICE
   a. Hands-on training in safe animal dissection using practical dissection models
   b. Determination of a positive result using standard rabies testing methods

III. DISCUSSION
   a. Exchange of opinions among local governments on the current status and issues of rabies measures based on the guidelines
   b. World Café-style workshop to clarify and strengthen practical approaches and preparedness for rabies outbreaks

Fig. 5.1. Overview of the national rabies training course. Photograph courtesy of S. Inoue.

for dealing with rabies outbreaks within their municipalities, in line with the three national guidelines. In addition, some of these municipalities have conducted their own technical training and tabletop exercises in case of a rabies outbreak.

To change perceptions among participating local governments and raise awareness of the need for ongoing efforts and preparedness in the event of a rabies outbreak (Yamada *et al.*, 2019), the World Café was implemented. This aims to have participatory workshops as a government–academia collaboration to strengthen participants' perceptions and raise awareness of the need for continuous efforts and preparedness in the event of a rabies outbreak. The workshops posed straightforward questions to stimulate ideas, encourage focused exploration and derive ideal conditions and new possibilities. This workshop method was also used in classes at undergraduate and graduate schools of veterinary medicine as a part of public health education.

5.2 Introducing a Participatory Workshop, the World Café, in the National Rabies Training Course

5.2.1 The aim of introducing the World Café targeting veterinary public health officers

The aim of introducing the World Café into the national rabies training course was to share experiences, lessons learned and challenges faced, and for participants from different municipalities to discuss solution strategies in a participatory workshop. All local governments were supposed to establish their practical action plans against rabies following the three national guidelines: (i) for risk management and rabies detection through a reporting system (MHLW of Japan, 2001); (ii) for the road map after an outbreak (MHLW of Japan, 2013); and (iii) for the surveillance of animals, including wildlife (MHLW of Japan, 2014). However, the progress of the preparation varied between municipalities. Some municipalities have established their own manuals for contingent outbreaks of rabies, and some held ongoing technical training for diagnosis and tabletop exercises for urgent responses to outbreaks. Meanwhile, others had difficulty in even developing the first steps of preparation, such as deciding where and how to perform post-mortems of animals suspected of having rabies as well as who should perform these post-mortems. The administrative officers in these municipalities did not sufficiently recognize the importance of preparedness for rabies outbreaks, and hence, no one had made even preliminary efforts to identify relevant authorities and stakeholders.

Therefore, the training course had to be designed to prompt public health veterinary officers to take the first steps towards rabies preparation by developing their awareness of the need even in this rabies-free country, which included efforts to create active networks among various players in their municipalities. For this purpose, we employed a participatory workshop called the World Café rather than a conventional lecture. This was because a workshop without spontaneous thinking seemed to be insufficient for changing awareness and ways of thinking about how to prepare for rabies before it occurs. Moreover, veterinary public health officers in neighbourhood municipalities having conversations and getting to know each other would all contribute to creating self-generated information exchange networks among municipalities that could work in an emergency. This attempt was introduced based on collaboration between the government – the Ministry and local governments – and academia (universities). Applying a novel educational method for maximizing teaching effectiveness contributed by universities' experienced educational efforts made it possible to share the latest scientific information on rabies and apply those novel ideas, knowledge and techniques to administrative operations in the local governments under the coordination of the Ministry.

5.2.2 The procedure of the World Café

The World Café is a participatory, interactive workshop that was developed by chance by Brown and Isaacs in 1995 (Brown *et al.*, 1999). It is now recognized as a powerful conversational process that enables groups to exchange ideas, share awareness and learn in relaxed, comfortable, café-style environments (Tan and Brown, 2005). Detailed information and its application are available on the World Café website (http://www.theworldcafe.com/ (accessed 2 February 2023)) and in various references (Steier *et al.*, 2015; Löhr *et al.*, 2020). The World Café consists of five components: (i) setting; (ii) welcome and introduction; (iii) small-group rounds; (iv) questions; and (v) harvest.

Our World Café discussion in the national rabies training course involved the following components: (i) setting: a special environment was created by providing small tables for group conversations, A0-size paper, coloured pens and pads of sticky notes for each table; (ii) welcome and introduction: the World Café began with a warm and casual atmosphere, including providing coffee, tea and snacks; (iii) small-group rounds: participants were organized into small groups of four or five persons; (iv) questions: specific subjects for discussion were presented in three rounds for 20–25 min each. After the first round, the host remained at the table, but each group member visited a different table. In the second round, mixed groups from other tables discussed either the first-round subject (though different ones were possible); mixing the tables was known as 'cross-pollination'. In the third round, the second-round group members returned to their original tables and exchanged what they had discussed in the second round to promote further cross-pollination. The final step was (v) harvest: this entailed all participants sharing insights and other outputs from their conversations using the ideas they had sketched out on paper (Fig. 5.2).

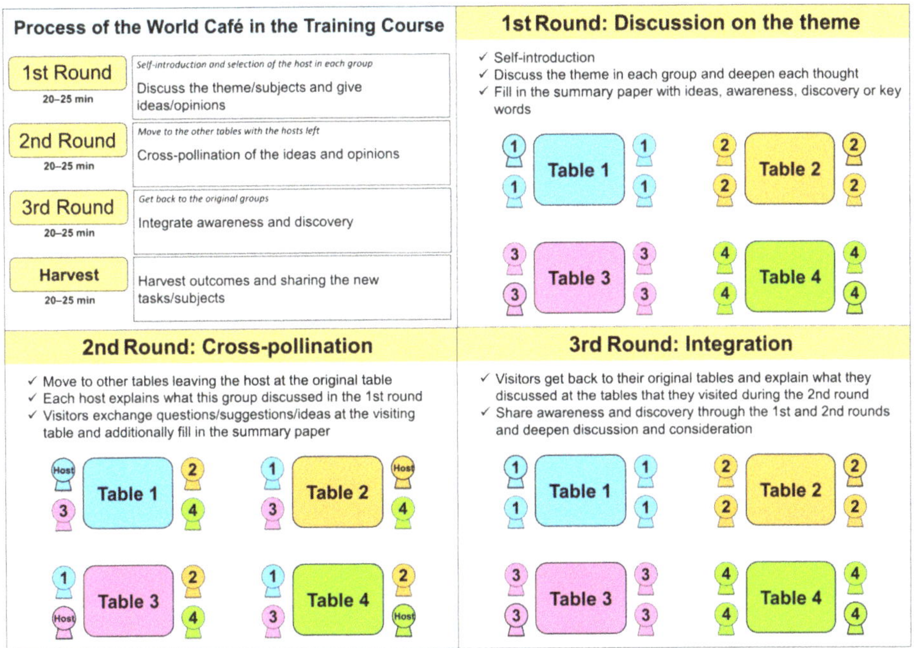

Fig. 5.2. Process of the World Café in the training course. Simili paper is a Japanese imitation parchment paper. Created by C. Kaneko, referring to Katori and Okawa, 2017.

5.3 The World Café at the National Rabies Training Course in the Kyushu-Okinawa Region

5.3.1 What was done

In the Kyushu-Okinawa region, the training course has been held annually in collaboration between Miyazaki prefecture and the University of Miyazaki using facilities at the Department of Veterinary Sciences, Faculty of Agriculture, University of Miyazaki. Approximately 20 participants join in total every year from various municipalities in the Kyushu-Okinawa region.

In the training course held between 2015 and 2017, designated the first phase, the World Café aimed to develop in participants an awareness of the need to be prepared, particularly to establish a system for detecting and diagnosing animals in Japan that are suspected of carrying rabies. The same World Café was also introduced in the public health curriculum for undergraduate veterinary students (described in Section 5.4).

The subject of the World Café in the first phase was 'Discuss the flow of countermeasures from the detection of a rabies-suspected case, survey, initial response and diagnosis when a dog-bite case in a human by a stray dog occurs in imaginary town A'. In the first round of the World Café, each group discussed the above mentioned subject having received some background information: (i) an owned dog was bitten by a stray dog and brought to the animal hospital in town A; (ii) a cattle farmer asked a veterinarian to see a calf that had been bitten by a stray dog in town A; and (iii) there have been increasing reports of dead raccoons on the road or near houses in town A.

In the second round, the shuffled members in each group shared what they had discussed in the first round. Additionally, they discussed 'What kinds of countermeasures should be taken when a new bite case caused by a wild raccoon occurs before a diagnosis of the above mentioned stray dog?'. In the third round, the participants returned to their original groups and shared what they had discussed during the second round. In addition, each group discussed the 'Flow of the countermeasures that should be taken after a definite diagnosis of the stray dog and the wild raccoon as rabies positive'. Finally, during the harvest, the participants in each group shared what they had learnt and exchanged during the session.

During the training course that began in 2019, the second phase, the World Café aimed to make participants aware of and reflect on the need for facilities that meet biosafety criteria for conducting post-mortems of suspected rabies-infected animals. To achieve this, we followed this process:

1. Participants were given a homework assignment before the training course date to depict floor plans of facilities to be used for post-mortem and laboratory diagnosis in the event of receiving suspected rabies-infected animals.
2. On the training course day, the participants were grouped so that each group contained members from different municipalities.
3. Then, each group was given one of the floor plans from the homework assignments completed by someone who was not in that same group.
4. All groups discussed how they could conduct post-mortem and laboratory diagnoses by observing biosafety protocols with each facility floor plan.
5. After discussion, each group shared what they had discussed.
6. Each participant confirmed their way of conducting diagnosis under limited circumstances and received ideas for improving their designs to ensure biosafety.

The subject of the World Café in the second phase of training was 'Discuss how you perform post-mortem and laboratory diagnosis in the existing facility in each municipality'. In the first round of the training course, each group discussed how they could conduct post-mortem and laboratory diagnoses in each existing facility using their assigned floor plans. This included drawing paths of travel on the floor plans to indicate where individuals would walk to transport a suspected rabies-infected animal into the laboratory and to dispose of the animal. In the second round, the shuffled group members shared what they had discussed in the first round and made revisions to the floor plans. In the third round, the participants returned to their original groups and once again shared what they had discussed in the second round, and then, we harvested reflections from the entire group.

5.3.2 Outcomes and tasks identified through the World Café

The World Café participatory workshops were highly successful for veterinary public health officers to share ideas and experiences and for heightening their awareness of the need to strengthen preparedness for a crisis management response to a rabies outbreak (including suspected cases) in Japan, where no rabies case has been reported for more than half a century. The World Café discussion style promoted a spontaneous exchange of experiences, including challenges (see Fig. 5.3). Furthermore, the workshops cultivated a sense of ownership in the participants, who were veterinary public health officers working on the front line of rabies detection and diagnosis.

In the World Café discussion in the first phase, participants discussed the need to develop a network system for detecting rabies in a suspected rabies-infected animal and for the flow of diagnosis and an emergency contact system between the relevant authorities. The group members discussed which facility should be responsible for capturing the suspected rabies-infected animal, observing the animal for 10–14 days, conducting the post-mortem, performing laboratory diagnoses and so on. The participants also discussed: (i) which department of which organization should contact victims of dog bites and collect epidemiological information; (ii) which department of which organization should be involved if the suspected rabies-infected animal is wild and has bitten pet animals and livestock; and (iii) which organizations should be networked and should take each emergency countermeasure once a suspected case is diagnosed as rabies positive.

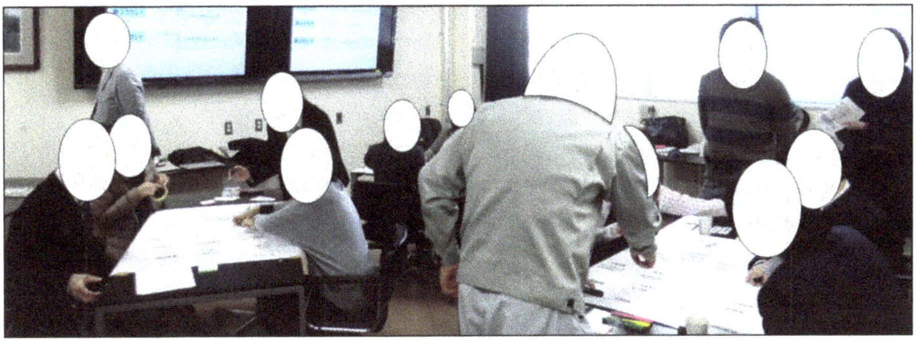

Fig. 5.3. The World Café discussion. Image courtesy of C. Kaneko.

After the training course, the participants gave their impressions of the World Café workshop discussion methodology. Comments included:

- 'It was valuable to understand the systems for rabies control and the progress of preparedness for rabies outbreaks in other municipalities through the World Café.'
- 'This activity was interesting because the content was concrete and specific according to the actual routine work in the organization that we each belong to.'
- 'This activity highlighted tasks to be undertaken in our municipality for the construction of countermeasures against a rabies outbreak and also shared the current situation and progress in other municipalities.'
- 'I became aware of how I (we) should cope with the construction and operation of a network/diagnosis system for rabies outbreaks in animals as an officer in the central government office in the municipality.'
- 'It was helpful to be assigned tasks because we usually do not have the opportunity to simulate cases of a rabies outbreak. I would introduce similar activities for municipalities into a rabies training course that is organized by our prefecture.'

In the second phase of World Café, participants shared their awareness of how they should perform a post-mortem and a laboratory diagnosis for a suspected rabies-infected animal by observing biosafety in the existing facility. The participants discussed: (i) where and how they should bring a suspected rabies-infected animal into the facility without encountering general clerks or visitors to the facility; (ii) where a post-mortem should be performed without contaminating the clean areas where other staff members worked; (iii) appropriate zoning for biosafety; (iv) how a post-mortem should be performed under BSL-2 (Biosafety Level 2) conditions in existing facilities; and (v) what kinds of materials would be needed in the room for a post-mortem, etc. The groups again discussed the movement lines for travel, from bringing a suspected rabies-infected animal for observation for 10–14 days to conducting the post-mortem, including diagnostic testing of the brain tissue.

After three rounds of discussion, each group presented what they had discussed. In particular, the representative participants of each facility whose floor plan was discussed added comments, including future facility development plans. Finally, rabies and biosafety specialists added their comments and advice, particularly regarding biosafety. After the training, the participants again shared their comments on the World Café process. For example: 'It was very interesting to have the opportunity to consider the established movement lines for post-mortem and laboratory diagnosis in each facility in each municipality.'

In our activities, the World Café successfully promoted the participating veterinary public health officers' awareness of the importance of beginning to prepare rabies countermeasures and gave them a sense of ownership that they were on the front line of rabies control. The World Café participatory workshop style encouraged participants' spontaneous thinking and exchange of ideas and experiences.

5.4 The World Café in Undergraduate Education

5.4.1 Current rabies education in veterinary medical schools in Japan

In Japan, rabies is an important topic in core curricula in undergraduate veterinary public health education. The goal of the course is to understand the characteristics of rabies, including its transmission cycle, clinical symptoms in humans and animals and the vaccine and treatment, as well as to learn skills for diagnosis in practice. The students study the relevant rabies laws: the Rabies Prevention Act and the Act on the Prevention of Infectious Diseases and Medical Care for Patients with Infectious Diseases. In the practical skills component of veterinary public health education, students should be trained to detect rabies virus antigens in the brains of animals suspected of infection. However, because the rabies virus is highly regulated by law in Japan, only a few veterinary universities possess live rabies virus and conduct practice for diagnosis.

Today, veterinary public health officers are taking on increasing importance in controlling zoonosis, including the rabies virus in Japan but, as mentioned above, many lack experience with rabies because no domestic cases have been reported in Japan for more than 60 years. In conjunction with the limited practice for rabies diagnosis, undergraduate students studying veterinary public health have difficulty simulating rabies control measures. To improve these conditions, the School of Veterinary Medicine, Hokkaido University, adopted the World Café discussion format to simulate rabies control measures used by veterinary public health officers in several local governments in cooperation with NIID.

5.4.2 World Café discussion at the School of Veterinary Medicine

A World Café was held with approximately 40 students in their fourth year who were taking a veterinary public health practice each year from 2017 to 2019 in the School of Veterinary Medicine, Hokkaido University. They received lectures and practical training on rabies diagnosis in advance and participated in six groups of six or seven students per group.

In the first round of discussions, the following scenario was presented to the students (Fig. 5.4a):

> You work for a local government as veterinary public health officers in Hokkaido, which is located on the northernmost point of Japan. You were informed that a stray dog had bitten a resident. You captured the dog that was behaving aggressively and seemed to walk unsteadily. How will you cope with this issue?

The students discussed the first steps, such as information gathering and diagnosis.

After 15 min of discussion, the groups were shuffled, and the second-round groups were given this additional information:

> In recent days, there have been several reports that companion dogs were bitten by a stray dog and taken to animal hospitals; a veterinarian reported a case that an animal had bitten a dairy farmer's calf, and several raccoon carcasses were detected near the area.

After another 15 min discussion, the participants returned to their original groups and were told that the stray dog and dead raccoons had been diagnosed with rabies infections. After confirming the suspected cases as definitive, the students discussed the changes and additional actions. They also considered administrative responses. During the discussion, each group made a poster and gave a presentation based on it (Fig. 5.4b), and all participants evaluated each presentation. After the World Café discussion, each student reconsidered the subject and submitted a report.

Because students were not familiar with the World Café discussion, they tended to show confusion and passive attitudes during the discussions at first, but soon, the discussion became animated with their curiosity and interest. In the first round, many students had the preconception that the stray dog had been infected with the rabies virus, and they did not discuss the possibility of a common bite case. In particular, the students did not consider epidemiological situations specific to Japan, such as how the rabies virus was introduced into the area. The participants' discussions focused mainly on responses to the captured dog and

(a) Subject

1st Round: Scenario

- You work for a local government as veterinary public health officers in Hokkaido. You were informed that a stray dog had bitten a resident. You captured the dog that was behaving aggressively and seemed to walk unsteadily.
  <u>How will you cope with this issue?</u>

2nd Round: Additional information

- In recent days, there have been several reports that companion dogs were bitten by a stray dog and taken to animal hospitals.
- A veterinarian reported a case that a dairy farmer's calf had been bitten by an animal.
- Several raccoon carcasses were detected near the area.

3rd Round: Definitive information

- The stray dog and dead raccoons had been diagnosed with rabies infections.

(b) Poster

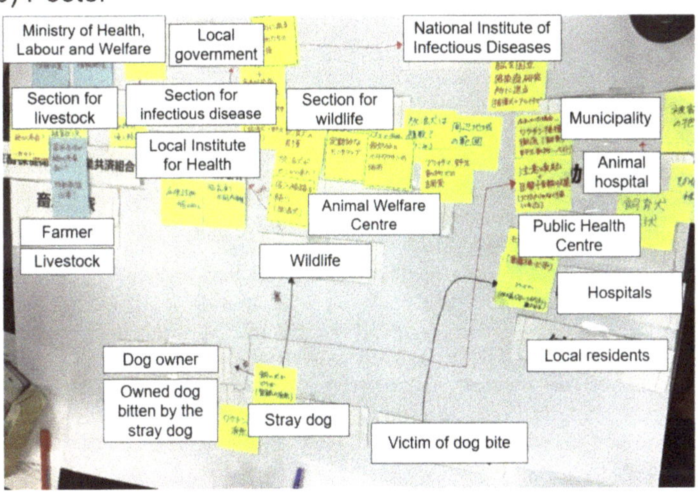

Fig. 5.4. World Café interactive workshop used in undergraduate education at the School of Veterinary Medicine, Hokkaido University, Japan. (a) Subject given for the World Café discussion; and (b) poster made by a group during the discussion. Photographs courtesy of K. Yoshii.

bitten resident that had been presented in the lectures and practical training in advance.

The students' discussions became more sophisticated after we presented the additional information in the second round. They began to consider the background information and understand the serious epidemiological situations in which the rabies virus might be widely endemic in Hokkaido, Japan. They discussed the surveillance to clarify the epidemiological situation of the rabies virus circulation in the area: where and what kinds of animals were infected with the rabies virus, and how the virus had been introduced and circulated in the area. After the students had received the definitive information in the third round, they discussed administrative responses to control the situation, including surveillance, disclosure of information, protective measures against rabies virus infection and control of rabies-infected animals. Then, they summarized their discussion and gave presentations.

The students' major discussion points were as follows. The first was how the rabies virus had been introduced into Japan after it had been eradicated more than 60 years ago. Students suggested that an infected dog had arrived on a foreign trading ship without quarantine and escaped into the suburban area. Next, they discussed how the virus could circulate in the area. Many wild animals in Hokkaido, such as foxes and raccoons, can be natural hosts of the rabies virus, and if infected animals enter the wild, an endemic focus could be established. Furthermore, because wild animals often stray into human living spaces and dairy industry areas, humans, companion animals and domestic animals could be exposed to the virus. Finally, the students discussed what should be done to control endemic rabies, and they determined that the following should be the requirements: (i) establishing a surveillance system; (ii) a large-scale vaccination campaign for humans and companion dogs; (iii) preparing PEP; and (iv) actions against infected wild animals, including stray dogs. All of these are the fundamental responses required for rabies control.

Compared with the veterinary public health officers, the undergraduate students lacked knowledge and experience in actual public health veterinary works, including relevant laws, administrative procedures and posts in local government. Therefore, their discussions tended to be rough and unrealistic, especially given limited budgets and hands, but they did give open-ended, novel ideas beyond preconceived understandings. The World Café small groups stimulated individual discussion and communication, and mixing the groups allowed all students to share ideas they had heard in the first groups. On a questionnaire regarding the World Café, more than 95% of the students answered that the discussions had increased their understanding of the importance of administrative responses for rabies control and that they took more interest in veterinary public health.

5.4.3 Outputs of the World Café discussion

Based on the 3 years of our implementation of the World Café discussion, we attempted to share our results to improve veterinary public health education. First, we conducted a World Café discussion with undergraduate students and veterinary public health officers in the Hokkaido area as postgraduate education. In 2019, the first postgraduate education programme in rabies control was held for veterinary public health officers who worked in the local Hokkaido government; the programme was nearly the same as that held in other prefectures, as described above. Students who attended a veterinary public health practice class participated in the programme, including the World Café discussion. The discussions with veterinary public health officers supplied them with practical viewpoints that they could not have acquired in class, especially actions based on laws, organizations and budgets. In turn, the students' opinions stimulated discussion among the veterinary public health officers because they represented the views of amateurs and the public. The interactions between the undergraduate students, as an undergraduate group, and veterinary public health officers, as a postgraduate group, led to more sophisticated discussions than would have been possible with two separate groups.

To disseminate the effectiveness of the World Café discussions in veterinary public health education, we presented our efforts at the Japanese Society of Veterinary Science annual meeting in 2019. After introducing the recent national guidelines for rabies surveillance, we presented our activities in the undergraduate education programme at Hokkaido University and in the postgraduate education programme for veterinary public health officers in the Kyushu-Okinawa region. After the presentation, several veterinary universities communicated with us to ask us to introduce the World Café participatory workshop methodology into their curricula. Because many veterinary public health faculties have experienced the difficulties that we discussed above, they were very interested in this new style of education.

Recently, undergraduate veterinary education has been calling for more practical education. Each university has introduced cooperation with public animal shelter centres and collaborations with faculties of veterinary medicine overseas. In combination with this practical training, World Café discussions can be effective for undergraduate education in veterinary public health.

5.5 Conclusion

Our attempt to introduce a participatory interactive workshop, the World Café, was successful in developing a sense of ownership and an awareness of the need to prepare countermeasures against rabies in Japan among veterinary public health officers. We also attributed this success to the government–academia collaboration. This method was also fruitful for educating undergraduate veterinary medicine students of public health on proper administrative actions and workflows in response to a rabies outbreak. The World Café can be applied to any subject, such as veterinary medicine, human medicine, public health and education.

Rabies control is sometimes hampered by a lack of coordination within and between departments in charge. The officials in charge and persons concerned should be prompted to apprise themselves in their respective situations to have a localized solution for effective and sustainable rabies control. The World Café will work as a tool to support such scenarios. Interactive conversation at various levels, such as conversations within a single department in charge, mixing multidisciplinary departments and beyond relevant organizations, will help share many-faceted views and promote the generation of novel solutions for achieving the common goal, Zero by 30. Many references are available to learn how to organize a World Café. Any topic can be a theme/subject and function well in a World Café. A more tangible theme/subject would be recommended in promoting rabies control rather than an abstract one. However, it can be arranged according to a real-life situation in an application. Besides, the facilitator will be expected to be neutral and non-judgemental to bring out various ideas and perceptions of participants. Participants will have to listen seriously to others' opinions even if they do not agree.

In this interactive conversation, the World Café successfully raised the rabies diagnosis/prevention framework standard in officials' minds and departments through our activities. This will be applicable to strengthen and promote measures to achieve Zero by 30 worldwide.

Acknowledgements

The Ministry of Health, Labour and Welfare (MHLW) zoonosis section of the Tuberculosis and Infectious Diseases Control Division in the Health Service Bureau has been promoting rabies control and prevention in Japan, and we are grateful for the opportunity to contribute to strengthening rabies countermeasures in local governments through government–academia collaboration. This work was supported by the Cooperative Research Program of CADIC, University of Miyazaki (CADIC, CRP2016).

Authors' Declaration

All authors declare that they have no conflict of interest.

All authors have approved this manuscript, agree with its submission, and share collective responsibility and accountability.

This manuscript has not been published or is not under review elsewhere.

References

Aguirre, A.A. (2017) Changing patterns of emerging zoonotic diseases in wildlife, domestic animals, and humans linked to biodiversity loss and globalization. *ILAR Journal* 58(3), 315–318. DOI: 10.1093/ilar/ilx035.

Allen, T., Murray, K.A., Zambrana-Torrelio, C., Morse, S.S., Rondinini, C. *et al.* (2017) Global hotspots and correlates of emerging zoonotic diseases. *Nature Communications* 8(1), 1124. DOI: 10.1038/s41467-017-00923-8.

Anonymous (1968) Japan. The rabies prevention law and regulations. Law No.247 of 1950. *Bulletin – Office International Des Épizooties* 69(3), 535–585.

Brown, J., Isaacs, D., Margulies, N. and Warhaftig, G. (1999) The World Café: catalyzing large-scale collective learning. *Leverage: News and Ideas for the Organizational Learner* 33, 1–2.

Hu, S.-C., Hsu, C.-L., Lee, M.-S., Tu, Y.-C., Chang, J.-C. *et al.* (2018) *Lyssavirus* in Japanese Pipistrelle, Taiwan. *Emerging Infectious Diseases* 24(4), 782–785. DOI: 10.3201/eid2404.171696.

Huang, A.S.-E., Chen, W.-C., Huang, W.-T., Huang, S.-T., Lo, Y.-C. *et al.* (2015) Public health responses to reemergence of animal rabies, Taiwan, July 16–December 28, 2013. *PLOS ONE* 10(7), e0132160. DOI: 10.1371/journal.pone.0132160.

Inoue, S., Kaneko, C., Yoshii, K., Hotta, A. and Fei, A.C.-Y. (2018) Current issues of rabies prevention in Japan. *Journal of Veterinary Epidemiology* 22(1), 8–14. DOI: 10.2743/jve.22.8.

Katori, K. and Okawa, K. (2017) [*Let's Do a World Café. New Edition: Conversations Connect, Worlds Connect.*] *World Café Wo Yarou. Shinpan: Kaiwa Ga Tsunagari, Sekai Ga Tsunagaru* [in Japanese]. Nikkei Business Publications, Tokyo.

Löhr, K., Weinhardt, M. and Sieber, S. (2020) The "World Café" as a participatory method for collecting qualitative data. *International Journal of Qualitative Methods* 19, 160940692091697. DOI: 10.1177/1609406920916976.

Mailles, A., Boisseleau, D., Dacheux, L., Michalewiscz, C., Gloaguen, C. *et al.* (2011) Rabid dog illegally imported to France from Morocco, August 2011. *Euro Surveillance* 16(33), 19946. DOI: 10.2807/ese.16.33.19946-en.

MHLW (Ministry of Health, Labour and Welfare) of Japan (2001) The rabies guidelines 2001 [in Japanese]. Available at: https://www.mhlw.go.jp/bunya/kenkou/kekkaku-kansenshou18/pdf/05-01.pdf (accessed 24 October 2022).

MHLW (Ministry of Health, Labour and Welfare) of Japan (2013) The rabies guidelines 2013 [in Japanese]. Available at: https://www.mhlw.go.jp/bunya/kenkou/kekkaku-kansenshou18/pdf/guideline2013.pdf (accessed 24 October 2022).

MHLW (Ministry of Health, Labour and Welfare) of Japan (2014) The guidelines for animal rabies survey [in Japanese]. Available at: https://www.mhlw.go.jp/bunya/kenkou/kekkaku-kansenshou10/dl/140826-01.pdf (accessed 24 October 2022).

Nakajima, K. (2005) Enhanced zoonotic infectious disease control measures subsequent to the revision of the infectious diseases control law. *Infectious Agents Surveillance Report* 26(8), 196–198. Available at: http://idsc.nih.go.jp/iasr/26/306/de3062.pdf (accessed 5 December 2022).

NIID (National Institute of Infectious Diseases) (1999) The national epidemiological surveillance of infectious diseases in compliance with the enforcement of the new infectious diseases control law. *Infectious Agents Surveillance Report* 20, 4. Available at: http://idsc.nih.go.jp/iasr/20/230/de2309.html (accessed 24 September 2021).

Nomura, T., Takahashi, H. and Takeda, Y. (2003) Changes in measures against infectious diseases in Japan and proposals for the future. *Journal of the Japan Medical Association* 46(9), 390–400.

Nosaki, Y., Maeda, K., Watanabe, M., Yokoi, T., Iwai, K. *et al.* (2021) Fourth imported rabies case since the eradication of rabies in Japan in 1957. *Journal of Travel Medicine* 28(8), taab151. DOI: 10.1093/jtm/taab151.

Shih, T.-H., Chiang, J.-T., Wu, H.-Y., Inoue, S., Tsai, C.-T. *et al.* (2018) Human exposure to ferret badger rabies in Taiwan. *International Journal of Environmental Research and Public Health* 15(7), 1347. DOI: 10.3390/ijerph15071347.

Smith, K.F., Sax, D.F., Gaines, S.D., Guernier, V. and Guégan, J.F. (2007) Globalization of human infectious disease. *Ecology* 88(8), 1903–1910. DOI: 10.1890/06-1052.1.

Steier, F., Brown, J. and Mesquita da, F. (2015) The World Café in action research settings. In: Bradbury, H. (ed.) *The SAGE Handbook of Action Research*, 3rd edn. SAGE Publications, London, pp. 211–219.

Available at: https://methods.sagepub.com/book/the-sage-handbook-of-action-research-3e (accessed 5 December 2022).

Takahashi-Omoe, H., Omoe, K. and Okabe, N. (2008) Regulatory systems for prevention and control of rabies, Japan. *Emerging Infectious Diseases* 14(9), 1368–1374. DOI: 10.3201/eid1409.070845.

Tan, S. and Brown, J. (2005) The World Café in Singapore: Creating a learning culture through dialogue. *The Journal of Applied Behavioral Science* 41(1), 83–90. DOI: 10.1177/0021886304272851.

Yamada, A., Makita, K., Kadowaki, H., Ito, N., Sugiyama, M. *et al.* (2019) A comparative review of prevention of rabies incursion between Japan and other rabies-free countries or regions. *Japanese Journal of Infectious Diseases* 72(4), 203–210. DOI: 10.7883/yoken.JJID.2018.431.

6 Animal Bite Treatment Centres in the Philippines: Functions and Challenges Experienced in the Delivery of Rabies Post-exposure Prophylaxis

Beatriz Puzon Quiambao[1]*, Timothy John Reyes Dizon[1], Nicole Perreras Grande[1] and Mary Elizabeth Gutierrez Miranda[2]

[1]Research Institute for Tropical Medicine, Muntinlupa, Philippines; [2]Field Epidemiology Training Program Alumni Foundation, Manila, Philippines

Abstract

Rabies post-exposure prophylaxis (PEP) consists of wound care and vaccination after exposure to a suspect rabid animal. Animal Bite Treatment Centres (ABTCs) are government-run clinics in the Philippines providing rabies PEP for free. In addition to PEP provision, ABTCs also perform recording and reporting, rabies awareness activities, and case investigation of suspect rabies cases. Insufficient and erratic vaccine/rabies immunoglobulin (RIG) supply, fast turnover of trained personnel, multiple reporting systems, and the coronavirus 2019 (COVID-19) pandemic have tested the capacity of the ABTCs to sustain essential services and have been addressed in various ways. ABTCs need to go beyond PEP provision and embrace other programme strategies such as education and advocacy, responsible pet ownership, and the implementation of an Integrated Bite Case Management (IBCM) programme together with animal health counterparts to guide careful and informed use of PEP. All sectors must work together to achieve the global goal of zero human rabies deaths by 2030.

6.1 Introduction

Rabies post-exposure prophylaxis (PEP) is administered after exposure to a suspect rabid animal. This consists of wound care and administration of rabies vaccines, with or without rabies immunoglobulin (RIG). PEP is a life-saving intervention and should be considered for all rabies exposures.

An animal bite is considered a medical emergency. Although many people are bitten by animals that are eventually determined to be non-rabid, this information is usually not known at the time of presentation. Thus, it is common in rabies endemic countries, including the Philippines, to treat every bite as a suspected rabies case and provide PEP to the patient without full knowledge of the biting animal's status. This translates into overutilization of scarce rabies biologicals. However, practitioners and patients must err on the side of caution and perform rabies prophylaxis given the fatal nature of the disease.

Rabies vaccination is administered using either of two routes: (i) intradermal (ID); or (ii) intramuscular (IM). On the one hand, the

*Corresponding author: QuiambaoBeych2019@gmail.com

 One Health for Dog-mediated Rabies Elimination in Asia (eds V. Slack *et al.*)

DOI: 10.1079/9781800622975.0006

ID regimen uses a smaller volume of vaccines, allowing sharing of a vial, thereby reducing the cost. This regimen is more cost-effective for high-volume clinics that see significant numbers of patients each day. On the other hand, the IM regimen uses about five times more vaccine volume, constituting the entire amount of the vial per administration. This regimen is more practical for clinics that see only a few patients per day.

Dog bite incidence in the Philippines is extremely high, exceeding 800 individuals per 100,000 annually, with the vast majority caused by healthy dogs. Over the past few years, the number of animal bite victims receiving PEP has doubled from around 500,000 in 2013 to more than 1.4 million in 2016, and the number has remained above 1 million since then (NRPCP, 2020). Because of awareness and availability of PEP, more bite victims have sought treatment, and so more dog bites are being reported (Amparo *et al.*, 2018a).

In the Philippines, rabies control activities of the human and animal health sides are mostly implemented separately. The current national surveillance detects only a fraction of animal rabies cases, and frequently, human cases are recorded with a lack of information regarding the animal source. The Animal Bite Treatment Centres (ABTCs) provide PEP with no further investigation of the biting animal except to ask the victim or dog owner to report the animal's status after an observation period of 14 days. This information is not shared with animal health workers and is used mainly for deciding whether the remaining vaccine doses will be given; victims of healthy animals are not given the rest of the doses.

6.1.1 One Health

One aspect that has been increasingly recognized as significant in rabies control is the coordination between human and animal health, the so-called One Health approach. Because rabies is a zoonosis, the animal vector needs to be considered alongside the human victim.

Integrated Bite Case Management (IBCM), advocated by the World Health Organization (WHO), incorporates risk assessments of bite victims to guide more careful and informed use of PEP (WHO, 2018). This is also important in unifying human and animal disease surveillance for a more coherent disease management programme.

To develop and implement a cost-effective, epidemiologically robust, enhanced surveillance and response package and contribute to rabies elimination, two pilot provinces in the Philippines implemented IBCM starting in 2019. The ABTC plays a central role in the field implementation of IBCM, where frontline human and animal healthcare workers are trained on protocols for exposure risk assessment, PEP administration, outbreak investigation and reporting. Where bite patients are first encountered, the ABTC staff assesses the risk of rabies exposure and decide on the need for immediate PEP and follow-up animal investigation. Peer support through social media group chat has been useful in increasing shared information between sectors, linking the ABTC to the animal health or agriculture worker who does the animal investigation, and possibly sample collection for confirmatory diagnosis.

A monthly epidemiologic report with trends in bite cases and animal investigations is generated and disseminated from these activities. These reports contain: (i) graphs comparing bite cases and animal investigations categorized by municipality and risk; (ii) graphs comparing the number of bite patients submitted per ABTC; (iii) maps depicting the municipalities of the suspect or confirmed rabid bite patients or biting animals; (iv) reports of suspect human rabies cases; and (v) confirmed animal rabies cases. Since the regular generation of monthly reports, improvements have been noted, including more open communications between animal and human health workers, more timely updates on recent suspect rabies cases, an increased number of animal investigations completed, and more consistent patient data submission from health workers (M.E. Miranda, Manila, 2021, personal communication). Although a reduction in the number of human rabies cases has not yet been clearly observed, the enhanced surveillance and monitoring have continued to identify high-risk localities for targeted remedial dog vaccination and community engagement. Continued implementation with the expansion of this strategy to other areas will foster a One Health mindset which is important in achieving zero human rabies deaths by 2030.

Initial efforts towards integrating the rabies control activities of the human and animal health sectors began with establishing the Philippine Inter-Agency Committee on Zoonoses in 2020 (Department of Health, Department of Agriculture and Department of Environment and Natural Resources, 2020). However, broader adoption at the level of the units implementing the rabies programme, such as the ABTC and the Municipal Agriculture Office, remains a challenge.

6.1.2 National Rabies Prevention and Control Program (NRPCP)

The Philippines' NRPCP is embodied in Republic Act 9482, otherwise known as the Republic of Philippines, 2007. This law mandates the creation of the National Rabies Prevention and Control Committee composed of the Departments of Agriculture, Health, Education, and Interior and Local Government and Local Government Units with the assistance of the Department of Environment and Natural Resources, non-governmental organizations and people's organizations.

Components of the control programme include: (i) mass vaccination of dogs; (ii) establishment of a central database for registered and unvaccinated dogs; (iii) impounding, field control and disposition of unregistered, stray and unvaccinated dogs; (iv) conduct of information and education campaigns on the prevention and control of rabies; (v) provision of pre-exposure treatment to high-risk personnel and post-exposure treatment to animal bite victims; (vi) provision of free routine immunization or pre-exposure prophylaxis (PrEP) of schoolchildren aged 5–14 years in areas where there is a high incidence of rabies; and (vii) encouragement of the practice of responsible pet ownership (Anti-Rabies Act of 2007, Section 4; Republic of Philippines, 2007).

The Anti-Rabies Act of 2007 provides the legal basis for PEP provision, administered mostly through ABTCs. ABTCs are clinics catering exclusively to victims of bites by rabies-prone animals (mainly dogs and cats). Across the Philippines, there are an estimated 750 ABTCs. They are government run and administer rabies vaccines for free using the intradermal regimen. RIG is infiltrated in and around the wound for severe bites and may be provided for free on a limited basis. Aside from the ABTCs, there are privately run bite clinics called Animal Bite Centres or Clinics (ABCs), providing the same service for a fee. In addition, hospital emergency rooms and some doctors in their clinics also provide PEP. However, because they see fewer patients, they often give the vaccine intramuscularly and RIG is either given via IM injection or infiltration depending on the expertise of the clinic staff.

6.2 Animal Bite Treatment Centres (ABTCs) in the Philippines

6.2.1 Functions of ABTCs

The functions of ABTCs are shown in Fig. 6.1 but are described in detail as follows.

6.2.1.1 PEP provision

The main function of ABTCs is PEP provision. ABTC staff assesses rabies risk by examining the bite and reviewing the bite incident. They perform wound care and administer rabies PEP, when indicated. The staff also provide one-on-one counselling and health education to patients regarding expected adverse reactions, correct first aid practices, completion of the vaccination schedule and observation of the biting animal.

6.2.1.2 Recording and reporting

Each patient is given a PEP card which serves as their personal vaccination record. ABTCs maintain a rabies exposure registry detailing each patient's personal information, exposure history, rabies vaccine/RIG received, including dates of receipt and type of biting animal. The data are used for the animal bite and cohort analysis reports which are submitted quarterly and annually to rabies coordinators at the provincial and regional levels and eventually to the NRPCP (NRPCP, 2019). Completing all vaccine doses is emphasized with a target PEP completion rate of 90% (NRPCP, 2020).

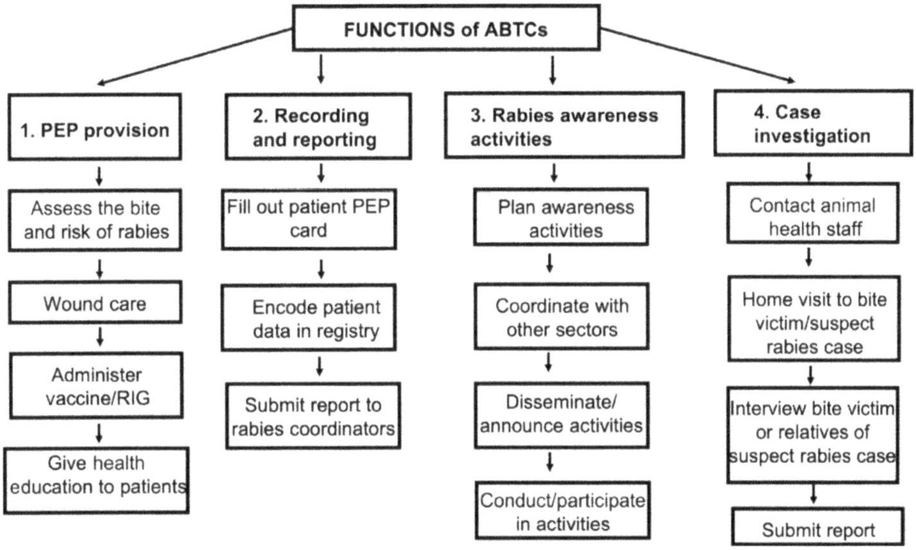

Fig. 6.1. Functions of Animal Bite Treatment Centres (ABTCs). PEP, post-exposure prophylaxis; RIG, rabies immunoglobulin. Created by Beatriz P. Quiambao.

6.2.1.3 Rabies Awareness

ABTCs also provide health education through posters, pamphlets and videos with an emphasis on responsible pet ownership, avoidance of bites, first aid for bites and wound care. ABTCs initiate or join a variety of activities in their localities such as: (i) lectures; (ii) mass dog vaccination; (iii) rabies trivia contests; (iv) motorcades (i.e. a parade of vehicles with posters and loudspeakers spreading rabies-related information); (v) poster-making contests; and (vi) PrEP for health workers in celebration of Rabies Awareness Month in March and World Rabies Day on 28 September.

6.2.1.4 Case Investigation

A few ABTCs also perform case investigations of high-risk bites and suspect human rabies cases.

6.2.2 Evolution of ABTCs

In the early 1990s, there was limited access to rabies PEP in the country. One had to join the long queues for vaccination at San Lazaro Hospital, a government-run infectious disease hospital in Metro Manila with the largest bite centre. A few hospital emergency rooms also administered rabies vaccine.

The Research Institute for Tropical Medicine (RITM), an infectious disease facility located in southern Metro Manila, pioneered the use of the ID regimen for rabies vaccination in the country. Since 1985, RITM had provided PEP to bite victims using the IM regimen as part of its emergency room services. Supported by the RITM's various research on ID rabies vaccination and the growing evidence for its use, RITM established its animal bite clinic, implementing ID rabies vaccination in 1993.

National implementation of ID rabies vaccination followed in 1997, enabling the country to discontinue the use of the locally manufactured nerve tissue vaccine (NTV) and replace them exclusively with modern tissue culture vaccines to provide safe, effective and affordable PEP. The ID regimen allowed the sharing of a vial of the vaccine among multiple patients, which meant that at least two patients should be seen at the clinic so that the remaining vaccine is not wasted. The Department of Health (DOH) established a network of ABTCs to improve access to these life-saving rabies biologicals nationwide.

The advantages of these clinics include: (i) the optimal use of the vaccine through ID vaccination; (ii) the simplification of logistic distribution since the supply of vaccines and RIG are allocated only to these clinics; and (iii) focused training of ABTC staff. Finally, consolidation of the PEP services into one clinic facilitates easy recollection in the event of a bite and eases the traceability of biting animals.

Several obstacles prevented the broad implementation of these rabies control efforts. Initially, many provinces were unaware of the need for establishing ABTCs, as the necessity for providing PEP was not fully recognized. There was no formal training available for managing animal bites, and doctors and nurses were not confident to manage bite victims despite the presence of guidelines on ID vaccine administration and RIG infiltration. Likewise, the supply of tissue culture vaccines and the budget for vaccine procurement was limited.

In the late 1990s, the NRPCP implemented massive health education and media campaigns, initiated discussions with local government officials, and integrated rabies education materials into the elementary school curricula, enabling a better understanding of the disease and the danger of animal bites. This led to an increase in consultation for animal bites. Consequently, regional and provincial rabies coordinators began establishing ABTCs in their areas (M. Vinluan, Manila, 22 August 2021, personal communication). In 2002, DOH formed a technical working group to update the national guidelines on animal bite management. This group has continued to revise the guidelines based on available international and local recommendations and local data. In 2008, the RITM started its first formal training course on animal bite management and, together with the DOH Centres for Health Development, continues to provide training to government doctors and nurses that operate the ABTCs, as well as private individuals who are interested in establishing an ABC. The budget of the national rabies programme gradually increased, mostly for the procurement of vaccines and immunoglobulins, which were distributed through the Expanded Programme on Immunization (EPI) cold chain facilities. With the obstacles slowly being addressed, the number of ABTCs gradually increased with the target of one ABTC per 100,000 population (NRPCP, 2020).

By July 2017, 513 ABTCs had been established all over the country. Of the 82 provinces (with the National Capital Region considered as a province), 70 (85%) provinces have at least one ABTC and 12 provinces have no ABTCs. The maximum number of ABTCs in any given province was 30. The number of ABTCs was not related to the income class of the locality (Amparo *et al.*, 2018a). Fig. 6.2 shows the nationwide distribution of 750 ABTCs as of August 2021, with only two provinces (Tawi-Tawi and Guimaras) having no ABTC. Thirty-four provinces (41%) have attained the target of one ABTC/100,000 population. Sixty-one provinces have ten or fewer ABTCs; 14 of these have a small population and have met the required number of ABTCs (R. Deray and H. Berano, Manila, 2021, personal communication). For the country to reach the goal of one ABTC/100,000 population, the NRPCP needs to establish an additional 355 ABTCs, approximately half of the current number.

6.3 Enablers of the ABTC system

To perform its functions efficiently, the ABTC requires a reliable reporting system, trained staff, updated management guidelines, a steady supply of rabies vaccines and RIG, and a way to sustain the operations financially.

6.3.1 Recording and reporting

Manual reporting of animal bites was done until 2014 when an electronic system for collecting, reporting and analysing rabies data called the National Rabies Information System (NaRIS) was implemented. It is an information and notification portal for the public, an online registry for rabies exposures for ABTC personnel, and a dashboard for decision makers (NRPCP, 2019). This is similar to the R36, a voluntary hospital-based reporting platform for rabies exposures in Thailand (Yurachai *et al.*, 2020).

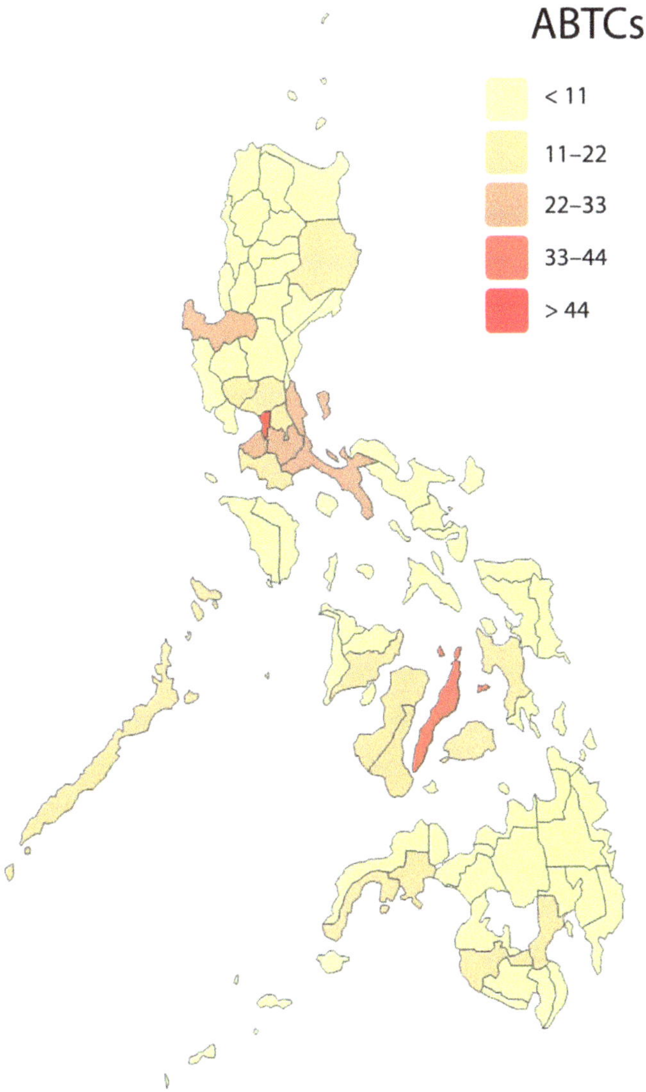

Fig. 6.2. Distribution of ABTCs in the Philippines, August 2021. Map is the property of Beatriz P. Quiambao.

6.3.2 ABTC certification

With the proliferation of ABTCs and ABCs, the DOH released an administrative order (Administrative Order 2013–004; Department of Health, 2013) in 2013 governing the certification of government ABTCs and private ABCs to standardize the operations and ensure the provision of affordable, quality and safe PEP. To pass the certification process, ABTCs/ABCs should comply with the DOH administrative orders on rabies and animal bite management using vaccines/RIG approved by the Philippine Food and Drug Administration. ABTCs should be adequately equipped with vaccines/RIG, emergency drugs, a dedicated refrigerator and adequate water supply, and should practise proper waste disposal. They should be operated by DOH-trained doctors and nurses who can categorize the exposures and provide correct clinical management. A standardized recording and reporting system and a two-way referral system had to be maintained (NRPCP, 2019).

The certification process consists of four cycles: (i) the Self-Assessment Cycle, where the ABTC/ABC assesses their status using a self-assessment form, with an option to request technical assistance; (ii) the Application Cycle where the self-assessment form is submitted; (iii) the Certification Cycle, where the certification visit is conducted; and (iv) the Registration and Issuance Cycle where the certificate is issued. The certification is good for 2 years (Department of Health, 2013).

6.3.3 Training

Only trained and licensed doctors and nurses are allowed to manage bite victims. Specific training on animal bite management, which includes a hands-on session on ID vaccination and RIG infiltration, has been ongoing since 2008. Regional and provincial coordinators are being capacitated to conduct the training themselves to address the increasing requests for training.

Since the COVID-19 pandemic, this necessary training course migrated to an online platform to address the restrictions of social distancing and infection control practices. The online animal bite management modules were launched in November 2021.

6.3.4 Vaccine supply

To provide free PEP to bite victims, rabies vaccines and RIG are allocated by the DOH to ABTCs with supplementation from the local government. In the early years when ABTCs were still being established, the budget for vaccines was limited, and there was no clear policy on how many doses should be given for free.

In 2005, WHO recommended the shortening of the ID regimen from 90 days (days 0, 3, 7, 28/30, 90) to 28 days (days 0, 3, 7, 28) by eliminating the day 90 dose and doubling the day 28 dose (WHO, 2005). The ID regimen now consisted of two ID doses on each of 4 days, for a total of eight ID doses. In the Philippines, the shortened ID regimen was implemented in 2007 (Department of Health, 2007). Due to limited funds, only two to four out of the eight ID doses of rabies vaccine were administered for free beginning in 2009 (Amparo *et al.*, 2018a). The remaining doses were an out-of-pocket expense for the patients. The number of free doses increased to four out of eight doses by 2013, six out of eight doses by 2015, and by 2016, all eight doses were given for free (Rabies Prevention and Control Program; Department of Health, n.d.). Implementing the recently recommended 1 week two-site PEP regimen called the Institut Pasteur du Cambodge (IPC) regimen (WHO (World Health Organization), 2018; Department of Health, 2018) will further reduce the number of doses needed, lessen clinic visits, and allow more patients to complete the vaccination regimen for free.

6.3.5 Animal bite management guidelines

One of the main requirements for certification was compliance with the latest DOH administrative order on rabies and animal bite management. Since 1996, these guidelines have been regularly updated based on the most current WHO recommendations with local modifications. The most recent version, DOH Administrative Order 2018–0013, dated 16 April 2018, recommended the IPC PEP regimen, a 1 week schedule on days 0, 3 and 7, without the day 28 dose (Department of Health, 2018). However, the rollout of the latest guidelines could not be immediately implemented because of the inability to train all ABTC staff and the shortage of WHO-prequalified vaccines. The COVID-19 pandemic further exacerbated these problems.

6.3.6 Financial sustainability

To subsidize the cost of PEP, the Philippine Health Insurance Corporation (PHIC) implemented the Animal Bite Treatment Package in 2012.

This allowed the reimbursement of PEP for category III bites (severe bites with bleeding) at a fixed rate of Php3000 per case (approximately US$60). Only ABTC/ABCs which have passed the certification process may avail of the Animal Bite Treatment Package. The package covered the cost of the rabies vaccine, RIG, wound care, tetanus prophylaxis and antibiotics (Philippine Health Insurance Corporation, 2012). As the amount is not sufficient to cover all costs, out-of-pocket expenses persisted.

By June 2016, there were 269 PHIC-accredited ABTCs and ABCs availing of the Animal Bite Treatment Package (Amparo *et al.*, 2018a). This number was reduced to 232 by 31 May 2021, since the lockdowns and reassignment of healthcare workers impeded the accreditation process. By 30 November 2022, the number had risen to 468 accredited ABTCs and ABCs (Philippine Health Insurance Corporation, 2022).

6.4 Challenges

The ABTC has many benefits including: (i) a physical infrastructure manned by trained personnel; (ii) implementation of updated management guidelines; (iii) vaccine supply provided by the government and distributed through the existing EPI cold chain structure; (iv) a reimbursement package for animal bites; (v) a standardized certification process; (vi) an available formal training programme; and (vii) a national reporting system. However, challenges continue to be experienced by the ABTCs.

6.4.1 Challenges before the COVID-19 pandemic

Although the DOH has successfully developed the ABTC network, the number of human rabies cases has not shown a progressive decline and continues to be more than 200 cases annually. A recent study in the Philippines has shown that only 45% of animal bite victims consult ABTCs for PEP (Amparo *et al.*, 2018b). The unvaccinated bite victims are the ones who eventually end up with symptoms of rabies and die.

The recurring shortage of WHO-prequalified rabies vaccines continues to be a major challenge to the programme, especially since these vaccines were the only ones recommended by the DOH. Consequently, the programme allowed the use of non-WHO-prequalified rabies vaccines, of which several are locally available. In addition, RIG supply has also been frequently inadequate, and, because of its high cost, only around 40% of bite victims receive it (NRPCP, 2019).

The rapid turnover of ABTC staff meant that training had to be done continuously to ensure adherence to management guidelines. However, only RITM and a few regional Centres for Health Development were accredited by DOH to provide the training, resulting in an accumulation of untrained ABTC staff. Delays in reporting due to Internet connectivity problems and intervening tasks hampered the collation of national data. Compartmentalized surveillance and reporting systems resulted in multiple reporting systems for rabies, such as the Philippine Animal Health Information System for animal rabies and the Philippine Integrated Disease Surveillance and Response, the Field Health Service Information System, and NaRIS, for human rabies. Data are not easily accessible for the different rabies stakeholders with challenges in data sharing, transparency and collaboration within and across health sectors (A.C. Amparo, Manila, 2021, personal communication).

Coordination with animal health personnel was also a challenge. It was observed that ABTC staff focused only on the management of human bite victims and did not communicate with the veterinary office regarding biting animals that were highly suspected of being rabid. This was seen as an additional task and was not considered a priority, especially if the ABTC managed many high-risk bite patients. Often, the responsibility of reporting suspect animals to the veterinary office fell on the bite victim (A.C. Amparo, Manila, 2021, personal communication).

ABTCs can certainly do more towards achieving the global goal of zero human rabies deaths by 2030. As part of its health education function, ABTCs should emphasize to their clients the importance of submission of samples from animals that die for laboratory confirmation of rabies. ABTCs can also reach out to the public through currently available social media platforms to send clear messages and clarify myths and misinformation on a daily basis, not only during World Rabies Day or Rabies Awareness Month celebrations. Prompt coordination with animal health workers in the event of a suspect biting animal should be implemented to remove rabid animals from the community and to conduct case investigation to identify additional bite victims needing PEP.

6.4.2 Challenges during the COVID-19 pandemic

Not least among all the challenges faced by the ABTCs was the COVID-19 pandemic. Just like all health programmes, rabies control activities were severely disrupted by the pandemic. Many hospital-based ABTCs either ceased operations or were relocated elsewhere since clinic space was used for COVID-19 response. In the National Capital Region (NCR), the pandemic's epicentre, 70% of ABTCs were closed, some permanently. RITM, one of the two largest ABTCs in NCR, catering for around 30,000–40,000 animal bite victims/year, closed its ABTC in March 2020 to focus on the COVID response. Animal bite victims were referred to nearby ABTCs and it is likely that some bite victims did not receive PEP at all. The RITM ABTC reopened in March 2022. In other regions, however, more than 90% of the ABTCs remained operational.

Health workers in many ABTCs were reassigned to COVID-related duties such as contact tracing, triage, clinical care and other roles, while animal/agriculture workers were tasked to do relief goods distribution, as the COVID-19 response was prioritized over rabies programme activities. Some of the staff were exposed to COVID-19 or contracted the infection and had to be isolated. This resulted in delays in many activities, including certification and re-certification of ABTCs/ABCs, case investigation of suspect human rabies cases, encoding in the NaRIS and report submission. Many ABTCs and ABCs had expired accreditation (J.C. Estioco *et al.*, Manila, 2020, personal communication; C. Andaya *et al.*, 2021, personal communication).

In addition, training was cancelled beginning in 2020, such that new rabies coordinators and ABTC staff were functioning with no formal training on animal bite management. The use of face masks and face shields for protection against COVID-19 hampered vision and made it more difficult to administer the injections. The prolonged lockdown periods severely affected people's mobility, thus preventing animal bite victims from accessing the ABTCs for consultation. Subsequent vaccine doses could not be given on schedule, and many patients could not complete the vaccination series. The lack of transport during lockdowns delayed vaccine delivery to the country and distribution to the regional vaccine warehouses and ABTCs.

On the national level, approximately 12% of funds for the rabies programme were reappropriated for the COVID-19 response, compounding the vaccine and RIG supply problem. Even the celebration of Rabies Awareness Month in March and World Rabies Day in September was cancelled (R. Quintana, Manila, 2020, personal communication).

For the IBCM project in the pilot provinces, the local government workforce also experienced increased workload or diversion to other duties, mobility restrictions and quarantine measures. Data from their ABTCs lagged, although animal health workers continued to initiate animal investigations when ABTCs prompted them (Philippine Council for Health Research and Development, 2021, unpublished).

Surprisingly, instead of the expected decrease in animal bites due to reduced mobility, there was a reported slight increase from 1,173,274 in 2019 to 1,211,892 in 2020. This may be due to increased bites by pet dogs and cats, as keeping a pet became popular during the lockdown. Nationwide, human rabies cases decreased from 243 in 2019, to 222 in 2020 and 195 in 2021 (R. Quintana, Manila, 2021, personal communication; R. Deray, Manila, 2022, personal communication). While it is hoped that this represents an actual decrease in human rabies cases, it can also be due to inadequate reporting. Achieving the Zero by 30 goal relies on accurate reporting.

6.4.3 Addressing the challenges brought about by the pandemic

Multiple measures were put in place to address these challenges. First, the referral system between ABTCs was strengthened such that patients were referred to functional ABTCs, and vaccine supplies from non-functioning ABTCs were reallocated or even donated to them. Second, online delivery of care through telemedicine, virtual case investigations of suspect human rabies cases, and patient follow up through social media were instituted. Third, infection control measures and minimum public health standards for protection against COVID-19 (e.g. the use of personal protective equipment, a no mask-no entry policy, physical distancing, temperature check and health declaration checklist) were strictly implemented. To avoid crowding, an appointment scheduling system was employed, and triaging was done to prioritize vulnerable individuals such as the elderly, pregnant women and children. Some innovative measures included installation of a plastic partition to separate the patient from the health worker and the use of a testing booth as a vaccination booth (shown in Fig. 6.3). Fourth, the operating hours of some ABTCs were adjusted to accommodate the reduction in personnel. Fifth, the training course on animal bite management shifted to an online platform. Sixth, the validity of ABTC and ABC certification was extended. Finally, virtual information campaigns were conducted through social media platforms (J.C. Estioco *et al.*, Manila, 2020, personal communication; C. Andaya *et al.*, Manila, 2020–2021, personal communication).

6.5 Learning from Other Countries

6.5.1 PEP provision in other Asian countries

There have been many experiences among Asian countries on PEP provision. The ID regimen of rabies vaccination originated in Thailand, and most Asian countries have adopted this regimen using cell culture vaccines (Khawplod *et al.*, 2006). NTV production and use ceased in Pakistan in 2015 (Ahmad *et al.*, 2021), while Myanmar continues to use NTV (Salahuddin *et al.*, 2016).

PEP provision in the Philippines has many similarities to that of other Asian countries. Like the ABTC in the Philippines, PEP is provided through community-based clinics such as the Public District Rabies Prevention and Control Centers in Bangladesh, Basic Health Units in Bhutan, and Vietnam's Provincial and District Preventative Medicine Centers, some of which are dedicated rabies clinics. There are also hospital-based rabies clinics in Sri Lanka and Thailand and specialized centres such as the Angkor Hospital for Children and the Institut Pasteur in Cambodia. PEP is likewise available in the private sector in most countries except for Bhutan. Cambodia, Sri Lanka and Bhutan implement follow ups of high-risk non-compliant patients. Like the Philippines, vaccine supply in many Asian countries is procured nationally and provided to patients for free, with out-of-pocket expenses accruing when out of stock. PEP is not free in China, Vietnam or Cambodia, while government subsidy is available in Thailand. RIG is in short supply in most Asian countries, even for countries that produce RIG such as Thailand and Vietnam. Reporting of animal bites ranges from no reporting to an electronic reporting system in Sri Lanka, Thailand and the Philippines. Especially noteworthy is the web-based voluntary, hospital-based reporting platform in Thailand, called R36, which collects information for all suspect rabies exposures and monitors adherence to PEP recommendations. Sri Lanka and Bhutan also report adverse events following immunization (Nguyen *et al.*, 2019; Li *et al.*, 2019; Sreenivasan *et al.*, 2019; Yurachai *et al.*, 2020).

The COVID-19 pandemic has also affected the rabies programme in other countries. In Thailand, curfew implementation resulted in the decline of outdoor activities, thus diminishing the risk of potential rabies exposure. Bite victims tended to not receive PEP because of the COVID-19 health risks associated with travel or clinic attendance (S. Purin, Manila, 2021, personal communication). While outdoor exposure also declined in the Philippines due to lockdowns, interaction with pets increased in the homes, resulting in a slight increase in reported bites. In Pakistan, PEP utilization

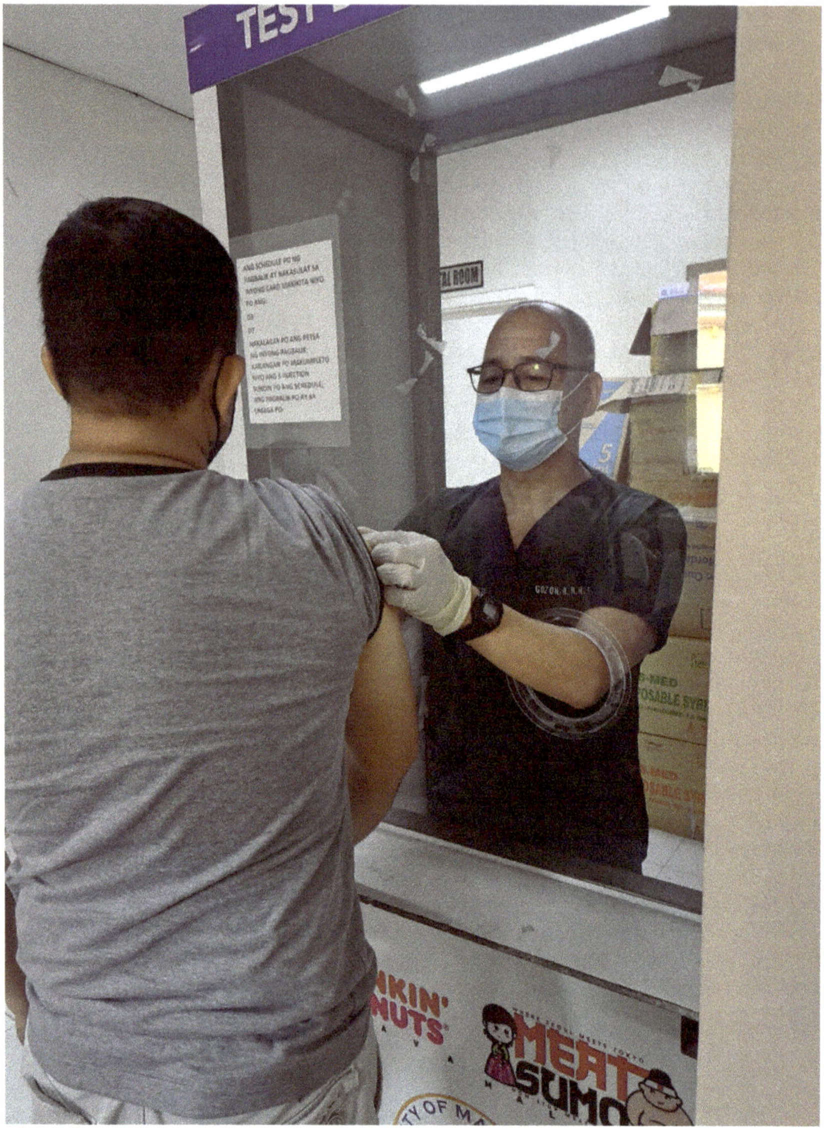

Fig. 6.3. Use of a plastic partition to separate the health worker from the patient. Photograph courtesy of Rex Gozon, 2022.

was reduced during the first few months of the pandemic but returned to pre-pandemic levels afterwards (S. Jamali, Manila, 2021, personal communication).

The Philippines can certainly learn from the successful practices of other countries. Since many of the bite victims are bitten by healthy pets, an assessment of rabies risk is important to determine the need for PEP. Follow up of non-compliant patients who are high risk or have a proven rabies exposure is a good strategy to ensure proper completion of the PEP regimen. A hospital-based reporting system for PEP provision may be very useful, while domestic production of RIG should help local supply.

6.5.2 Sharing best practices

The Philippines has a lot of best practices that can be shared with other countries. First is the provision of a specific training programme for healthcare workers on animal bite management. Refresher training is required whenever the guidelines are updated. Second is a certification process for the ABTCs, which includes the facility, the trained personnel, and reporting and compliance to the guidelines. However, it lacks a system to regularly monitor compliance with guidelines. The third is the presence of a national online surveillance system specific to rabies. While the Thai R36 surveillance is based in hospitals, the NaRIS covers ABTCs in the communities and hospitals. Attempts to include private clinics/hospitals have also been made. However, prompt encoding and regular analysis of the NaRIS data need to be addressed. Fourth is the specific animal bite management package from the PHIC, where ABTCs can get reimbursed for the treatment of bite patients. This is in addition to the provision of all ID doses for free by the national government.

6.6 Conclusion

The ABTC is the main avenue for PEP provision and therefore plays an integral role in human rabies prevention in the Philippines. Challenges before and during the COVID-19 pandemic have tested the system's resilience and have been addressed in various ways.

The Philippines has a long way to go to attain the Zero by 30 goal. The COVID pandemic has set back the programme and now the country has to work double time to achieve the goal. ABTCs need to go beyond PEP provision and embrace other programme strategies such as education and advocacy, responsible pet ownership and an integrated approach to animal bites together with animal health counterparts as part of One Health. There is no need for more policies. Rather, the focus should be on policy implementation to achieve the goal of a world free from dog-mediated rabies by 2030.

Acknowledgements

The authors would like to acknowledge the data shared by Dr Raffy Deray, Dr Ronald Quintana, Dr Minerva Vinluan and Mr Hipolito Berano of the National Rabies Prevention and Control Program (NRPCP); Ms Anna Charina Amparo of the Global Alliance for Rabies Control; Dr Suda Purin of the Queen Saovabha Memorial Institute, The Thai Red Cross Society, Thailand; Dr Seemin Jamali of the Jinnah Postgraduate Medical Center, Pakistan; and Department of Health Regional Rabies Coordinators (Jeff Carl Estioco, Odielon Filomeno, Christy Andaya, Ricardo Cabigas, Angelito Cuevas, Nieto Fernandez, Aida Ruby Jeremias and Gina Manlapig). We also acknowledge the assistance of Ana Dominique Pablo for the references and map; Rex Gozon for the photograph and Steve Scholand for review and editing of the manuscript.

Authors' Declaration

All authors declare that they have no conflict of interest.

All authors have approved this manuscript, agree with its submission, and share collective responsibility and accountability.

This manuscript has not been published or is not under review elsewhere.

References

Ahmad, W., Naeem, M.A., Akram, Q., Ahmad, S. and Younus, M. (2021) Exploring rabies endemicity in Pakistan: major constraints & possible solutions. *Acta Tropica* 221, 1–10. DOI: 10.1016/j.actatropica.2021.106011.

Amparo, A.C., Jayme, S., Roces, M.C., Quizon, M.C., Villalon, E.E.S. *et al.* (2018a) The evaluation of operating Animal Bite Treatment Centers in the Philippines from a health provider perspective. *PLOS ONE* 13(7), 1–25. DOI: 10.1371/journal.pone.0199186.

Amparo, A.C., Jayme, S., Roces, M.C., Quizon, M.C., Mercado, M. *et al.* (2018b) The evaluation of operating Animal Bite Treatment Centers in the Philippines from a patient perspective. *PLOS ONE* 13(7), 1–17. DOI: 10.1371/journal.pone.0200873.

Department of Health (2007) Administrative Order No.2007-0029 revised guidelines on management of animal bite patients. Department of Health, Republic of the Philippines. [No longer available online as superseded by recent guidelines.].

Department of Health (2013) Administrative Order No.2013-0004 implementing guidelines on the conduct of Animal Bite Treatment Centers (ABTC) and Animal Bite Centers (ABC) certification of the national rabies prevention and control program. Republic of the Philippines. Available at: https://dmas.doh.gov.ph:8083/Rest/GetFile?id=336924 (accessed 10 September 2021).

Department of Health (2018) Administrative Order No.2018-0013 revised guidelines on the management of rabies exposure. Republic of the Philippines. Available at: www.psmid.org/wp-content/uploads/2020/03/CPG-rabies-AO-2018-0013.pdf (accessed 12 October 2021).

Department of Health (n.d.) Rabies prevention and control program. Republic of the Philippines. Available at: https://doh.gov.ph/national-rabies-prevention-and-control-program (accessed 1 September 2021).

Department of Health, Department of Agriculture and Department of Environment and Natural Resources (2020) Joint Department Administrative Order No.2020-02 guidelines to operationalize the Philippine Inter-Agency Committee on Zoonoses (PhilCZ). Republic of the Philippines. Available at: www.da.gov.ph/wp-content/uploads/2020/11/doh_da_denr_s2020.pdf (accessed 12 October 2021).

Khawplod, P., Wilde, H., Sirikwin, S., Benjawongkulchai, M., Limusanno, S. *et al.* (2006) Revision of the Thai Red Cross intradermal rabies post-exposure regimen by eliminating the 90-day booster injection. *Vaccine* 24(16), 3084–3086. DOI: 10.1016/j.vaccine.2006.01.051.

Li, A.J., Sreenivasan, N., Siddiqi, U.R., Tahmina, S., Penjor, K. *et al.* (2019) Descriptive assessment of rabies post-exposure prophylaxis procurement, distribution, monitoring, and reporting in four Asian countries: Bangladesh, Bhutan,Cambodia, and Sri Lanka, 2017–2018. *Vaccine* 39, A14–A19. DOI: 10.1016/j.vaccine.2018.10.011.

Nguyen, H.T.T., Le, N.D.T., Pham, T.N., Urabe, M.I., Afriyie, D.O. *et al.* (2019) Evaluation of Vietnam's post-exposure prophylaxis delivery system, 2017. *Vaccine* 37, A20–A27. DOI: 10.1016/j.vaccine.2019.02.041.

NRPCP (National Rabies Prevention and Control Program) (2019) National rabies prevention and control program manual of procedures (2019). Available at: https://doh.gov.ph/sites/default/files/publications/Rabies%20Manual_MOP_2019%20nov28.pdf (accessed 16 September 2021).

NRPCP (National Rabies Prevention and Control Program) (2020) National rabies prevention and control program strategic plan 2020–2025. Available at: https://doh.gov.ph/sites/default/files/publications/FINAL%20MTP%20RABIES%20%28nov%2028%29%202.0_ed.pdf (accessed 12 September 2021).

Philippine Health Insurance Corporation (2012) Circular No.015, s. 2012 PhilHealth for Animal Bite Package (Rabies Post-exposure Prophylaxis). Available at: www.philhealth.gov.ph/circulars/2012/circ15_2012.pdf (accessed 20 August 2021).

Philippine Health Insurance Corporation (2022) List of accredited animal bite package providers. Available at: www.philhealth.gov.ph/partners/providers/institutional/accredited/ABPP_113022.pdf (accessed 29 October 2022).

Republic of Philippines (2007) 'Anti-Rabies Act of 2007'. Republic Act no.9482: an Act providing for the control and elimination of human and animal rabies. Available at: www.officialgazette.gov.ph/2007/05/25/republic-act-no-9482/ (accessed 13 December 2021).

Salahuddin, N., Gohar, M.A. and Baig-Ansari, N. (2016) Reducing cost of rabies post-exposure prophylaxis: experiences of a tertiary care hospital in Pakistan. *PLOS Neglected Tropical Diseases* 10(2), e0004448. DOI: 10.1371/journal.pntd.0004448.

Sreenivasan, N., Li, A., Shiferaw, M., Tran, C.H., Wallace, R. *et al.* (2019) Overview of rabies post-exposure prophylaxis access, procurement and distribution in selected countries in Asia and Africa, 2017–2018. *Vaccine* 37, A6–A13. DOI: 10.1016/j.vaccine.2019.04.024.

WHO (World Health Organization) (2005) *WHO Expert Consultation on Rabies*, First Report. WHO Technical Report series 931. WHO, Geneva, Switzerland. Available at: https://apps.who.int/iris/bitstream/handle/10665/43262/WHO_TRS_931_eng.pdf?sequence=1&isAllowed=y (accessed 15 August 2021).

WHO (World Health Organization) (2018) *WHO Expert Consultation on Rabies*, Third Report. WHO Technical Report Series 1012. WHO, Geneva, Switzerland. Available at: https://apps.who.int/iris/bitstream/handle/10665/272364/9789241210218-eng.pdf (accessed 15 August 2021).

Yurachai, O., Hinjoy, S. and Wallace, R.M. (2020) An epidemiological study of suspected rabies exposures and adherence to rabies post-exposure prophylaxis in Eastern Thailand, 2015. *PLOS Neglected Tropical Diseases* 14(2), 1–17. DOI: 10.1371/journal.pntd.0007248.

7 The 'Pooling Strategy' in Himachal Pradesh, India: An Innovation for Rabies Post-exposure Prophylaxis During Crisis of Shortage of Life-saving Biologicals

Vanessa Slack[1] and Omesh Kumar Bharti[2]*
[1]Texas Department of State Health Services, Zoonosis Control Branch, Texas, USA; [2]State Institute of Health and Family Welfare, Department of Health & Family Welfare, Shimla, India

Abstract
Resource-limited clinics and countries often face ethical dilemmas when treating patients exposed to rabies during rabies vaccine and rabies immunoglobulin (RIG) shortages. This chapter discusses an innovative 'pooling strategy' that was developed during shortage crises of life-saving biologicals in Himachal Pradesh, India. Patients received a fraction of a rabies vaccine vial administered via the intradermal route. RIG was prioritized and only given to high-risk patients with severe exposures. RIG was only infiltrated into wounds to provide immediate virus neutralization, and the dosage was determined by the size and number of wounds (as opposed to systemic administration of RIG intramuscularly using a dosage calculated by patient body weight). Residual drops of vaccine and RIG were consolidated, or 'pooled', to the next vial for the next patient – preventing even a single drop of waste. To date, this protocol has been > 99.9% effective and has saved considerable costs, rabies biologicals and human lives.

7.1 Introduction

Dogs (*Canis familiaris*) are the foremost reservoir for the rabies virus throughout Asia and Africa, accounting for more than 99% of all human rabies deaths. Among the estimated 59,000 human rabies deaths that occur worldwide each year, 20,800 fatalities occur in India alone (Hampson *et al.*, 2015). After exposure, the rabies virus travels to the central nervous system, where it is undetectable by the host's immune system (Fooks *et al.*, 2017). The incubation period can be long, and the symptoms can be highly variable (Wilson *et al.*, 2019). A person exposed to the rabies virus must receive timely and effective post-exposure prophylaxis (PEP) to neutralize the virus and prevent the onset of clinical signs and symptoms, or else death inevitably follows within days of symptom onset.

The World Health Organization (WHO) recommends that any person exposed to the rabies virus receives rabies PEP, but the scheduled regimen may vary depending on the type or severity of exposure, the resources available, and the preference of the treating clinic. A post-exposure regimen generally consists of

*Corresponding author: bhartiomesh@yahoo.com

 One Health for Dog-mediated Rabies Elimination in Asia (eds V. Slack *et al.*)
DOI: 10.1079/9781800622975.0007

thorough wound washing, a single dose of rabies immunoglobulin (RIG) upon presentation, and a series of rabies vaccinations over 1 week or 1 month (WHO, 2018b).

RIG is a vital component of rabies PEP because it provides passive immunity to a person, while the rabies vaccine has time to induce a humoral immune response against the virus (Anderson, 2007). When infiltrated into and around bite wounds, RIG immediately neutralizes the virus at the wound site, preventing it from accessing the central nervous system (Wu *et al.*, 2017). RIG can be either human-derived (hRIG) or equine-derived (eRIG). Although eRIG carries a risk (albeit extremely rare) for severe allergic reactions in some patients, it is often preferential to hRIG which can be unaffordable and unavailable in many countries (Anderson, 2007).

Rabies vaccines can induce a detectable immune response after 10–14 days, but that delay can be significant when incubation periods are reduced due to proximity to the nervous system or an increased viral load from significant wounds (Anderson, 2007). Of greatest concern are severe Category III exposures and wounds to innervated areas such as the face, head, neck and hands (Wilson *et al.*, 2019). Rabies vaccines can be administered either intradermally (ID) or intramuscularly (IM) and require multiple doses to ensure adequate protection. Administering the vaccine via the ID route costs five times less and uses less product than the IM route, yet the immunogenicity and vaccine efficacy are equivalent for both routes of vaccine administration (Denis *et al.*, 2019). There are several approved schedules for subsequent vaccinations, but vaccines are typically administered immediately, and then 3, 7, and between 14 and 28 days after exposure (Warrell, 2019).

While modern rabies biologicals are safe and extremely effective, they are not readily available in many regions where rabies is endemic and poorly controlled. Rabies biologicals are often in high demand, in short supply, and must maintain a specific temperature-controlled supply chain to be effective – which can be difficult in resource-limited areas. Such regions suffer disparities that further hinder the ability to maintain a sufficient supply, such as poor infrastructure, limited treatment facilities, inadequate distribution systems, and frequent shortages of RIG and vaccines (Hampson *et al.*, 2015). Healthcare centres that are able to stock and provide rabies PEP to patients often find themselves short of personnel or medical supplies (such as syringes needed to administer the vaccines). Patients face barriers that obstruct access to life-saving PEP, including: (i) unaffordable costs of PEP; (ii) reluctance to seek medical care; (iii) productivity and income losses while seeking PEP; (iv) high travel costs; and (v) travelling long distances multiple times to complete the vaccination series.

Areas with inadequate resources could prioritize RIG and vaccines for patients with the greatest risk of rabies exposure. To ensure limited supply can meet demand, clinics could implement innovative strategies to ensure that every person who needs rabies PEP has access to – and can afford – these biologicals. This chapter explores various evidence-based innovations, including a pooling strategy, which is now practised in many hospitals across India.

7.2 The Context

One specific region suffering from frequent supply shortages of rabies biologicals was Himachal Pradesh, a northern state in India situated in the Western Himalayas. This mountainous state is predominantly rural, and the extreme landscape complicates the ability of some citizens to seek medical care (Bharti *et al.*, 2015). Patients in Himachal Pradesh would be required to search multiple hospitals and clinics, running 'from pillar to post' to find a facility that stocked rabies biologicals. The state also suffered from poor compliance with the PEP regimen among lower-income patients who could not afford these life-saving biologicals nor return to the clinic on subsequent days to complete the rabies vaccine series (Bharti *et al.*, 2015). In addition, a lack of government funding to cover the cost of PEP for low-income patients further hindered the state's ability to stock clinics with rabies PEP biologicals, making it unavailable to even those that could afford it.

With only a finite supply that could not meet demand, the Deen Dayal Upadhyaya (DDU) Hospital in Shimla (the capital of Himachal Pradesh) recognized the need for unconventional

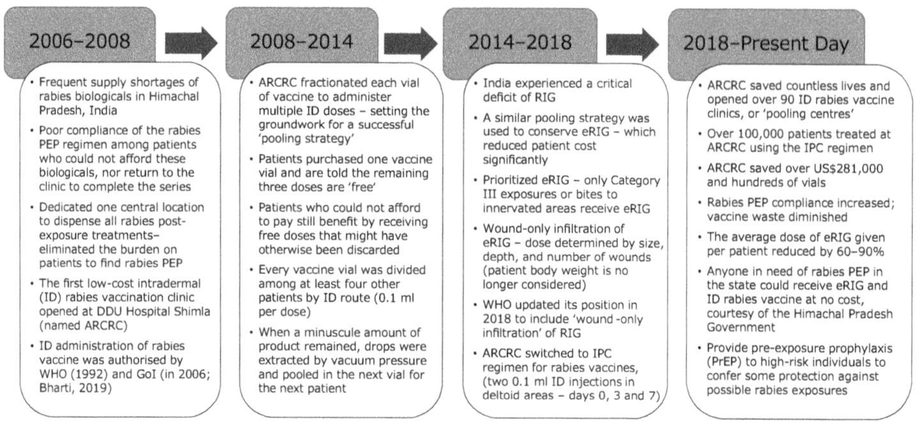

Fig. 7.1. The timeline of the development, implementation, challenges and outcomes of the 'pooling strategy' in Himachal Pradesh, India. ARCRC, Anti-Rabies Clinic and Research Centre; DDU, Deen Dayal Upadhyaya; eRIG, equine rabies immunoglobulin; GoI, Government of India; ID, intradermal; IPC, Institut Pasteur du Cambodge; PEP, post-exposure prophylaxis; RIG, rabies immunoglobulin; WHO, World Health Organization. Image created by V. Slack.

clinical interventions to ensure all persons exposed to rabies could receive adequate post-exposure treatment, without compromising patient safety or rabies PEP efficacy. After extensive literature reviews and advocacy sessions, in August 2008, the first low-cost ID rabies vaccination clinic, named Anti-Rabies Clinic and Research Centre (ARCRC), opened at DDU Hospital Shimla, amid a rabies vaccine shortage (Bharti *et al.*, 2015). Since 2008, ARCRC has saved countless human lives and substantial costs by opening over 90 ID rabies vaccine clinics (called 'pooling centres') in the state. The timeline for this progress is shown in Fig. 7.1.

Overcoming logistical and ethical challenges, DDU Hospital Shimla implemented the following innovations and modifications to the rabies PEP protocol which revolutionized patient care in a rabies-endemic region and reduced the number of human rabies deaths:

- the infrastructure:
  - centralize rabies PEP care for *all* patients to one facility (ARCRC at DDU Hospital Shimla);
  - patients purchase one vaccine vial (which covers the cost of the entire four-dose PEP series) and are told their remaining three doses are 'free';
  - create a pooling strategy by administering partial doses from opened vials while maintaining the sterility and temperature of the vials;
  - residual drops are amalgamated to the next vial or allocated to the poorest patients – ensuring no waste;
- rabies vaccine
  - administer vaccines via the ID route, using a fraction of the vial compared to the IM route;
  - modify rabies PEP schedule, so patients require fewer visits;
- RIG:
  - provide eRIG, which is more affordable and as effective as the alternative, hRIG;
  - infiltrate wounds with eRIG instead of systemic injection via the IM route;
  - calculate eRIG dosage depending on severity and number of wounds rather than patient body weight;
  - prioritize eRIG for patients with severe animal bites (Category III exposures);
- evidence-based innovations and One Health partnerships:

- provide pre-exposure prophylaxis (PrEP) to high-risk individuals at no cost;
- discontinue PEP administration when transmission of rabies is unlikely (i.e. consumption of milk from rabid bovines);
- conduct canine and wildlife rabies surveillance; and
- administer rabies vaccine via the ID route to canines and livestock in high-risk areas.

7.3 The Infrastructure

The first challenge was to dedicate one central location to dispense all rabies post-exposure treatments and eliminate the burden on patients to find rabies PEP. All healthcare facilities in Shimla Municipality were asked to provide wound washing and first aid to animal bite patients before referring them to DDU Hospital Shimla for rabies PEP (Bharti *et al.*, 2012). ARCRC could fractionate each vial of vaccine (and later eRIG) to administer multiple doses – this would set the groundwork for a successful 'pooling strategy' and reduce vaccine waste. Vaccines would be drawn up from one vial at a time, but once a vial was opened, any unused product must be discarded after 6–8 hour (WHO, 2018a).

When a patient presented to DDU Hospital with possible rabies exposure, they would be asked to purchase one vial of rabies vaccine (approximately US$5–7.50) during the first visit. Patients were told that the subsequent doses would be 'free' to incentivize them to return for the remaining doses and complete the PEP series (Bharti *et al.*, 2012). Although a patient paid for an entire vial, they would receive doses from a partial vial that was opened earlier that day and shared among other patients. Every vial was divided up among at least four other patients by ID route, at 0.1 ml per dose on each deltoid, leading to a fivefold reduction of vaccine use compared to the IM route. All precautions were taken to maintain the sterility and temperature of the vaccines, while partial vials were discarded at the end of the day.

When only a minuscule amount of product remained in a vial, the drops were extracted by vacuum pressure and pooled in the next vial for the next patient (Bharti *et al.*, 2015). This pooling strategy was especially advantageous for the poorest patients who could not afford to pay for the vial of vaccine, yet could still benefit from this system by receiving doses that might have otherwise been discarded (Bharti *et al.*, 2012). Since this pooling strategy came to fruition, hundreds of vials have been saved by this economic strategy and vaccine waste has diminished (Bharti *et al.*, 2015). A similar pooling strategy was later used for eRIG to conserve every drop of immunoglobulin for patients in need (Bharti *et al.*, 2019).

From 2008 to 2015, over 12,000 animal bite victims were treated under this pooling strategy at ARCRC in DDU Hospital. As a result of these low-cost interventions, over US$281,000 was saved, and the rabies PEP compliance rate increased by 2.8 times (Bharti *et al.*, 2015). Implementing this pooling strategy and prioritizing patient care significantly lowered the cost of rabies PEP in the state. The vaccine became so affordable that other clinics in Himachal Pradesh could supply the rabies vaccine by implementing similar pooling strategies, working towards zero human rabies deaths in the region (Bharti *et al.*, 2015). Presently, 90 such pooling centres are in operation throughout the state of Himachal Pradesh.

Beginning in May 2018, every person in need of rabies PEP in the state could receive eRIG and ID-administered rabies vaccine at no cost, courtesy of the Himachal Pradesh Government (Bharti, 2019).

7.4 Rabies Vaccines

7.4.1 Preliminary phase

The recommended potency for rabies vaccine is ≥2.5 IU per dose, regardless of the route of injection (IM or ID). If administered IM, each dose requires an entire vial (0.5 ml or 1.0 ml, depending on the vaccine brand). If given ID, however, the volume of vaccine needed is just a fraction of that – only 0.1 ml per injection (Denis *et al.*, 2019). Vaccine effectiveness is equivalent for either route of vaccine administration. Therefore, a single vial of vaccine can be fractionated to provide five (or ten) doses when administered ID, rather than one dose administered IM (Warrell, 2019). In India, the estimated cost of one complete course of rabies vaccines administered IM route was US$35–44.50, while the cost for the same vaccine given ID route was US$5–7.50 (Bharti *et al.*, 2012).

Multiple-dose schedules would still be required, but ID administration would significantly lower the cost of vaccinations, making it more affordable to patients who would be more likely to return, purchase and complete the PEP series. Unfortunately, vaccine manufacturers often marketed rabies vaccine vials as 'IM use' only and administering rabies vaccines through an IM route became standard practice. WHO endorsed the administration of rabies vaccines through the ID route in 1992 (WHO, 1992). However, the Government of India (GoI) did not fully authorize ID administration until 2006, despite higher costs, frequent shortages and increased demand for the vaccine (Bharti, 2019).

7.4.2 Strategy development and implementation

Many physicians were hesitant to transition to ID administration due to a lack of training and a reluctance to administer vaccines both off-label and in fractional doses (Bharti *et al.*, 2015). Therefore, the first barrier to optimizing care at DDU Hospital in Shimla was to obtain rabies vaccine vials that were licensed and labelled 'for IM/ID use' by specifically requesting this from vaccine manufacturers. Medical staff at DDU Hospital were subsequently trained to administer vaccines via the ID route and in fractionated doses.

In 2008 DDU Hospital Shimla opened ARCRC and began ID administration as a low-cost solution to ration rabies vaccines. The clinic implemented the vaccine 'pooling strategy' by providing vaccinations from an opened vial under aseptic conditions. The residual drops were pooled in the next opened vial, significantly reducing vaccine waste.

Initially, the DDU Hospital was practising a WHO-recommended rabies PEP schedule called the ID two-site Thai Red Cross (TRC) regimen which required two 0.1 ml ID injections on four visits: first upon presentation and then on days 3, 7 and 28. Many patients in the state found this month-long regimen difficult to complete so DDU Hospital sought alternate options approved by WHO with fewer visits. In May 2018, ARCRC started using the PEP vaccine schedule called the Institut Pasteur du Cambodge (IPC) regimen, or the 'ID two-site 1 week IPC' regimen (Bharti *et al.*, 2019). This schedule requires two 0.1 ml ID injections of rabies vaccine in the deltoid areas during three visits on days 0, 3 and 7 (Warrell, 2019).

7.4.3 Outcomes and challenges

A reduction of patient visits from four (within a month) to three (within a week) benefited patients in Himachal Pradesh and increased PEP compliance. The financial burden was alleviated among patients who could not be spared from lost wages at work or time from home duties while seeking care. Medical resources and staff time were preserved for other patients.

Since the implementation of the three-visit IPC regimen, over 100,000 patients have been treated at DDU Hospital using this pooling strategy. The only confirmed human rabies deaths in Himachal Pradesh have been those that could not acquire timely rabies post-exposure treatment (vaccines and immunoglobulin) (Bharti, 2019). Although the number of visits for rabies PEP had been reduced, some patients still struggled to obtain or afford this life-saving treatment after exposure to rabies. DDU Hospital continued to consider other strategies to reduce PEP costs and eliminate human rabies deaths in the state.

7.5 Rabies immunoglobulin (RIG)

7.5.1 Preliminary phase

Until recently, the WHO position for rabies PEP indicated that any confirmed or suspected exposure to rabies should receive a dose of RIG as determined by the patient's body weight. As much RIG as possible would be infiltrated into or around the wound(s), while any remaining RIG would be systemically administered IM in a site distal to the rabies vaccine (SAGE Working Group on Rabies Vaccines and Immunoglobulins and WHO Secretariat, 2017). However, several studies and clinical interventions have found that this practice of providing a systemic injection of excess RIG after wound infiltration fails to produce additional benefits to immunization efficiency (Anderson, 2007; Madhusudana *et al.*, 2013; Wu *et al.*, 2017). Controlled animal studies have demonstrated no basis for calculating the dose of immunoglobulin based on body weight. Modern rabies vaccines are highly effective and immunogenic, so the only benefits of RIG seem to be immediate and localized – by neutralizing the virus at the wound site and preventing it from entering the nervous system (Anderson, 2007).

Many facilities in the state that carried rabies vaccines would not stock RIG, so patients would not receive immediate protection from immunoglobulins. In fact, there have been documented cases of patients in Himachal Pradesh who were not given RIG, but only received the rabies vaccine series, and subsequently died from rabies (Bharti *et al.*, 2017).

India experienced a critical deficit of RIG in 2014 and minimal supply was available to hospitals and clinics (Bharti *et al.*, 2016). As a result, DDU Hospital Shimla was again constrained to make RIG affordable and available to those who needed it – even if it defied the latest WHO rabies PEP guidelines. ARCRC at DDU Hospital took a few approaches to reduce the cost and amount of product injected per patient without compromising patient safety or RIG immunogenicity.

7.5.2 Strategy development and implementation

The first step towards a more cost-effective approach for rabies PEP was to exclusively use the equine-derived eRIG in DDU Hospital, which would significantly reduce the cost to patients. In Himachal Pradesh, the maximum dose of eRIG was 40 IU/kg body weight and could cost around US$9–20 per patient. When human-derived hRIG was given, the maximum dose was 20 IU/kg body weight and could cost nearly US$400–500 per patient (Bharti, 2019).

Administering eRIG carries the rare risk of severe allergic reactions in some patients, so clinics would often test a small dose on every patient and monitor for concerning symptoms before administering the full dose. Such a reaction could occur at any stage of administration of eRIG, and the product was in such limited supply, so staff at ARCRC discontinued this step from patient care. Instead, the staff prepared to treat any patient who might experience an adverse reaction to save precious biological for other patients in need. By 2018, WHO discouraged skin testing of eRIG before administration because it was unreliable to predict adverse effects (WHO, 2018b).

After switching to the more affordable eRIG option of immunoglobulins, DDU Hospital had to draft emergency clinical intervention protocols to guarantee that eRIG would be available to anyone who needed it. In 2014, the clinic decided that only wounds would be infiltrated with eRIG and the dose would be determined by size, depth, and the number of wounds – patient body weight would no longer be considered (Bharti *et al.*, 2016). An ethics committee approved this protocol for emergency use (despite not yet being a WHO-approved treatment), and Government Health Services authorized the use of wound-only infiltration of eRIG for patients with severe bite wounds (Bharti *et al.*, 2017).

Every patient seen for possible rabies exposure received meticulous wound washing and a complete series of rabies vaccination. However, eRIG had to be further prioritized so that only those with severe wounds (Category III exposures) or animal bites to innervated areas would receive eRIG. This prioritization was based on WHO Category of Exposure recommendations; thus, RIG was not indicated for patients who were licked by or scratched by an animal (Category I or Category II exposures, respectively) (WHO, 2018b).

While this protocol was not officially condoned by WHO until 2018, physicians in rabies-endemic regions with a lack of PEP resources were faced with an ethical predicament: either provide care to fewer patients using a dose of RIG determined by their body weight or provide care to more patients using a fraction of the dose – based on animal studies that supported wound-only infiltration. At ARCRC, physicians opted for the latter as a clinical intervention to save lives (Bharti *et al.*, 2016). This was not a novel concept but an intervention that could not be tested until it was implemented out of desperation during a RIG shortage crisis (Anderson, 2007).

7.5.3 Outcomes and challenges

From June 2014 to June 2018, ARCRC saw 10,830 patients for rabies exposures. Among those, 7506 patients were deemed Category III exposure and received wound-only infiltration of eRIG and the rabies vaccine series. The average dose of eRIG given per patient was reduced by 60–90% of what would have been given based on patient body weight (what used to be about 10 ml of eRIG per patient averaged 1.26 ml in 2014 with the new protocol; that dose decreased further to 0.87 ml in 2017 due to clinical nurses' increased expertise in wound infiltration over time) (Bharti, 2019). Such a reduction in the volume of RIG could significantly reduce costs worldwide, saving countless human lives.

Over 70% of patients receiving wound-only infiltration of eRIG required a volume of <0.5 ml, which cost patients, on average, only US$0.75 (Bharti *et al.*, 2019). In the first year, 269 patients with Category III exposures were treated with this protocol, and rather than 363 vials of eRIG being given IM, the clinic only used 42 vials for local infiltration instead. ARCRC was able to follow up with over 80% of patients for 9 months and none developed signs of rabies infection (Bharti *et al.*, 2016).

In addition, DDU Hospital implemented the vaccine pooling strategy for RIG, ensuring that multiple patients could share eRIG vials, and every drop of eRIG was being utilized to eliminate human rabies deaths in the region. This low-dose, low-cost model of eRIG wound infiltration transformed rabies post-exposure care for patients in Himachal Pradesh. As of April 2018, WHO updated the previous position so that as much RIG as feasible be infiltrated into and around wounds; no longer should excess RIG be systemically injected IM at a site distant from the wound (WHO, 2018a). This recommendation of localized RIG infiltration could save countless human lives worldwide by improving accessibility and affordability, especially in areas where canine rabies is endemic.

Clinical studies have demonstrated successful outcomes (i.e. protective rabies titre levels) after wound-only infiltration and dosage reduction of RIG in guinea pigs, monkeys and mice (Anderson, 2007; Madhusudana *et al.*, 2013; Wu *et al.*, 2017). Field studies have demonstrated similar outcomes in domestic bovines who were bitten by rabid dogs or mongooses and received wound-only infiltration of eRIG (Bharti *et al.*, 2018). More studies must be done to substantiate the safest reduction of RIG volume needed to infiltrate wounds.

The limitation of this, of course, is that such a clinical study on humans would be impossibly unethical to conduct in a research setting. The exact amount of RIG needed per patient should consider many factors before reducing virus-neutralizing immunoglobulins for patient care (e.g. location and severity of the bite, and viral load inoculated into a wound) (Bharti *et al.*, 2017).

7.6 Evidence-based Innovations and One Health Partnerships

The first pooling clinic at ARCRC, DDU Hospital Shimla, revolutionized rabies post-exposure healthcare by significantly reducing the cost and dosage of rabies PEP, without compromising patient safety and survival. While these pooling strategies and collaborative clinical interventions ensure that more patients can receive rabies PEP (both vaccines and immunoglobulins), many patients still face obstacles in complying with and affording this treatment. Although ARCRC had reduced the number of visits for a patient seeking rabies PEP, it was still difficult for some citizens to complete the vaccination series after potential exposure.

One possible intervention would be to vaccinate high-risk individuals before they were exposed to rabies. Such patients would require fewer visits to complete the post-exposure series. In addition, there was a growing concern regarding the number of patients coming to ARCRC for rabies PEP following exposure to rabid livestock and wildlife species. Once again, DDU Hospital sought solutions for its citizens.

7.6.1 Pre-exposure prophylaxis (PrEP) for high-risk cohorts

Although rabies PEP failures are rare, many rabies experts advocate the practice of rabies PrEP among populations of individuals at a higher risk of being exposed to rabies. This would include children in canine rabies-endemic regions and individuals whose jobs or lifestyles put them at risk for exposure (Fooks *et al.*, 2017). Pre-emptive protection against rabies in these cohorts will help significantly reduce costs, use fewer biologicals, and eliminate human rabies deaths. Fewer vaccine doses are required for a previously vaccinated person seeking post-exposure treatment. Furthermore, RIG is not indicated for a previously vaccinated person and can be conserved for other patients (WHO, 2018a).

With the success of the pooling strategy at DDU Hospital, staff was able to give high-risk individuals, such as ragpickers, PrEP vaccines from remnants of vials that would have otherwise been discarded (Bharti *et al.*, 2015). WHO recommends a two-site ID vaccine administered on two visits for a complete PrEP schedule (on days 0 and 7). In case of constraints, WHO has determined that a one-visit PrEP will 'likely confer some protection' (WHO, 2018b, p. 217). A one-visit PrEP schedule could be crucial among the poorest citizens who may not be available for subsequent doses or, in case of actual rabies exposure, for post-exposure vaccines. For vulnerable patients receiving PrEP at DDU Hospital, staff provided four 0.1 ml vaccinations on both deltoids and suprascapular skin areas during a single visit (Bharti *et al.*, 2015). One study has shown this schedule demonstrated an acceptable immunogenic response when the alternative is no protection (Bharti, 2015). More studies are required to validate the efficacy of such a schedule, but for vulnerable populations living in precarious environments, this intervention may be invaluable for eliminating human rabies deaths in this region.

7.6.2 Implementation of One Health partnerships for non-canine exposures

As an additional way to preserve limited resources, ARCRC discontinued the practice of administering rabies PEP to patients who consumed raw milk or butter of a bovine that had died of rabies (Bharti *et al.*, 2015). At the start of the ARCRC ID clinic, staff would give rabies vaccines to approximately 40–50 cases every year (primarily to schoolchildren) who had consumed raw dairy products of a rabid cow or buffalo. DDU Hospital attempted to vaccinate these cohorts, but there were high dropout rates and a loss of follow up to complete the PEP schedule (Bharti *et al.*, 2015).

Ingesting unpasteurized milk from a rabid cow is theoretically a possible route of exposure, although, to date, the infectious rabies virus has not been isolated from these products (Wilson *et al.*, 2019). The 2018 WHO position no longer recommends PEP to individuals who consume milk from a rabid animal; therefore, ARCRC has ceased this practice, saving the clinic hundreds of rabies vaccine doses annually (WHO, 2018a).

While the primary concern in this region is canine-rabies exposures, a significant number of citizens have also been bitten or scratched by wildlife species susceptible to rabies infection. Wild animals are not the primary concern for rabies transmission in canine-rabies endemic regions, as they may be considered 'dead-end' infections and are not likely to encounter or infect humans, canines or livestock (Fooks *et al.*, 2017). However, villages in Himachal Pradesh border forested regions, and wildlife may adapt and live close to humans, canines and livestock, thus creating a potential for human rabies transmission.

Rabies control measures had to be implemented using One Health partnerships to further reduce the number of patients seen at DDU Hospital Shimla. The first step was to conduct surveillance and identify wildlife species infected with the rabies virus in Himachal Pradesh. In 2015, Shimla Municipality collected animals found deceased and tested the brain samples for rabies virus using the fluorescent antibody test at Central Research Institute, Kasauli, Himachal Pradesh (Bharti *et al.*, 2015). The municipality confirmed rabies-positive results in rhesus macaques, leaf-eating monkeys, palm civets, mongooses, squirrels and domestic cats (Bharti, 2016). While it was not disputed these species were capable of being infected with or transmitting rabies, there was little evidence in that region until this surveillance effort confirmed it. Therefore, exposure to these animals must be considered a possible risk for rabies transmission and protective measures should be implemented to save lives. Of most concern were the numerous patients referred to ARCRC with scratches and/or bites from wild monkeys which were unobtainable for a rabies confirmation test. DDU Hospital now follows the same protocols for treating patients with wildlife exposures as they do for canine exposures.

DDU Hospital is coordinating with forestry and animal husbandry departments to identify rabies focus areas, based on the locations of confirmed rabid animals or human rabies deaths. These rabies focal points can be mapped and compared to nearby villages where rabies vaccination campaigns could be conducted throughout the state. Using this strategy, efforts to prophylactically vaccinate dogs, livestock or wildlife with oral rabies vaccine baits can be orchestrated with permission from the GoI (Bharti *et al.*, 2015).

Following the protocols used by ARCRC, veterinarians in Himachal Pradesh have adopted a post-exposure protocol to infiltrate eRIG into wounds of bovines, horses and dogs after bite exposure from a rabid animal. In addition, veterinarians can implement a similar PrEP protocol for domestic animals and livestock by reducing the volume of vaccine per dose if given via the ID route, further saving costs and precious biologicals (Sharma *et al.*, 2022). Implementation of these protocols has been successful, and one study demonstrates the survival of livestock when PrEP or PEP was administered using the Himachal Pradesh model (Bharti *et al.*, 2018).

7.7 Conclusion

From a global public health perspective, access to life-saving rabies biologicals must be improved and prioritized in rabies-endemic regions. Implementing low-cost innovations in resource-limited regions can ameliorate community hardships, reduce the financial burden for patients, spare valuable resources in clinics and save human lives. In times of vaccine shortages, physicians face ethical barriers with little guidance from expert organizations; thus, creative local clinical interventions, such as vaccine pooling strategies, are needed. Innovative control programmes should be considered in regions where rabies is ubiquitous and transmission of the virus is not species specific.

The success of the Himachal Pradesh pooling strategy model has allowed over 90 pooling centres to open in the state. These low-cost ID rabies vaccine clinics have substantially reduced the volume of rabies biologicals administered per patient without compromising immunogenicity. To date, there have been no human rabies deaths among the patients receiving rabies PEP at DDU Hospital Shimla. Rabies vaccines and immunoglobulins were so affordable that in 2018, Himachal Pradesh Government made them free for all in the state, while the GoI is working towards making PEP free for all in the country. The most recent WHO position has independently approved these strategies for rabies PrEP and PEP, citing the success of the pooling strategy in Himachal Pradesh, among other studies. These strategies should be adopted globally to help achieve zero human rabies deaths by 2030. While these innovations have proved indispensable in clinical settings in rabies-endemic regions, more studies must be done in laboratory settings to ensure adequate titre levels are detected in pre-exposure and post-exposure treatments.

While most patients presenting to ARCRC are treated for dog bites, a multi-pronged One Health approach is needed to control rabies in regions like Himachal Pradesh, where wildlife and canines cohabitate near humans and livestock. Partnerships are needed across all sectors to control rabies: physicians and clinics, veterinarians and canine vaccination teams, and wildlife and forestry departments – all serve to interrupt rabies transmission to humans, canines and wildlife, respectively.

Authors' Declaration

All authors declare that they have no conflict of interest.

All authors have approved this manuscript, agree with its submission, and share collective responsibility and accountability.

This manuscript has not been published or is not under review elsewhere.

References

Anderson, D.C. (2007) WHO guidelines dealing with immunoglobulin use impede rabies prevention. *Asian Biomedicine* 1(1), 103–107.

Bharti, O.K. (2015) Immunizing vulnerable populations like rag pickers, garbage collectors, municipality workers and newspaper hawkers against rabies in Shimla Municipality, HP, India. *World Journal of Vaccines* 05(1), 19–24. DOI: 10.4236/wjv.2015.51003.

Bharti, O.K. (2016) Human rabies in monkey (*Macaca mulatta*) bite patients a reality in India now! *Journal of Travel Medicine* 23(4), taw028. DOI: 10.1093/jtm/taw028.

Bharti, O.K. (2019) Using a modified rabies immunoglobulin protocol in a crisis of unavailability: ethical and practical challenges. *Indian Journal of Medical Ethics* 4(2), 139–143. DOI: 10.20529/IJME.2019.022.

Bharti, O.K., Madhusudana, S.N., Gaunta, P.L. and Belludi, A.Y. (2016) Local infiltration of rabies immunoglobulins without systemic intramuscular administration: an alternative cost effective approach for passive immunization against rabies. *Human Vaccines & Immunotherapeutics* 12(3), 837–842. DOI: 10.1080/21645515.2015.1085142.

Bharti, O.K., Madhusudana, S.N., Kale, A., Gaunta, P.L., Chaudhry, L.S. *et al.* (2015) Success story of a low cost Intra-Dermal Rabies Vaccination (IDRV) clinic-lessons learnt over five years of 12,000 patient vaccinations "without failure" at DDU Hospital Shimla, Himachal Pradesh, India: "saving a drop of rabies vaccine and immunoglobulins" 12 innovations to make Himachal Pradesh rabies free state by 2020. *World Journal of Vaccines* 5(3), 129–139. DOI: 10.4236/wjv.2015.53014.

Bharti, O.K., Van Damme, W., Decoster, K., Isaakidis, P., Appelmans, A. *et al.* (2012) Breaking the barriers to access a low cost intra-dermal rabies vaccine through innovative 'pooling strategy.' *World Journal of Vaccines* 2(3), 121–124. DOI: 10.4236/wjv.2012.23016.

Bharti, O.K., Madhusudana, S.N. and Wilde, H. (2017) Injecting rabies immunoglobulin (RIG) into wounds only: a significant saving of lives and costly RIG. *Human Vaccines & Immunotherapeutics* 13(4), 762–765. DOI: 10.1080/21645515.2016.1255834.

Bharti, O.K., Thakur, B. and Rao, R. (2019) Wound-only injection of rabies immunoglobulin (RIG) saves lives and costs less than a dollar per patient by 'pooling strategy.' *Vaccine* 37(1), A128–A131. DOI: 10.1016/j.vaccine.2019.07.087.

Bharti, O.K., Sharma, U.K., Kumar, A. and Phull, A. (2018) Exploring the feasibility of a new low cost intra-dermal pre & post exposure rabies prophylaxis protocol in domestic bovine in Jawali Veterinary Hospital, District Kangra, Himachal Pradesh, India. *World Journal of Vaccines* 8(1), 8–20. DOI: 10.4236/wjv.2018.81002.

Denis, M., Knezevic, I., Wilde, H., Hemachudha, T., Briggs, D. *et al.* (2019) An overview of the immunogenicity and effectiveness of current human rabies vaccines administered by intradermal route. *Vaccine* 37(1), A99–A106. DOI: 10.1016/j.vaccine.2018.11.072.

Fooks, A.R., Cliquet, F., Finke, S., Freuling, C., Hemachudha, T. *et al.* (2017) Rabies. *Nature Reviews Disease Primers* 3(17091), 1–19. DOI: 10.1038/nrdp.2017.91.

Hampson, K., Coudeville, L., Lembo, T., Sambo, M., Kieffer, A. *et al.* (2015) Estimating the global burden of endemic canine rabies. *PLoS Neglected Tropical Diseases* 9(4), e0003709. DOI: 10.1371/journal.pntd.0003786.

Madhusudana, S.N., Ashwin, B.Y. and Sudarshan, S. (2013) Feasibility of reducing rabies immunoglobulin dosage for passive immunization against rabies: results of *in vitro* and *in vivo* studies. *Human Vaccines & Immunotherapeutics* 9(9), 1914–1917. DOI: 10.4161/hv.25431.

SAGE Working Group on Rabies Vaccines and Immunoglobulins and World Health Organization (WHO) Secretariat (2017) Background paper: proposed revision of the policy on rabies vaccines and rabies immunoglobulins. Available at: www.who.int/publications/i/item/WHO-CDS-NTD-NZD-2018.04 (accessed 22 March 2022).

Sharma, A.K., Bharti, O.K., Prakash, V.C., Lakshman, D., Isloor, S. *et al.* (2022) Survival of 21 dogs after post-exposure prophylaxis using intra-dermal rabies vaccine in the pre-scapular region and emergency use of expired equine rabies immunoglobulin for local wound infiltration in victims of a rabid dog bite using One Health approach. *Epidemiology International* 7(4), 12–18. DOI: 10.24321/2455.7048.202215.

Warrell, M.J. (2019) Rabies post-exposure vaccination in 2 visits within a week: a 4-site intradermal regimen. *Vaccine* 37(9), 1131–1136. DOI: 10.1016/j.vaccine.2019.01.019.

Wilson, P.J., Rohde, R.E., Oertli, E.H. and Willoughby, R.E. (2019) *Rabies: Clinical Considerations and Exposure Evaluations*. Elsevier, St Louis, Missouri.

WHO (World Health Organization) (1992) *WHO Expert Committee on Rabies, Eighth Report*. WHO, Geneva, Switzerland.

WHO (World Health Organization) (2018a) *WHO Expert Consultation on Rabies,* Third Report. WHO Technical Report Series, No.1012. WHO, Geneva, Switzerland.

WHO (World Health Organization) (2018b) Weekly epidemiological record. Rabies vaccines: WHO position paper – April 2018. *World Health Organization – Technical Report Series* 931, 201–220.

Wu, W., Liu, S., Yu, P., Tao, X., Lu, X. *et al.* (2017) Role of systemic injection of rabies immunoglobulin in rabies vaccination. *Archives of Virology* 162(6), 1701–1703. DOI: 10.1007/s00705-017-3263-y.

8 Investigation of Rabies Control in Free-roaming Dogs: A Mathematical Modelling Approach From Bangladesh

Md Hamidul Islam[1]* and M.A. Masud[2]
[1]University of Rajshahi, Rajshahi, Bangladesh; [2]Natural Product Informatics Research Center, KIST Gangneung Institute of Natural Products, Gangneung-si, South Korea

Abstract

This chapter addresses the issues relevant to the transmission of rabies in free-roaming dogs (FRDs) and its control by using mathematical epidemic models. Compared to well-supervised dogs, there are some additional challenges to controlling rabies in FRDs which arise due to a number of factors, such as: (i) the lack of a regular population census; (ii) very little or no control over dog movement; (iii) uncontrolled access to food waste around markets and garbage dumps; and (iv) uncontrolled mating. We present a series of mathematical models, such as an ordinary differential equation model, a pulse vaccination model and an optimal vaccination model. In addition to vaccination, controlling carrying capacity is considered to be another control measure. The analysis shows that vaccination alone cannot be effective enough in containing rabies in all possible transmission scenarios. Rather, it is found to be successful in containing rabies in all possible transmission scenarios when accompanied by carrying capacity control.

8.1 Introduction

Rabies is one of the oldest and most terrifying diseases taking a heavy toll on human lives every year around the world. Around 55,000 people are reported dead due to rabies every year around the world (Pieracci *et al.*, 2019). Dog-mediated human rabies has been declared eliminated from countries in Europe and America including both north and south continents (Singh *et al.*, 2017), but it remains a major public health challenge in many Asian and African countries (Acharya *et al.*, 2020). Human rabies deaths in Asia account for 59.6% of global human rabies cases (Hampson *et al.*, 2015). Among the Asian countries where rabies is found highly prevalent are Bangladesh and India. In contrast, rabies is found relatively less prevalent in Nepal, Myanmar, Bhutan, Thailand and Indonesia (Acharya *et al.*, 2020). Every affected country is fighting against rabies in its own way. To coordinate these fragmented efforts, the world called for action by setting a global goal of eliminating dog-mediated human rabies deaths by 2030 (Fahrion *et al.*, 2017). Later, the World Health Organization (WHO), the World Organization for Animal Health (OIE), the Food and Agriculture Organization of the United Nations (FAO) and the Global Alliance for

*Corresponding author: hamidul.islam@ru.ac.bd

 One Health for Dog-mediated Rabies Elimination in Asia (eds V. Slack *et al.*)
DOI: 10.1079/9781800622975.0008

Rabies Control (GARC) decided to work together on this action plan to reach the goal.

Well-supervised dogs are vaccinated regularly and their mating is also controlled. However, dog-keeping practices vary greatly with the cultural and economic conditions of a society. Existing studies show that rabies is highly prevalent where the majority of dogs are poorly supervised or not supervised at all. For example, a major portion of the total dog population in Bangladesh and India is free roaming (Ruan, 2017). The reason why it is harder to control rabies in free-roaming dogs (FRDs) is explained later in this section. Therefore, the action plan against rabies needs to be designed by considering the following: (i) the economic conditions of the society; (ii) dog-keeping practices; and (iii) cultural practices of communities towards poorly supervised community dogs or FRDs. The global collaboration against rabies is thus planned to help countries by empowering their existing tools and expertise with renewed international support to act efficiently to save human lives from this preventable disease (Fahrion *et al.*, 2017).

More than 99% of human deaths due to rabies are attributed to dog-mediated rabies (WHO, 2021). About eight in ten deaths are reported in rural areas where about half of the deaths are children, which gives a clear indication that rabies affects mostly vulnerable people around the world. Therefore, it could be expected that increasing the social awareness of rabies and access to affordable post-exposure prophylaxis for people may significantly reduce rabies-induced fatalities. In addition, making the dog population resistant to rabies through vaccination and stopping the spread of the disease at its source could be considered as one of the most effective ways to reduce human rabies deaths significantly. Unfortunately, these fragmented efforts have proved not to be effective enough in stopping the spread of the disease. Prevention and control of an epidemic disease like rabies requires a more meticulous plan based on the complex interaction between human and animal health and the environment. Therefore, the collaborative, multisectoral and transdisciplinary One Health approach has been considered the best possible way to deal with rabies in the most affected countries of Asia (Acharya *et al.*, 2020).

According to a data-driven empirical study conducted in 2008, rabies killed about 2100 people every year in Bangladesh (Hossain *et al.*, 2012), which was 1.35 times higher than the WHO prediction. So, rabies has been considered a national priority in Bangladesh, with the aim of eliminating the disease through the One Health approach by 2020 (Acharya *et al.*, 2020). However, controlling rabies in a country like Bangladesh is quite a challenging task where a major portion of the total dog population is free roaming (Ruan, 2017). Following the classifications of dogs given by Wandeler (1985) the dog population of Bangladesh can be classified into two major categories: (i) community dogs which are owned but very poorly supervised; and (ii) FRDs which are fully independent and unrestricted (Ruan, 2017). In this study, we refer to these categories of dogs as FRDs. Study shows that FRDs are mostly responsible for the transmission of several zoonotic pathogens (Pastoret *et al.*, 2014).

In order to control rabies, Bangladesh introduced mass dog culling in its major towns in early 2012, but this approach failed, and rabies infections remained endemic (Ghosh *et al.*, 2020). The literature shows that multiple factors are associated with the failure of the culling method (Hossain *et al.*, 2012; Tenzin *et al.*, 2015; Taylor *et al.*, 2017). For example, culling is unable to distinguish between vaccinated and non-vaccinated dogs. Consequently, culling can result in a reduction in the number of vaccinated dogs and increase the proportion of susceptible dogs above the threshold for herd immunity (Nunes *et al.*, 2008). As a result, Bangladesh stopped mass culling in 2012 (Tenzin *et al.*, 2015). Shortly after that, another programme was launched by the Obhoyaronno Bangladesh Animal Welfare Foundation (OBAWF) in Dhaka city with the objective of controlling rabies using the catch-neuter-vaccinate-release (CNVR) strategy. There are many challenges to implementing this programme successfully. For example, this programme requires substantial financial support. In addition, the success of this programme requires a solid knowledge of the dog population size, which is a very challenging task in the case of FRDs (Tenzin *et al.*, 2015).

During the CNVR programme in 2012 and 2013, a FRD population estimation study was conducted in Dhaka, the capital city of Bangladesh, where the ratio of FRDs to humans was reported to be 1:828 (Tenzin *et al.*, 2015). However, the report of Choudhari (2016) showed that this ratio increased to 1:193, suggesting a fourfold increase in just 3 years. FRDs have access to environmental resources such as waste food around markets and garbage dumps, and their mating is not controlled. As a result, their reproduction rate is higher than that of fully controlled dogs. Further, these poorly managed FRDs are not regularly vaccinated and their lifespan is comparatively short (Pal, 2001). Therefore, rabies in this dog population follows different dynamics compared to well-supervised dogs and demands particular attention (Czupryna *et al.*, 2016). A substantial amount of research has been carried out investigating rabies transmission and control management in well-supervised dogs (Ruan, 2017; Taylor *et al.*, 2017). In contrast, rabies transmission in FRDs and the idea of controlling the disease is relatively less understood. Existing literature suggests that controlling rabies in FRDs is technically more difficult and requires further attention from the scientific and policy-making communities (Baquero *et al.*, 2015; Leung and Davis, 2017).

Regular immunization is a widely used technique to control the transmission of rabies among dogs. Regular immunization is expected to be more efficient in controlling rabies if the birth rate of dogs is controlled, as suggested by Ruan (2017). However, the regular immunization concept in the case of FRDs does not seem to be feasible, as it incurs major costs and continuous efforts as the FRD population is quickly renewed due to high birth rates and short lifespans. Therefore, following the One Health approach, the possible cheap and sustainable way to control the FRD population in such settings could involve controlling dogs' access to food waste and garbage dumps, as well as improving food waste management (Taylor *et al.*, 2017). Baquero *et al.* (2015) suggested that reduction of the environmental carrying capacity through food waste management is the most effective way to modify the dog population dynamics of FRDs. Therefore, the use of vaccinations as an additional control in conjunction with waste management may help to achieve the desired outcome in controlling rabies in FRDs.

Mathematical models are often used to acquire quantitative insights into the epidemiological characteristics of infectious diseases and to develop cost-effective control strategies. There are a large number of mathematical models focusing on the transmission dynamics of rabies in wildlife such as foxes and raccoons (Bingham, 2005; Keller *et al.*, 2013). Several other mathematical models considering the dog–human transmission of rabies have been proposed in recent years (Zhang *et al.*, 2011; Leung and Davis, 2017; Ruan, 2017). These studies have focused on the transmission of rabies among dogs and to humans and suggested different control strategies to limit this transmission. Zhang *et al.* (2011) studied the transmission of rabies within domestic dogs and from domestic dogs to humans in China, and the authors suggested that reducing the dog birth rate and increasing dog immunization coverage may be effective ways to control rabies. Another mathematical model focusing on domestic and FRDs suggested that rabies incidence can be reduced by increasing the vaccination rate of domestic dogs and reducing the FRD population (Hou *et al.*, 2012). In another recent sophisticated study conducted by Ruan (2017), the author introduced a rabies transmission model with increasing complexity.

Existing practices concerning data-driven studies of epidemics and population dynamics are based on time series data of infection prevalence (Kim *et al.*, 2020; Masud *et al.*, 2020b). However, no such data are available for rabies in FRDs. So, using mathematical modelling tools, we inferred key parameters of the dog population and its rabies dynamics. We examine the transmission dynamics of rabies among dogs in the presence of different vaccination and sterilization strategies to investigate the effectiveness of vaccination in stopping the endemic spread. Besides the CNVR programme, we also consider food waste management as another control measure to reduce the environmental carrying capacity. We then analyse the model to investigate whether these combined control measures could be effective in stopping the spread or significantly reducing the incidence of rabies which is mainly caused by FRDs.

8.2 What Was Done

8.2.1 Basic model

We divide the dog population into three compartments: (i) susceptible (S); (ii) exposed (E); and (iii) infectious (I). We assume the logistic growth of the dog population relates to the growth rate with carrying capacity. We assume that only susceptible dogs give birth to susceptible newborns at a rate r. The carrying capacity is D. The growth term is modelled in such a way that the number of newborns depends on susceptible dogs only, while all dogs contribute to the limitations in available resources. The per-capita transition rate, β, is a product of the contact rate and transmission probability at each contact. In addition, homogeneous mixing of dogs is considered, so the probability that a susceptible dog will meet an infected dog is considered to be $\frac{I(t)}{N(t)}$. Therefore, the transition rate of dogs from the susceptible to the exposed compartment is $\beta S(t)\frac{I(t)}{N(t)}$. Dogs remain exposed for a period of $1/\sigma$ years on average, where σ is the mean period taken for an exposed dog to develop infection. Following the model proposed by Zhang *et al.* (2011), we assume that γ fraction of the exposed dogs (E) become infectious at the rate $\gamma\sigma$, while the rest do not become infectious and instead return to the susceptible compartment at the rate $(1-\gamma)\sigma$. The parameters m and μ represent the normal and disease-related death rates, respectively. With these assumptions, the rabies transmission model is given by:

$$\begin{aligned}\frac{dS(t)}{dt} &= rS(t)\left(1-\frac{N(t)}{D}\right)-\beta S(t)\frac{I(t)}{N(t)}\\ &\quad+\sigma(1-\gamma)E(t)\\ \frac{dE(t)}{dt} &= \beta S(t)\frac{I(t)}{N(t)}-(\sigma+m)E(t)\\ \frac{dI(t)}{dt} &= \sigma\gamma E(t)-(\mu+m)I(t)\end{aligned} \tag{8.1}$$

where $N(t) = S(t) + E(t) + I(t)$, here t represents time, which is measured in years. The model shown in Eqn 8.1 has a disease-free equilibrium (DFE), $\varepsilon^0 = (S^0, E^0, I^0) = (D, 0, 0)$. The basic reproduction number, R_0, is the average number of new rabid dogs due to a single rabid dog introduced into the DFE. We use the next-generation matrix method and follow the notations developed by van den Driessche and Watmough (2002) in calculating the basic reproduction number. According to this method, the matrix of new infection terms $\mathcal{F}$ and the matrix of transition terms $\mathcal{V}$ are given by:

$$\mathcal{F} = \begin{pmatrix}\beta S\frac{I}{N}\\ 0\\ 0\end{pmatrix};$$

$$\mathcal{V} = \begin{pmatrix}(\sigma+m)E\\ -\gamma\sigma E+(m+\mu)I\\ -rS\left(1-\frac{N}{D}\right)+\beta S\frac{I}{N}-\sigma(1-\gamma)E\end{pmatrix}$$

Calculating the derivatives of $\mathcal{F}$ and $\mathcal{V}$ at ε^0 yields $F_{\varepsilon^0} = \begin{pmatrix}0&\beta&0\\0&0&0\\0&0&0\end{pmatrix}$ and

$$V_{\varepsilon^0} = \begin{pmatrix}\sigma+m & 0 & 0\\ -\gamma\sigma & m+\mu & 0\\ r-\sigma(1-\gamma) & r+\beta & r\end{pmatrix}$$

The expression $F_{\varepsilon^0}V_{\varepsilon^0}^{-1}$ termed as the next-generation matrix gives the expected number of secondarily infected dogs in each compartment produced by a single rabid dog. The spectral radius of $F_{\varepsilon^0}V_{\varepsilon^0}^{-1}$ is the basic reproduction number $R_0 = \frac{\beta\sigma\gamma}{(\mu+m)(m+\sigma)}$, which is reasonable from the perspective of an epidemic due to the following:

$$\begin{aligned}R_0 &= \frac{\beta\sigma\gamma}{(\mu+m)(m+\sigma)}\\ &\approx \beta\times\frac{1}{\mu+m}\times\frac{D-1}{D}\times\frac{\sigma\gamma}{m+\sigma}\\ &= \text{infectious contact rate}\times\text{infectious lifetime}\\ &\times\text{probability of contact with susceptible dog}\\ &\times\text{probability of developing infection}\end{aligned}$$

The eigenvalues of the Jacobian matrix of the system shown in Eqn 8.1 at DFE ε^0 are $-r$ and

$$\frac{-(2m+\mu+\sigma)\pm\sqrt{(2m+\mu+\sigma)^2+4(R_0-1)(m+\mu)(m+\sigma)}}{2}$$

Here, the first eigenvalue is clearly negative, while the other two are negative if $R_0 < 1$, and at least one is positive if $R_0 > 1$. Therefore, ε^0 is locally asymptotically stable for $R_0 < 1$ and unstable for $R_0 > 1$. In addition, the model in Eqn 8.1 has an endemic equilibrium:

$$\varepsilon^* = \left(\frac{N^*}{R_0}, N^*\left(1-\frac{1}{R_0}\right)\frac{\mu+m}{\sigma\gamma+\mu+m}, N^*\left(1-\frac{1}{R_0}\right)\frac{\sigma\gamma}{\sigma\gamma+\mu+m}\right)$$

which is biologically relevant only when $R_0 > 1$. Here,

$$N^* = D\left[1-(R_0-1)\frac{(\mu+m)(m+\sigma\gamma)}{r(\sigma\gamma+\mu+m)}\right]$$

The expression of N^* shows that when $R_0 = 1$, the dog population reaches the carrying capacity; however, if $R_0 > 1$, then the short lifespan of infected dogs hinders the total dog population to reach the carrying capacity. If $R_0 > 1 + \frac{r(\sigma\gamma+\mu+m)}{(\mu+m)(m+\sigma\gamma)}$, then the dog population will become extinct. A similar indication can also be observed from the stability investigation of ε^*.

The stability investigation of ε^* requires tedious mathematical analysis; therefore, we used numerical software MATCONT (Dhooge *et al.*, 2003) to investigate the stability of this equilibrium. The parameters β and γ are chosen as bifurcation parameters due to their importance in epidemic outbreaks.

The bifurcation analysis provides an integrated view of the dynamics of the model (Eqn 8.1). It reveals that ε^* is stable and rabies persists for $R_0 > 1$ (see Fig. 8.1). The figure shows several ranges of values for β and γ for a given r. The endemic equilibrium ε^* that is stable in region II bifurcates on the Hopf curve emitting a stable limit cycle through supercritical Hopf bifurcation, resulting in the formation of region III. As a result, the system exhibits periodic oscillations when the values of β and

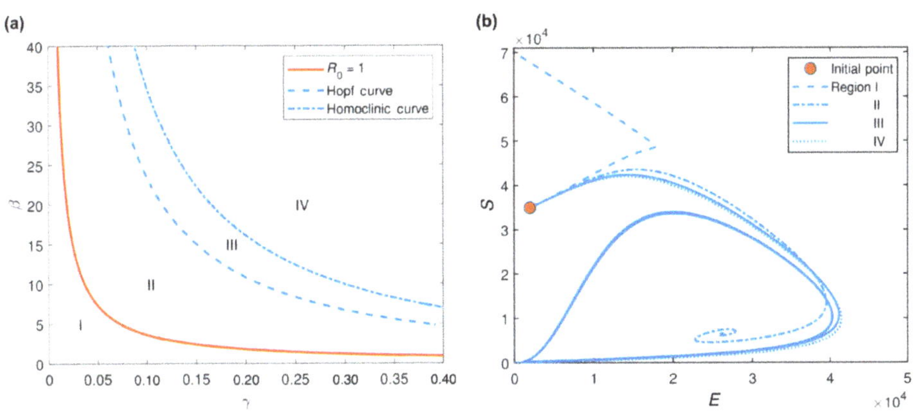

Fig. 8.1. (a) (β, γ) -parameter plane for = 1. Below the solid curve (region I), the system remains disease free, whereas rabies persists beyond this curve. Region III represents the ranges for **β** and γ, which lead to periodic oscillatory solutions. The dog population becomes extinct in region IV. (b) The trajectories corresponding to the four different regions of the bifurcation diagram. R_0, basic reproduction number; *E*, part of dog population that is exposed to rabies infection; *S*, part of dog population that is susceptible to rabies. Images created by the authors.

γ are chosen from this region. The size of the limit cycle that emerged from the Hopf bifurcation increases in the direction of increasing γ, which finally disappears via a saddle homoclinic bifurcation on the homoclinic curve, and thus, region IV is born. As a result, the values of β and γ chosen from region IV lead to the extinction of the dog population, suggesting that the values of β and γ corresponding to this region are unreasonably large for the given r.

8.2.2 Controlling rabies

The above analysis shows that the dynamics of the model shown in Eqn 8.1 are largely controlled by the parameter r in association with β and γ. Therefore, one can speculate that the spread of rabies can be hindered by controlling the birth rate in dog populations, or by securing the newborns, for example through vaccination. We experiment with three different vaccination strategies. The first one is continuous vaccination, in which the susceptible dog population is vaccinated at a constant rate. The next one is pulse vaccination, which refers to vaccinating a fraction of the entire susceptible dog population in a single pulse applied every θ year. The last one is time-dependent vaccination accompanied by carrying capacity control.

8.2.2.1 Continuous vaccination model

Here, we assume that dogs are vaccinated and neutered at a rate k and move to the recovered class. There are vaccines that need a booster dose after 2–3 years, while the lifespan of FRDs is approximately 3 years. Therefore, we assume that the vaccine provides permanent immunity. However, this may not be true in general as the duration of protection afforded by vaccines depends on the efficacy of the vaccine, the size of the dose, and various other factors. We, therefore, acknowledge that this is a limitation of the constant vaccination model.

$$
\begin{aligned}
\frac{dS(t)}{dt} &= rS(t)\left(1 - \frac{N(t)}{D}\right) - \beta S(t)\frac{I(t)}{N(t)} \\
&\quad + \sigma(1-\gamma)E(t) - kS(t) \\
\frac{dE(t)}{dt} &= \beta S(t)\frac{I(t)}{N(t)} - (\sigma + m)E(t) \\
\frac{dI(t)}{dt} &= \sigma\gamma E(t) - (\mu + m)I(t) \\
\frac{dR(t)}{dt} &= -kS(t) - mR(t)
\end{aligned}
\qquad (8.2)
$$

The basic reproduction number, R_0, reduces to:

$$R_0 = \frac{\beta\sigma\gamma m}{(\mu+m)(m+\sigma)(m+k)}$$

The above expression shows that R_0 increases with β and decreases with k. For $R_0 < 1$, we can easily find that, $k > \beta\frac{\sigma\gamma m}{(\mu+m)(m+\sigma)} - m$ (i.e. the threshold for k is a linear function of β). This shows that vaccination coverage should be high in the case of a high transmission rate.

8.2.2.2 Pulse vaccination model

Here, we present the pulse vaccination model derived from the model (Eqn 8.1). According to this strategy, a fraction $k' = k\theta$ of the entire susceptible dog population is vaccinated in a single pulse every θ year (Yongzhen *et al.*, 2011), where the value of k is such that $0 < k\theta < 1$.

$t \neq n\theta$

$$
\begin{aligned}
\frac{dS(t)}{dt} &= rS(t)\left(1 - \frac{N(t)}{D}\right) - \beta S(t)\frac{I(t)}{N(t)} \\
&\quad + \sigma(1-\gamma)E(t) \\
\frac{dE(t)}{dt} &= \beta S(t)\frac{I(t)}{N(t)} - (\sigma + m)E(t) \\
\frac{dI(t)}{dt} &= \sigma\gamma E(t) - (\mu + m)I(t) \\
\frac{dR(t)}{dt} &= -mR(t)
\end{aligned}
\qquad (8.3)
$$

$t^+ = n\theta,\ n \in \mathbb{Z}_+$

$$
\begin{aligned}
S(t^+) &= (1 - k')S(t) \\
E(t^+) &= E(t) \\
I(t^+) &= I(t) \\
R(t^+) &= k'S(t) + R(t)
\end{aligned}
$$

8.2.2.3 Optimal control model

In this section, we intend to explore strategies apart from vaccination that can be used independently or in combination with vaccination to reduce virus transmission among the dog population effectively. We consider vaccination and moderated growth rate through waste management as two potentially effective control measures. As we intend to limit the growth rate by reducing the carrying capacity, we aim to enhance the second-order term in the logistic growth model, since a decrease in carrying capacity can alternatively be achieved by increasing this term. We assume u_1 can increase the second-order term ρ times at maximum, where $u_1 \in [0, 1]$ and ρ are constants that depend on how much stress can be applied to suppress the growth. In our simulation, we assume $\rho = 10$, which means the impact of limiting resources (second order term in the logistic growth) can be increased to a maximum of ten times by implementing carrying capacity control.

We assume that the vaccine is provided at a rate u_2, where $u_2 \in [0, 1]$. Taking the above two controls into consideration, the model (Eqn 8.1) takes the following form:

$$\begin{aligned}\frac{dS(t)}{dt} &= rS(t)\left(1 - \frac{N(t)}{D}(1 + \rho u_1(t))\right)\\ &\quad -\beta S(t)\frac{I(t)}{N(t)} + \sigma(1-\gamma)E(t)\\ &\quad -u_2 S(t) \qquad (8.4)\\ \frac{dI(t)}{dt} &= \sigma\gamma E(t) - (\mu + m)I(t)\\ \frac{dR(t)}{dt} &= u_2(t)S(t) - mR(t)\end{aligned}$$

Though our objective is to reduce the number of rabid dogs, it is neither possible to reduce and maintain the growth rate at zero, nor to vaccinate all dogs. So, we construct a cost functional as follows to balance between the competing aims of reducing rabid dogs and implementing control measures.

$$\begin{aligned}J(u_1(t), u_2(t)) &= \int_0^T \left(CI(t) + \tfrac{1}{2}B_1 u_1(t)^2 + \tfrac{1}{2}B_2 u_2(t)^2\right)dt,\\ &\quad 0 \le u_1(t), u_2(t) \le 1 \qquad (8.5)\end{aligned}$$

Here, T is the final time; C is a constant representing the cost associated with each infected dog and plausible human infections resulting from the infected dog; and B_1 and B_2 are costs associated with the implementation of controls u_1 and u_2, respectively. Our aim is to identify the time-dependent piece-wise continuous controls $u_1^*(t)$ and $u_2^*(t)$ that minimize the cost $J(u_1(t), u_2(t))$ subject to the model shown in Eqn 8.4. Two time-dependent controls govern the dynamics of the system to minimize the number of infected dogs as well as the cost associated with the implementation of the controls.

Mathematically speaking, the aim is to determine the optimal controls $u_1^*(t)$ and $u_2^*(t)$ such that:

$$J(u_1^*(t), u_2^*(t)) = min\{J(u_1, u_2) \mid u_i : [0, T] \rightarrow [0, 1] \text{ is piece-wise continuous, } i = 1, 2\}$$

Such an optimal control pair $(u_1^*(t), u_2^*(t))$ exists (Masud *et al.*, 2020a) and the necessary conditions to be satisfied by the optimal controls are derived from Pontryagin's Maximum Principle (Fleming and Rishel, 1975). We construct the Hamiltonian, $\mathcal{H}$, given by:

$$\begin{aligned}\mathcal{H} := {} & CI(t) + \tfrac{1}{2}B_1 u_1(t)^2 + \tfrac{1}{2}B_2 u_2(t)^2\\ &+\lambda_1\left[rS(t)\left(1 - \tfrac{N(t)}{D}(1 + \rho u_1(t))\right)\right.\\ &\left. -\beta S(t)\tfrac{I(t)}{N(t)} + \sigma(1-\gamma)E(t) - u_2 S(t)\right]\\ &+\lambda_2\left[\beta S(t)\tfrac{I(t)}{N(t)} - (\sigma + m)E(t)\right]\\ &+\lambda_3[\sigma\gamma E(t) - (\mu + m)I(t)]\\ &+\lambda_4[u_2(t)S(t) - mR(t)] \qquad (8.6)\end{aligned}$$

Here λ_1, λ_2, λ_3 and λ_4 are called adjoin variables and given by the following system of ordinary differential equations:

$$\begin{aligned}\lambda_1'(t) = {} & -\lambda_1(t)r + \lambda_1 r(1 + \rho u_1(t))\tfrac{N(t)+S(t)}{D}\\ &+\beta I(t)(\lambda_1(t) - \lambda_2(t))\tfrac{N(t)-S(t)}{N^2(t)}\\ &+(\lambda_1(t) - \lambda_4(t))u_2(t)\end{aligned}$$

$$\lambda_2'(t) = \lambda_1 r(1 + \rho u_1(t))\frac{S(t)}{D} + \lambda_2(t)m - \beta(\lambda_1(t) - \lambda_2(t))\frac{S(t)I(t)}{N(t)^2} - \sigma(\lambda_1(t) - \lambda_2(t)) + \sigma\gamma(\lambda_1(t) - \lambda_3(t))$$

$$\lambda_3'(t) = -C + \lambda_1 r(1 + \rho u_1(t))\frac{S(t)}{D} + \lambda_3(t)(m + \mu) + \beta(\lambda_1(t) - \lambda_2(t))\frac{S(t)(N(t) - I(t))}{N(t)^2}$$

$$\lambda_4'(t) = \lambda_1 r(1 + \rho u_1(t))\frac{S(t)}{D} + \lambda_4(t)m - \beta(\lambda_1(t) - \lambda_2(t))\frac{S(t)I(t)}{N(t)^2}$$

with transversality conditions,

$$\lambda_i\left(t\right) = 0; i = 1, 2, 3, 4 \quad (8.7)$$

Finally, the optimal controls $u_1^*(t)$, $u_2^*(t)$ are characterized by $\frac{\partial \mathcal{H}}{\partial u_1} = 0$ and $\frac{\partial \mathcal{H}}{\partial u_2} = 0$, which give:

$$u_1^*(t) = min\left\{1, max\left\{0, \frac{r\rho\lambda_1^*(t)S^*(t)N^*(t)}{DB_1}\right\}\right\}$$

$$u_2^*(t) = min\left\{1, max\left\{0, S^*(t)(\lambda_1^*(t) - \lambda_4^*(t))\frac{1}{B_2}\right\}\right\}$$

In order to numerically solve the optimal control system, we use the Forward—Backward Sweep method (Lenhart and Workman, 2007). We assume $C = 1$, and $B_1 = B_2 = 10$. We set $B_1 = B_2$ in order to understand the relative importance of controlling carrying capacity and vaccination. The controls cannot be 100% effective (i.e. as we can neither reduce the growth rate to zero through waste food management, nor vaccinate all of the dogs at once). We assume that u_1 does not exceed 0.8. u_2 refers to the continuous vaccination rate, which has been reported to be 0.09 in Zhang *et al.* (2011) and 0.24 in Kitala *et al.* (2002). As this study deals with FRDs, we assume it does not exceed 0.3. Here, we present the details of a simulation to observe the optimal control strategies for 5 years. All the numerical simulations presented in this chapter are carried out using matlab 2018 a (matlab version 9.4.0 (R2018a), The MathWorks Inc., Natick, Massachusetts).

8.3 Parameterization

The population dynamics of unsupervised FRDs are very different from that of well-supervised, confined dogs. The lifespan of a FRD is about 3 years (Hampson *et al.*, 2007), which is much shorter than that of a domestic dog (Zhang *et al.*, 2011). The lifespan of domestic dogs varies depending on breeds, different patterns of breeding, body size, and responsible ownership that varies from community to community (Yordy *et al.*, 2020). However, Inoue *et al.* (2018) estimated the overall life expectancy of domestic dogs to be 13.7 (with 95% confidence interval (CI): 13.7–13.8) years. So, we assume the natural mortality rate $m = 0.33$/year. Yearly dog per-capita birth rate is assumed to be $0.0027 \times 365 - m = 0.6555$ (Leung and Davis, 2017). We assume a 2 month incubation period of the rabies virus in an infected dog (Zhang *et al.*, 2011) (i.e. the reciprocal of the incubation period $\sigma = 6$). The probability of clinical outcome is $\gamma = 0.4$. The duration of infectivity of an infected dog is assumed to be 5.7 days (Hampson *et al.*, 2007), which gives a disease-related death rate $\mu = 1/(5.7/365) \approx 64$. Rabies-related parameters in FRDs are described in Table 8.1. We now infer the carrying capacity D, and transmission rate, β, as well as R_0 in Dhaka.

From the data, we observe that the dog population almost doubled just in 3 years from 2013 to 2016 (i.e. according to our model, $1 < R_0 < 1.249$). We are now especially interested in which value of R_0 and D lead to that particular growth over 3 years. Assuming the dog population was at equilibrium in 2013, we simulated our model for $1 < R_0 < 1.249$, with different values of D. It should be noted that the initial conditions in 2013 were adapted according to the CNVR campaign outcome before the simulation (Tenzin *et al.*, 2015). The whole process is summarized as follows:

Table 8.1. Parameter values for numerical simulation. Table created by authors.

Notation	Value per year	Source
Dog per-capita birth rate, r	0.6555	Leung and Davis (2017)
Dog vaccination rate, k	≤ 0.3	Kitala *et al.* (2002), Zhang *et al.* (2011)
Dog-to-dog transmission rate, β	[169, 211]	Estimated
Risk of clinical outcome of exposed dogs, γ	0.4	Zhang *et al.* (2011)
The reciprocal of the incubation period, σ	6	Zhang *et al.* (2011)
Natural mortality rate, m	0.33	Hampson *et al.* (2007)
Disease-related death rate, μ	64	Hampson *et al.* (2007)

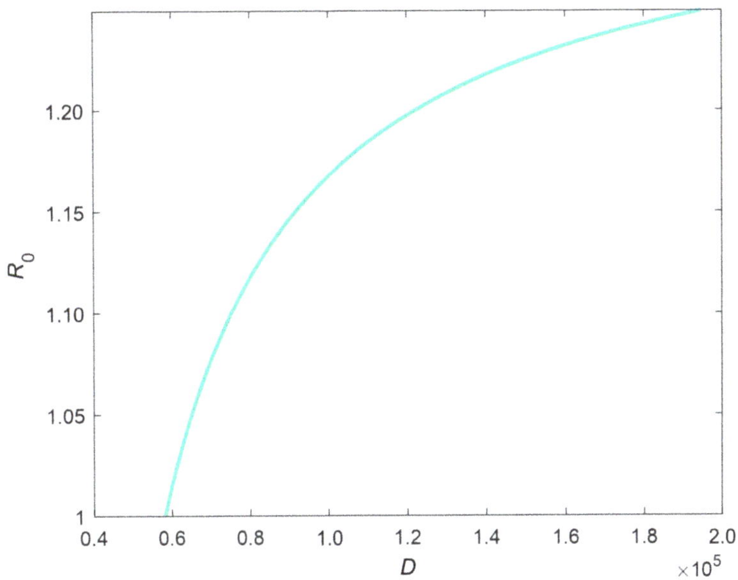

Fig. 8.2. Dog population dynamics in Dhaka 2013–2016. The curve shows the $R_0 - D$ pairs that explain dog population growth between the years 2013 and 2016 in Dhaka city, where R_0 is the basic reproduction number and D is the environmental carrying capacity of dogs. Image created by the authors.

- *Step 1*: Find all possible values of endemic equilibrium for $N^* = 18{,}585$ and $1 < R_0 < 1.249$ to be used as the initial condition.
- *Step 2*: Update the initial condition as $(S^*, E^*, I^*, R^*)+(-6665, 0, 0, 6665)$. Here, 6665 is the number of dogs vaccinated and neutered during the 2002–2003 CNVR programme (Tenzin *et al.*, 2015).
- *Step 3*: Run the model for each R_0 and associate the initial condition found in Step 2 to find the value of D that produces 37,009 dogs in 2016 (Choudhari, 2016).

The above algorithm gives us plausible $R_0 - D$ pairs to explain the rabies epidemic and dog population dynamics in Dhaka from 2013 to 2016, which is portrayed in Fig. 8.2.

8.4 What Was Found Out

The model considered in this study is concerned with rabies transmission among FRDs. The analyses show that a rabies outbreak is not possible if $R_0 < 1$. For $R_0 > 1$, an epidemic may take off. We consider reducing the dog carrying capacity as a means of reproduction control in addition to vaccination as plausible effective control measures. The effect of carrying capacity on the dog population has already been established. Reducing shelter and food sources through waste management could be a cheap way to reduce dog turnover. Here, we examine theoretically whether this method is effective in controlling rabies transmission.

For each $R_0 - D$ pair shown in Fig. 8.2, we find a distribution of the dog population for four different compartments for the year 2016. In further analysis, we use this as the initial condition and simulate the model from 2016. It is worth mentioning here that there was another CNVR programme in 2016, in which 26% of the dogs in the north and 13% of the dogs in the south of Dhaka were vaccinated (Hossain *et al.*, 2012; Ghosh *et al.*, 2020). We updated the initial conditions considering the vaccination that was administered in 2016. We then used these updated initial conditions to simulate the model with constant vaccination, the pulse vaccinating model, and the model that introduced optimal vaccination along with carrying capacity control.

The red dotted and solid lines in Fig. 8.3 show the number of infected dogs in a population that is obtained from the constant vaccination model with and without vaccination. The blue lines show the number of infected dogs in a population for pulse vaccination at three different rates, and the green lines show this dog population with optimal controls (viz. carrying capacity control, u_1 and vaccination, u_2). We simulated the models for a period of 5 years, where the value of R_0 was chosen as 1.1 and 1.23 along with two types of vaccination coverage ($0 < u_2 < 0.1$ and $0 < u_2 < 0.3$). The optimal control model is simulated under three different scenarios: (i) when both $u_1$and u_2 are implemented; (ii) when only u_1 is implemented; and (iii) when only u_2 is implemented. We also simulated the model for no vaccination coverage. This allows us to make a comparison between dog rabies scenarios in controlled and uncontrolled cases. The results are presented in Fig. 8.3(a–d). The corresponding values of the time-dependent controls are shown on the right panel. In our simulations, we refer to $R_0 = 1.1$ as a low transmission setting and to $R_0 = 1.23$ as a high transmission setting. Similarly, we refer to $k = 0.1$ as a low vaccination rate and to $k = 0.3$ as a high vaccination rate.

We notice that the implications of different controls used in this study may provide a significantly better dog rabies scenario compared to no control at all; however, when we look further into the details, we see that one produces better results in reducing the number of rabid dogs than the other. Note that the performance depends on the transmission settings as well as the vaccination rates.

We find that both the constant and the pulse vaccination corresponding to a low vaccination rate reduces the number of rabid dogs greatly; however, the optimal control model where the vaccination is accompanied by carrying capacity control produces an even better result. This is true both in low and in high transmission settings (see Fig. 8.3a and b). It is worth mentioning that carrying capacity control alone (when only u_1 is implemented) is shown to be more promising compared to constant and pulse vaccinations in high transmission settings, as opposed to low transmission. This indicates that the carrying capacity control alone may not be a reliable strategy to control rabies in general, and signifies the simultaneous implementation of vaccination and carrying capacity control. In contrast, the pulse vaccination model is shown to be more promising compared to other control strategies in the case of a high vaccination rate (see Fig. 8.3c and d). It is true both in low and in high transmission settings; however, it should be noted that vaccinating FRDs at the rate of 30% every year is a quite challenging task.

8.5 Conclusion

We studied a dog population model of rabies transmission dynamics to investigate the role of vaccination and dog population management in controlling rabies in FRDs. The analysis of the basic model showed that the birth rate of dogs significantly affects dog dynamics and rabies transmission in FRDs.

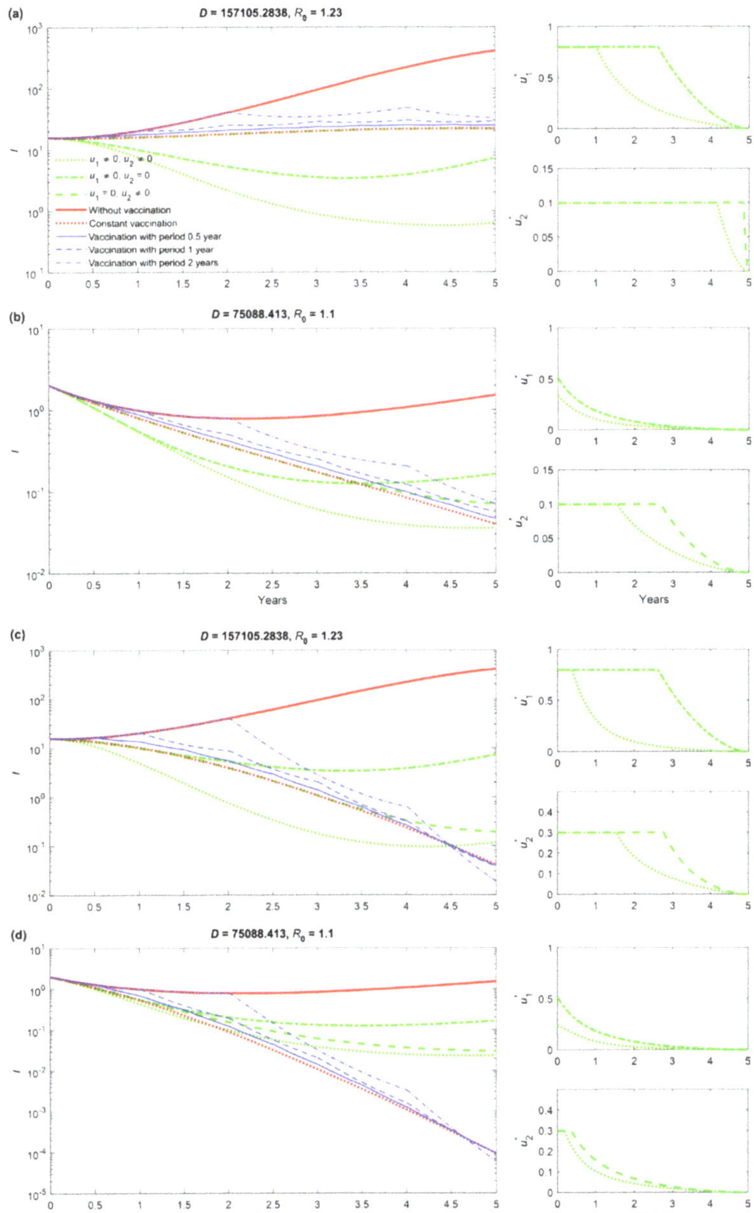

Fig. 8.3. Comparison between different types of control strategies (a–d). (a) Vaccination coverage, and basic reproduction number, R_0= 1.23; (b) k = 0.1, R_0= 1.1; (c) k = 0.3, R_0= 1.23 ; and (d) k = 0.3, R_0= 1.1. D is the environmental carrying capacity of dogs; I is the number of infected dogs; and u_1 and u_2 are the carrying capacity control and the vaccination rate control, respectively. The corresponding optimal control strategies are presented in the right-hand column of each graph (the asterisk '*' above 'u_1' and 'u_2' indicates the optimal value of 'u_1' and 'u_2' for which the cost function is minimized). Images created by the authors.

This quantity is directly associated with an upper limit of the transmission coefficient (in other words, an upper limit of the basic reproduction number at which the dog population becomes extinct); therefore, it was primarily expected that reducing this quantity may help to control or reduce the spread of rabies. The idea was validated by the results of the advanced model, where we found that carrying capacity reduction significantly reduces the number of rabid dogs. However, this control measure alone was not effective enough to control rabies, and it showed different performances under different transmission settings and vaccination rates as follows. Corresponding to a low vaccination rate, the vaccination was found to dominate other control measures in low transmission settings, whereas the carrying capacity reduction was found to play a dominant role in the case of high transmission settings; therefore, the carrying capacity control was expected to be a good supplementary control measure. It leads us to conclude that simultaneous implementation of vaccination and carrying capacity control could be a reliable approach to controlling rabies in FRDs. This notion is supported by the results corresponding to the low vaccination rate, as shown in Fig. 8.3(a and b). In contrast, an alternative scenario was observed in the case of a high vaccination rate, where the pulse vaccination was found to dominate other control measures both in low and in high transmission settings. However, vaccination at the proposed rate of 30% every year is a quite challenging task for FRDs, as the maximum vaccination rate in Dhaka was 26% in 2016. Therefore, we recommend the simultaneous use of carrying capacity control and vaccination as control measures.

The results of this study should encourage community awareness against dumping food waste in undesignated areas, leaving food garbage open for dogs to access, or feeding FRDs. Finally, it is worth mentioning that the model parameters were estimated on a trial basis due to a shortage of data. These parameters could be estimated more accurately when there are sufficient data and hence, lead to better accuracy of the predicted results. A One Health approach would therefore be the best possible way to inhibit the transmission of rabies in FRDs in Bangladesh and hence, achieve the Zero by 30 goal. The control strategies proposed in this study are expected to be equally effective in other Asian countries, as FRDs share the major portion of the total dog population, the food waste sites are poorly managed, and the community has the habit of dumping food in undesignated areas and feeding FRDs.

Authors' Declaration

All authors declare they have no conflict of interest.

All authors have approved this manuscript, agree with its submission, and share collective responsibility and accountability.

Most of the content of this chapter overlaps with our previously published work in Masud *et al.* (2020a).

References

Acharya, K.P., Acharya, N., Phuyal, S., Upadhyaya, M. and Lasee, S. (2020) One-health approach: a best possible way to control rabies. *One Health* 10, 100161. DOI: 10.1016/j.onehlt.2020.100161.

Baquero, O.S., Akamine, L.A., Amaku, M. and Ferreira, F. (2015) Defining priorities for dog population management through mathematical modeling. *Preventive Veterinary Medicine* 123, 121–127. DOI: 10.1016/j.prevetmed.2015.11.009.

Bingham, J. (2005) Canine rabies ecology in southern Africa. *Emerging Infectious Diseases* 11(9), 1337–1342. DOI: 10.3201/eid1109.050172.

Choudhari, A. (2016) Report of roaming dog survey of Dhaka City Corporation, Bangladesh. *Stray and Feral Animal Populations Collection* 5, 1–11.

Czupryna, A.M., Brown, J.S., Bigambo, M.A., Whelan, C.J., Mehta, S.D. *et al.* (2016) Ecology and demography of free-roaming domestic dogs in rural villages near Serengeti National Park in Tanzania. *PLOS ONE* 11(11), e0167092. DOI: 10.1371/journal.pone.0167092.

Dhooge, A., Govaerts, W. and Kuznetsov, Y.A. (2003) MATCONT: a MATLAB package for numerical bifurcation analysis of ODEs. *ACM Transactions in Mathematical Software (TOMS)* 29(20), 141–164.

Fahrion, A.S., Taylor, L.H., Torres, G., Müller, T., Dürr, S. *et al.* (2017) The road to dog rabies control and elimination – What keeps us from moving faster? *Frontiers in Public Health* 5, 103. DOI: 10.3389/fpubh.2017.00103.

Fleming, W.H. and Rishel, R.W. (1975) *Deterministic and Stochastic Optimal Control*, Vol. 1. Springer Science & Business Media, New York. DOI: 10.1007/978-1-4612-6380-7.

Hampson, K., Dushoff, J., Bingham, J., Brückner, G., Ali, Y.H. *et al.* (2007) Synchronous cycles of domestic dog rabies in sub-Saharan Africa and the impact of control efforts. *Proceedings of the National Academy of Sciences of the United States of America* 104(18), 7717–7722. DOI: 10.1073/pnas.0609122104.

Ghosh, S., Rana, M.S., Islam, M., Chowdhury, S., Haider, N., *et al.* (2020) Trends and clinico-epidemiological features of human rabies cases in Bangladesh, 2006–2018. *Scientic Reports* 10(1), 1–11. DOI: 10.1017/ice.2020.1072.

Hampson, K., Coudeville, L., Lembo, T., Sambo, M., Kieffer, A. *et al.* (2015) Estimating the global burden of endemic canine rabies. *PLoS Neglected Tropical Diseases* 9(4), e0003709. DOI: 10.1371/journal.pntd.0003709.

Hossain, M., Ahmed, K., Bulbul, T., Hossain, S., Rahman, A. *et al.* (2012) Human rabies in rural Bangladesh. *Epidemiology and Infection* 140(11), 1964–1971. DOI: 10.1017/S095026881100272X.

Hou, Q., Jin, Z. and Ruan, S. (2012) Dynamics of rabies epidemics and the impact of control efforts in Guangdong Province, China. *Journal of Theoretical Biology* 300, 39–47. DOI: 10.1016/j.jtbi.2012.01.006.

Inoue, M., Kwan, N.C.L. and Sugiura, K. (2018) Estimating the life expectancy of companion dogs in Japan using pet cemetery data. *The Journal of Veterinary Medical Science* 80(7), 1153–1158. DOI: 10.1292/jvms.17-0384.

Keller, J.P., Gerardo-Giorda, L. and Veneziani, A. (2013) Numerical simulation of a susceptible-exposed-infectious space-continuous model for the spread of rabies in raccoons across a realistic landscape. *Journal of Biological Dynamics* 7, 31–46. DOI: 10.1080/17513758.2012.742578.

Kim, S., Seo, Y.B. and Jung, E. (2020) Prediction of COVID-19 transmission dynamics using a mathematical model considering behavior changes in Korea. *Epidemiology and Health* 42, e2020026. DOI: 10.4178/epih.e2020026.

Kitala, P.M., McDermott, J.J., Coleman, P.G. and Dye, C. (2002) Comparison of vaccination strategies for the control of dog rabies in Machakos District, Kenya. *Epidemiology and Infection* 129(1), 215–222. DOI: 10.1017/s0950268802006957.

Lenhart, S. and Workman, J.T. (2007) *Optimal Control Applied to Biological Models*. Chapman and Hall, London. DOI: 10.1201/9781420011418.

Leung, T. and Davis, S.A. (2017) Rabies vaccination targets for stray dog populations. *Frontiers in Veterinary Science* 4, 52. DOI: 10.3389/fvets.2017.00052.

Masud, M.A., Islam, M.H., Adnan, M.I. and Oh, C. (2020a) Dog rabies in Dhaka, Bangladesh, and implications for control. *Processes* 8(11), 1–15. DOI: 10.3390/pr8111513.

Masud, M.A., Islam, M.H., Mamun, K.A., Kim, B.N. and Kim, S. (2020b) COVID-19 transmission: Bangladesh perspective. *Mathematics* 8(10), 1–19.

Nunes, C.M., de Lima, V.M.F., de Paula, H.B., Perri, S.H.V., de Andrade, A.M. *et al.* (2008) Dog culling and replacement in an area endemic for visceral leishmaniasis in Brazil. *Veterinary Parasitology* 153(1–2), 19–23. DOI: 10.1016/j.vetpar.2008.01.005.

Pal, S.K. (2001) Population ecology of free-ranging urban dogs in West Bengal, India. *Acta Theriologica* 46(1), 69–78. DOI: 10.1007/BF03192418.

Pastoret, P., Van Gucht, S. and Brochier, B. (2014) Eradicating rabies at source. *Revue Scientifique et Technique Office International des Epizooties (OIE)* 33(2), 509–519.

Pieracci, E.G., Pearson, C.M., Wallace, R.M., Blanton, J.D., Whitehouse, E.R., *et al.* (2019) Vital signs: trends in human rabies deaths and exposures – United States. *Morbidity and Mortality Weekly Report* 524, 1938–2018.

Ruan, S. (2017) Modeling the transmission dynamics and control of rabies in China. *Mathematical Biosciences* 286, 65–93. DOI: 10.1016/j.mbs.2017.02.005.

Singh, R., Singh, K.P., Cherian, S., Saminathan, M., Kapoor, S. *et al.* (2017) Rabies - epidemiology, pathogenesis, public health concerns and advances in diagnosis and control: a comprehensive review. *The Veterinary Quarterly* 37(1), 212–251. DOI: 10.1080/01652176.2017.1343516.

Taylor, L.H., Wallace, R.M., Balaram, D., Lindenmayer, J.M., Eckery, D.C. *et al.* (2017) The role of dog population management in rabies elimination – A Review of Current Approaches and Future Opportunities. *Frontiers in Veterinary Science* 4, 109. DOI: 10.3389/fvets.2017.00109.

Tenzin, T., Ahmed, R., Debnath, N.C., Ahmed, G. and Yamage, M. (2015) Free-roaming dog population estimation and status of the dog population management and rabies control program in Dhaka City, Bangladesh. *PLoS Neglected Tropical Diseases* 9(5), e0003784. DOI: 10.1371/journal.pntd.0003784.

van den Driessche, P. and Watmough, J. (2002) Reproduction numbers and sub-threshold endemic equilibria for compartmental models of disease transmission. *Mathematical Biosciences* 180, 29–48. DOI: 10.1016/s0025-5564(02)00108-6.

Wandeler, A.I. (1985) Ecological and epidemiological data requirements for the planning of dog rabies control. In: Kuwert, E., Mérieux, C., Koprowski, H. and Bögel, K. (eds) *Rabies in the Tropics*. Springer, Berlin, pp. 657–661.

WHO (World Health Organization) (2021) New global framework to eliminate rabies. Available at: https://www.who.int/news/item/10-12-2015-new-global-framework-to-eliminate-rabies (accessed 26 September 2021).

Yongzhen, P., Shuping, L., Changguo, L. and Chen, S. (2011) The effect of constant and pulse vaccination on an SIR epidemic model with infectious period. *Applied Mathematical Modelling* 35(8), 3866–3878. DOI: 10.1016/j.apm.2011.02.012.

Yordy, J., Kraus, C., Hayward, J.J., White, M.E., Shannon, L.M. *et al.* (2020) Body size, inbreeding, and lifespan in domestic dogs. *Conservation Genetics* 21(1), 137–148. DOI: 10.1007/s10592-019-01240-x.

Zhang, J., Jin, Z., Sun, G.-Q., Zhou, T. and Ruan, S. (2011) Analysis of rabies in China: transmission dynamics and control. *PLoS ONE* 6(7), e20891. DOI: 10.1371/journal.pone.0020891.

9 Rabies in the Greater Manila Area and Region IV-B of the Philippines and the Potential Impact of Age-targeted Dog Vaccination

Mirava Yuson[1]*, Daria Llenaresas Manalo[2], Mary Elizabeth Gutierrez Miranda[1], Katie Hampson[3] and Shynie Telmo[4]

[1]Field Epidemiology Training Program Alumni Foundation, Inc., Quezon City, Philippines; [2]Research Institute of Tropical Medicine, Muntinlupa City, Philippines; [3]Institute of Health & Wellbeing, University of Glasgow, Glasgow, UK; [4]Regional Animal Disease Diagnostic Laboratory IV-B, Naujan, Oriental Mindoro, Philippines

Abstract

Rabies is endemic to the Philippines, causing more than 300 human deaths nationwide/year. Locally, its mode of transmission is through the bites of dogs, which commonly roam provincial and city streets. Using rabies surveillance data from animal samples submitted between 2011 and 2020 to laboratories in the Greater Manila Area and Region IV-B, this study modelled the impact of age-targeted vaccination on dogs. A rabies transmission model was developed, with an emphasis on separate puppy and adult age classes. Puppies less than 3 months of age are normally exempt from rabies vaccination, but the model showed that mass vaccinations with a vaccine coverage of 50%, that included puppies, led to eventual elimination of the disease. This study emphasizes the importance of immunizing not just adult dogs but puppies as well, while suggesting the need for stricter population control methods to reduce the incidence of rabies in the Philippines.

9.1 Introduction

Hundreds of people die annually from rabies in the Philippines, wherein the disease remains endemic (Tohma *et al.*, 2016). Dogs are the main vector of *Lyssavirus* locally, shedding it in the saliva and thus spreading it to other mammals primarily through biting. When infected, a dog undergoes an incubation period ranging from 2 days to 2 months, before developing neurological signs very rapidly and dying within 2 weeks. The closer the bite wound to the brain, the shorter the distance for the virus to travel – thus, bite location may greatly decrease rabies' incubation period. In order to diagnose rabies in dogs, brain tissue must be brought to a diagnostic laboratory, wherein the viral antigen is confirmed using the fluorescent antibody test (FAT), which has a sensitivity of 98% (WHO, 2018).

Metro Manila is officially known as the National Capital Region (NCR), but when paired with bordering provinces Bulacan, Laguna, Cavite and Rizal, they are altogether referred to as the Greater Manila Area. Collectively,

*Corresponding author: myuson@fetpafi.org

 One Health for Dog-mediated Rabies Elimination in Asia (eds V. Slack *et al.*)
DOI: 10.1079/9781800622975.0009

these provinces were responsible for 23.61% of human rabies cases nationwide in 2018 (Department of Health, 2018). South of the NCR are the island provinces making up Region IV-B of the Philippines, also termed 'Mimaropa', comprising Mindoro, Marinduque, Romblon and Palawan. This region was responsible for 1.3% of human rabies cases nationwide in 2018. Province-wise, Mindoro and Palawan have consistently exhibited higher incidence rates of human rabies compared to Marinduque (which remained rabies free until 2018) and Romblon (rabies free as of mid-2019). For an area in the Philippines to be declared 'rabies free', it must have had no human or animal rabies case for at least 2 years, while maintaining an adequate rabies surveillance system and import policy (Department of Health, 2012). A map of the regions can be found in Fig. 9.1.

Stray dogs are common in the Philippines, and even dogs considered as 'owned' dogs blur the line between strays and pets, as many can be seen wandering the streets (Lapiz *et al.*, 2012). Demographic data on dogs is limited, but government efforts to eliminate the disease partly by combating dog population growth have included mass spay and neuter programmes and culling of strays. Having adopted the Zero by 30 global strategic plan, the Philippines' measures to curb the disease include nationwide mass dog

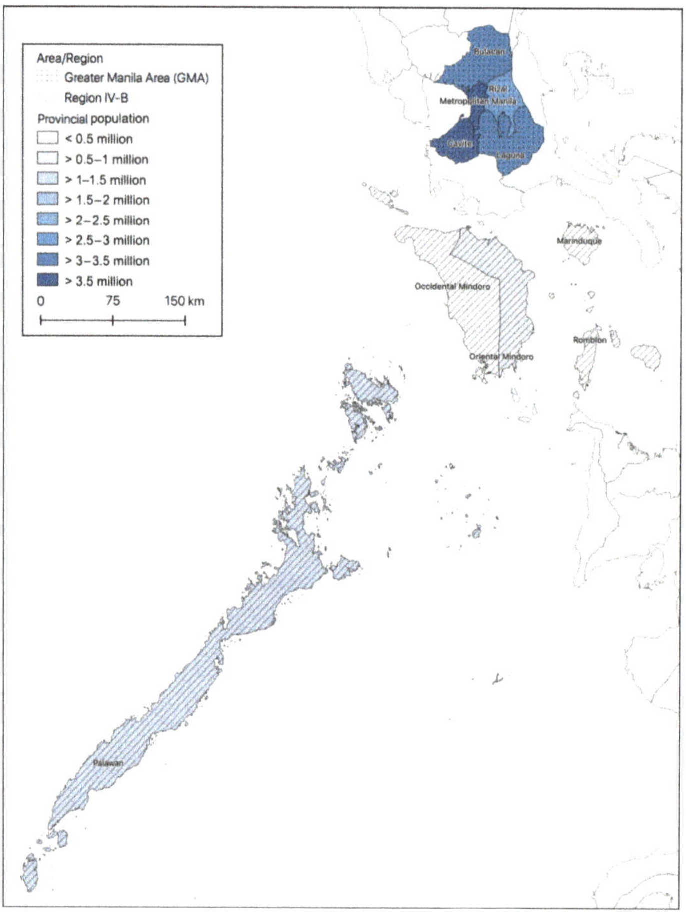

Fig. 9.1. Map of the Philippines depicting the provincial populations of the Greater Manila Area and Region IV-B. Figure created by Mirava Yuson.

vaccinations, which aim for an annual 70% vaccination coverage at minimum. But despite the aforementioned practices, dogs still persist as the primary cause for the majority of human rabies cases as of 2021. Nevertheless, the Philippines as a whole has aimed to follow the rabies guidelines set by World Health Organization (WHO) in 2008, and public awareness has been steadily increasing due to information drives and the establishment of over 500 animal bite treatment centres (ABTCs) throughout the country (Amparo *et al.*, 2018).

In accordance with national guidelines, the rabies vaccine is administered only to dogs that are at least 3 months old, although WHO recommends that it also be given to puppies regardless of age (WHO, 2018). The primary reason for the discrepancy is that vaccine manufacturers recommend against immunizing puppies, as immune responses are hypothesized to be hampered at that age due to their incompletely developed immune systems and circulating maternal antibodies. However, there is substantial evidence that inactivated rabies vaccines are actually effective in young puppies (Morters *et al.*, 2015). Previous studies show that up to 40% of rabies-positive puppies in the Philippines are 3 months old or younger (Domingo and Mananggit, 2014). Efforts have been made to determine what further actions can be taken to eliminate the disease locally. If mass vaccination efforts were to include puppies, incidents of rabies exposure in humans could potentially decrease – one among many expected positive results stemming from a transdisciplinary partnership with the animal health sector, that is performed congruently with the One Health approach.

In the event of a biting incident, bite victims are advised to undergo the rabies post-exposure prophylaxis regimen, while the biting animal is confined and observed for 10–14 days for signs of illness or sudden death. The FAT is only available for confirming rabies diagnosis in animals in select laboratories in the Philippines. The Research Institute of Tropical Medicine (RITM), located in Muntinlupa City, Metro Manila, receives the majority of animal rabies samples from the Greater Manila Area, while also performing genomic sequencing for rabies samples collected nationwide. The Regional Animal Disease Diagnostic Laboratory (RADDL) in Calapan City, Oriental Mindoro, plays a similar role in Region IV-B, but forwards samples to RITM for genomic sequencing.

Mathematical modelling is a tool for exploring epidemic scenarios and is commonly used to estimate the impact of different control measures such as mass dog vaccination. Modelling is also useful for determining which parameters have the biggest effect on the incidence of a disease. The objective of this study was to create a model that incorporates age classes and mass vaccination and examine whether age-targeted vaccination of puppies that are 3 months old or younger would decrease the incidence of rabies in dogs.

9.2 Materials and Methods

9.2.1 Assessing rabies incidence

The study was conducted using dog rabies samples submitted to RITM and RADDL from provinces within the Greater Manila Area and Region IV-B, respectively, in conjunction with the SPEEDIER project (managed by the Field Epidemiology Training Program Alumni Foundation, Inc. and the University of Glasgow). Brain tissue specimens from dogs suspected of rabies were brought to the diagnostic laboratory of the Veterinary Research Division in RITM or RADDL for confirmation via FAT. Information regarding each case included species, age, colour, municipality, and vaccination and ownership status. Samples were collected over a span of 10 years, from 2011 to 2020.

Rabies incidence was calculated for both human and dog populations in a given year on a regional and provincial level. We then explored the temporal trends in rabies cases across the two regions and compared the spatial distribution patterns in these cases.

9.2.2 Structuring an SEIV canine rabies model

A deterministic compartmental model consisting of differential equations and two age structures was developed to simulate the transmission dynamics of the rabies virus in the dog population of the two regions. The SEIV model comprised compartments Susceptible (S), Exposed (E), Infectious (I) and Vaccinated (V) for each age class (puppy ($_p$) and adult ($_a$)), with random mixing between dogs regardless of age assumed to take place and puppies referring to dogs of less than 3 months old, as shown in Fig. 9.2.

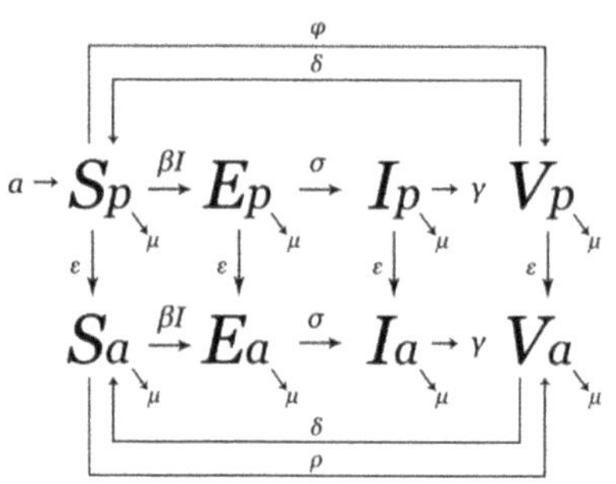

Fig. 9.2. Age-structured SEIV (**S**usceptible, **E**xposed, **I**nfectious and **V**accinated) canine rabies model. A compartmental diagram depicting the transmission of canine rabies, further categorized by the inclusion of two age classes, puppy ($_p$) and adult ($_a$). *a*, birth rate; *βI*, the effective infectious contact rate, or the force of infection; *σ*, incubation period; *γ*, infectious period; *μ*, natural mortality rate; *ε*, ageing rate; *φ*, puppy vaccination rate; *ρ*, adult vaccination rate; *δ*, duration of immunity. For full explanation of the flow diagram see text. Diagram created by Mirava Yuson.

The flow diagram of the age-structured model shown in Fig. 9.2 depicts the transmission of rabies among dogs with the inclusion of birth (*a*) and death (*μ*) rates. Although the same alphabet letter is used for both, the parameter for birth rate (*a*) is unrelated to the adult compartments (which use 'a' as a subscript, shown as $_a$). In this model, dogs were born as susceptible puppies (*Sp*), and could eventually become exposed (*Ep*) to rabies, get infected (*Ip*), then die. Following vaccination, susceptible puppies are moved on to the vaccinated puppies (*Vp*) compartment. After a period of 1-year, vaccine-induced immunity is assumed to wane according to national guidelines (Bureau of Animal Industry, 2019a), therefore dogs could become susceptible once more. Hence, vaccinated puppies could return to becoming susceptible puppies as shown in the diagram (in accordance with the model's differential equations) but factoring in the 1 year duration of vaccine immunity will always cancel out the number of puppies doing that, as they will have aged into adults instead. Susceptible puppies could also transition into susceptible adults (*Sa*); the same condition applies to exposed puppies who would grow up into exposed adults (*Ea*), and so forth. Similarly, vaccinated adults (*Va*) could lose their immunity and return to being susceptible adults. To see all the parameters and their definitions, please refer to Table 9.1.

The following system of differential equations was developed for the model, and

Table 9.1. Parameters described in the SEIV model.

Parameter	Value	Source
Birth rate (*a*)	0.38/year	Ferguson *et al.* (2015)
Natural mortality rate (1/*μ*)	3 years	Leung and Davis (2017)
Mean habitat carrying capacity (*K*)	337.44 dogs/km	Estimated
Growth rate (*r*)	0.38/year	Estimated
Ageing rate (1/*ε*)	90 days	Estimated
Incubation period (1/*σ*)	10 days	Department of Health (2012)
Infectious period (1/*γ*)	7 days	Department of Health (2012)
Total population (initial *N*)	1,250,000 dogs	Estimated
Puppy vaccination rate (*φ*)	0/year	Estimated
Adult vaccination rate (*ρ*)	0.53/year	Bureau of Animal Industry (2019b)
Duration of immunity (*δ*)	1 year	Estimated

implemented in the programming language R (RStudio Team, 2020), using the package deSolve (Soetaert *et al.*, 2010):

$$N = S_p + E_p + I_p + V_p + S_a + E_a + I_a + V_a \quad (9.1)$$

$$dS_p/dt = a(S_a + E_a + I_a + V_a) - \beta S_p(I_p + I_a) - \mu S_p - \varphi S_p - (rN/K)S_p - \varepsilon S_p + \delta S_p \quad (9.2)$$

$$dE_p/dt = \beta S_p(I_p + I_a) - \sigma E_p - \mu E_p - (rN/K)E_p - \varepsilon E_p \quad (9.3)$$

$$dI_p/dt = \sigma E_p - \gamma I_p - \mu I_p - (rN/K)I_p - \varepsilon I_p \quad (9.4)$$

$$dV_p/dt = -\mu V_p + \varphi S_p - (rN/K)V_p - \varepsilon V_p - \delta V_p \quad (9.5)$$

$$dS_a/dt = -\beta S_a(I_p + I_a) - \mu S_a - \rho S_a - (rN/K)S_a + \varepsilon S_p + \delta S_a \quad (9.6)$$

$$dE_a/dt = \beta S_a(I_p + I_a) - \sigma E_a - \mu E_a - (rN/K)E_a + \varepsilon E_p \quad (9.7)$$

$$dI_a/dt = \sigma E_a - \gamma I_a - \mu I_a - (rN/K)I_a + \varepsilon I_p \quad (9.8)$$

$$dV_a/dt = \rho S_a - \mu V_a - \delta V_a - (rN/K)V_p + \varepsilon V_p \quad (9.9)$$

9.2.3 Parameterization

A summarized description of the parameters used is listed in Table 9.1. Due to the limited availability of dog demographic data in the Philippines, the birth rate (a) and natural mortality rate ($1/\mu$) were taken from other literature. The authors believed that the following studies' findings can be generalized and applied to this study: the birth rate was based on that of a previous study conducted in Region VI (Ferguson *et al.*, 2015), while the natural mortality rate was based on the life expectancy of owned free-roaming dogs in sub-Saharan Africa (Leung and Davis, 2017). The growth rate (r) was the calculated difference between the birth rate and natural mortality rate. Since the model assumes logistic growth, the growth rate indicates the rate at which the population will reach equilibrium at a given carrying capacity. The ageing rate was derived from the number of days dogs spent as puppies (90) before being qualified to receive the rabies vaccine as 'adults'. Random mixing was assumed since, despite confinement, household dogs are often given outside access by their owners and can still interact with stray dogs, maintaining similar home ranges. Regardless of urban morphology, dog population density and number of contacts of puppies and adult dogs were presumed to be the same, as was the total force of infection. Human-mediated transfer of dogs in and out of the regions was not taken into account.

The effective contact rate (β) represents the probability of transmitting rabies per contact with another dog. Carrying capacity (K) was incorporated to restrict population growth, with the assumption that in the model, the value remained the same throughout the study duration. The following equation was used to calculate the carrying capacity (Kitala *et al.*, 2002):

$$K = (a + \sigma)(a + \gamma)/\beta\sigma \quad (9.10)$$

The dog population (1,250,000 dogs) was estimated by multiplying the mean number of Philippine dogs per household according to a study from the province of Bicol (Barroga *et al.*, 2018) with the number of households in the areas from which samples were collected (Greater Manila Area and Region IV-B). The number of households was calculated by means of dividing the total recorded population of 2020 by the average number of persons per household, based on a government census conducted in 2015 (Philippine Statistics Authority, 2016).

The basic reproduction number (R_0) was calculated using the following equation (Tohma *et al.*, 2016):

$$R_0 = \sigma\beta K/[(\sigma + a)(\gamma + a)] \quad (9.11)$$

Puppy vaccination rate (φ) was maintained at zero since dogs that are younger than 3 months old are considered exempt from vaccination in the Philippines. The adult vaccination rate (ρ) was set at 53.3%, which is the estimated overall percentage of dogs that were vaccinated in the Philippines in 2018 (Bureau of Animal Industry, 2019b). Vaccine efficacy was assumed to be 100%. Because dogs are recommended for yearly vaccination in the Philippines, the vaccine's duration of immunity was assumed to be 1 year.

The incubation period ($1/\sigma$) was estimated at 10 days, in accordance with the Department of Health (2012)'s data showing the minimum duration in which the virus incubates in dogs. Because clinical symptoms can manifest for up to 7 days before the dog falls into a coma, that aforementioned value was defined as the infectious period ($1/\gamma$).

9.2.4 Modelling scenarios

The model was initialized with 100 cases in the dog population split into puppy and adult infections based on the proportion in each age class, with vaccination coverage initially set to zero for both puppies and adults to mimic a rabies outbreak situation taking place in an entirely susceptible population. Then, to simulate the current situation, wherein rabies is endemic, alteration of parameters of the model was applied to create a curve resembling that of the data collected between 2011 and 2020. Vaccination was initiated at the defined rate, assumed to have been implemented consistently over the years beginning with the model's initialization.

The proportion of puppies to adults in the total population was adjusted from 10 to 80% to see whether dog demographics had a large impact on the transmission of rabies. To determine the effect of immunizing puppies, the vaccination rates for adults and puppies were adjusted separately between 0 and 70% and the effect on rabies incidence was observed. Taking into account scenarios depicting different proportions of puppies to adults, the resulting number of rabies cases per age group was calculated and compared. Since infected dogs remain in the modelled population until they die, cases at the end of the year in the model were summed and their proportion to the dog population was estimated as a measure of the disease's prevalence. These results were then divided by the length of the infectious period in order to produce the incidence rate.

Calibration of the carrying capacity and vaccination rate by experimenting with slight reductions and increases were performed in an attempt to achieve a better fit. Sensitivity analyses were conducted in order to determine the quantitative effect of each given parameter on the total number of rabies cases.

9.3 Results

9.3.1 Demographic age groups of rabid dogs

Ages of rabid dogs within the Greater Manila Area were recorded based on interviews with the person who submitted the carcass, who was – more often than not – also the dog's owner. Age-wise, 495 (73%) were adults, 14 (2.1%) were puppies of less than 3 months of age, and 169 (24.9%) were dogs of unknown age. Throughout the same duration, Region IV-B's data on confirmed rabid dogs were sparse, with not more than three cases confirmed annually, and these were based solely on specimens submitted to RITM.

The precise ages of Region IV-B dogs were often listed as unknown between 2011 and 2013, while no rabid dogs were recorded in 2014. A rabid puppy was recorded as early as 2012 in Oriental Mindoro. Out of a total of 50 dog specimens reported over 10 years, ten (20%) were specified to be puppies, 16 (32%) as adults, and the remaining 24 (48%) had no given ages. The limited data from Region IV-B does not allow for much inference. However, generalized data compiled for the Greater Manila Area show some notable patterns: among adult dogs with specified ages listed, 339 (69.8%) were between 1 and 4 years old, 99 (20.4%) were between 5 and 8 years old and the remaining 48 (9.9%) were 9 years old or above, which is considered a senior age for dogs. When contrasted with the aforementioned subdivided age categories, puppies still made up the smallest group.

Generally, it was more common for the age of the dog to be included: 512 dogs' ages were recorded in their respective files, compared to 166 that were listed as unknown. The oldest rabid dog found was listed as 36 years, though this could very well have been an inaccurate

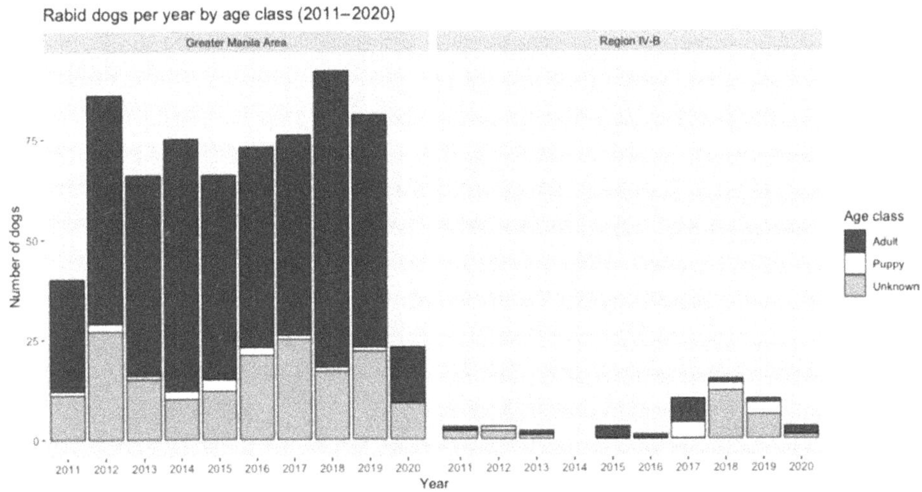

Fig. 9.3. Varying age classes of dogs that tested positive for rabies in the Greater Manila Area and Region IV-B between 2011 and 2020. Image created by Mirava Yuson.

answer from the owner. For owned and confined dogs, the majority of their ages (87.6%) were listed, and 84.6% of free-roaming but owned dogs' ages were known. For strays, 65.3% of dogs had ages listed as unknown. A summary of the data can be found in Fig. 9.3.

9.3.2 Comparisons between the Greater Manila Area and region IV-B

From 2011 to 2020, 678 dog specimens submitted to RITM from the Greater Manila Area tested positive for rabies. A bulk (279) of these animal rabies cases originated from Metro Manila – where RITM is located – which is the most densely populated region in the Philippines. For the four provinces bordering Metro Manila, 237 rabid dogs were confirmed in Cavite province over the 10 year span. For the remaining provinces, 144 cases originated from Laguna, 13 came from Rizal and five were from Bulacan.

Ten-year data on rabid dogs paints a somewhat incomplete picture of the geographic distribution of the disease in the Greater Manila Area. Overall, Metro Manila experienced the most rabies cases over the 10 years, but the annual incidence varied, with the provinces of Cavite and Laguna having most cases in some years.

With an average of 67.8 rabid dogs/year, 2018 was the year with the highest number of confirmed animal rabies cases in the Greater Manila Area, with 92 cases, while 2020 had the lowest at 23. Coronavirus 2019 (COVID-19)-imposed restrictions remain the probable reason for all 23 samples in 2020 originating strictly from Metro Manila.

Although Cavite had fewer annual canine rabies cases on average than Metro Manila and, in general, has a much smaller human population, its 0.018 average incidence rate of canine rabies (18 cases for every 1000 dogs) is more than three times that of the Metro's (which has five cases for every 1000 dogs) – perhaps an alarming consequence of concentrating most dog population control and mass vaccination efforts in the nation's capital. The high number of positive canine samples indicates that rabies may have a higher prevalence in Cavite, which has lower vaccination coverage than Manila, or that cases in Manila are still under-reported.

Canine rabies deaths in Region IV-B are mostly concentrated in the province of Oriental Mindoro, where 29 rabid dogs have been recorded, with eight each from Palawan and Occidental Mindoro (a province separated only from Oriental Mindoro by land borders). Having implemented adequate rabies surveillance measures and registering zero animal or human rabies cases, Marinduque was declared a rabies-free province in 2012. Romblon was also deemed a rabies-free province in 2019, although two suspect human rabies cases in Romblon were identified in 2020 (Romblon representative, Regional Epidemiology and Surveillance Unit, 12 January 2021, personal communication). Since there are seven canine rabies cases for every 1000 dogs in Region IV-B, the region's incidence rate (0.007) slightly exceeds that of Metro Manila's, although grouping the provinces of the Greater Manila Area together produces the same incidence rate as Region IV-B's.

Reported canine rabies incidence is exceptionally low in Region IV-B compared to the incidence rate of human rabies, primarily due to under-reporting of animal rabies cases as only a few specimens are sampled yearly. In terms of human rabies cases, the Greater Manila Area sees 2.9 deaths per 1,000,000 people every year, which is exceeded by Region IV-B's, wherein for every 1,000,000 people, 3.3 human rabies cases occur. Within Greater Manila itself, provincial incidence per million population in Cavite shows a rate of 7.7 human rabies deaths, compared to 2.4 human rabies cases for every 1,000,000 people in Metro Manila.

9.3.3 Vaccination and ownership status among rabid dogs

Greater Manila Area data show that 207 rabid dogs (30.5%) had an unknown vaccination status, while 57 (8.4%) dogs had previously been vaccinated for rabies compared to the 414 (61.1%) who were not. Of the vaccinated dogs, 38 (66.6%) of them had specific vaccination dates listed. Due to the high efficacy of the vaccine, it is likely that the rabid dogs' vaccinated status was due to one of the following: (i) being vaccinated after infection; (ii) low protective titre levels from having been vaccinated more than a year prior; or (iii) wrong information given by the owner.

Among vaccinated dogs, 21 (36.8%) were vaccinated for rabies less than 1 year prior to their deaths. Ten of those dogs had been vaccinated within 10 days of their demise. All such dogs were owned, but only one was described as 'owned but free roaming'. Up to 14 dogs had been vaccinated more than 1 year before their death, whereas 22 dogs lacked information regarding their death date or date of vaccination.

Ownership status varied greatly among rabid dogs: (i) 355 (52.4%) of rabid dogs were owned and confined to the house; (ii) 169 (24.9%) dogs' ownership status was listed as 'owned but free roaming'; (iii) 124 (18.3%) were strays with no known owner; and (iv) the remaining 30 (4.4%) had an unknown ownership status. For 'owned but free roaming' dogs, 127 were vaccinated, while 16 had been vaccinated and 26 had an unknown vaccination status.

The vaccination status of a stray dog is difficult to ascertain – none of the stray dogs were listed as vaccinated, although some may have been runaways, previously owned or owned but free roaming, having died without the owner's knowledge. Ultimately, only a titre test would confirm the vaccination status of a stray dog, but performing one as part of a necropsy is not considered to be common practice.

In terms of age, 20.8% of rabid senior dogs were previously vaccinated. Only one senior dog was listed as having been vaccinated more than 1 year prior compared to the 11 others with specified vaccination dates. However, 59.6% of previously vaccinated dogs were adult dogs between 1 and 4 years old.

In Region IV-B, of the 15 confirmed rabid puppies, six were free roaming but owned, compared to two that were restricted to the owner's home but otherwise had regular household contact with people. Three puppies were listed as owned, two were strays and one had an unknown vaccination status. All recorded rabid puppies were either unvaccinated (11) or had an unknown vaccination status (3), and their age group remains the only one wherein no vaccinated dog got rabies.

Records of rabid dogs' sexes showed that 351 (51.8%) of rabid dogs were male, whereas 240 (35.4%) were female and 87 (12.8%) did not have their sex listed. Male dogs were usually free roaming and owned (100 free roaming in contrast to 131 owned) whereas only 36.6% of females were free roaming. For rabid dogs in the Greater Manila Area, 79.3% were found dead and the remainder had been euthanized, save for one, whose details were left blank. Animals tended to show signs and symptoms of rabies between 1 and 7 days before death.

Because Region IV-B's data pre-2017 lacks details on vaccination and ownership status, a comparison cannot be made with the Greater Manila Area regarding the aforementioned variables.

9.3.4 Model scenarios

The model illustrated the transmission dynamics of rabies among puppies aged 3 months and below, and adult dogs. Culling was not taken into account, as well as purposeful breeding or sterilization (spaying/neutering) of pets. The model also did not cover the potential spread of the disease through wildlife vectors. Similarly, movement of dogs in and out of the Greater Manila Area was not incorporated into the model.

Different scenarios were modelled, using various vaccination coverage levels in puppies and adults. The number of infected puppies and infected adults were summed, and the result divided by the infectious period in order to produce the incidence rate.

An epidemic scenario was modelled initially: upon introducing one infected adult, an oscillatory behaviour was evident assuming no vaccination in adults or puppies, with yearly time lags between peaks of each epidemic cycle during the first 3 years. These time lags would gradually increase to the point that an epidemic cycle occurred only once every 3 years. The higher the number of infected dogs, the smaller the time lag between successive peaks. There was no difference whether the epidemic originated from a puppy or an adult.

In an endemic situation, rabies cases in dogs dropped dramatically to a certain point with increasing vaccination coverage in adults. For vaccination coverage at 10%, the number of cases followed a downward slope – rather than a parabolic curve if no vaccination was done – leading to a 12% decrease annually, and an overall 40% decrease in cases after 7 years. A 50% adult vaccination coverage was associated with a 14% decrease in rabies cases annually, peaking with a total reduction of cases by 60% after 10 years. However, any vaccination rate upwards of 50% showed little difference in decreasing rabies incidence, and vaccinating all adults exclusively was not shown to eliminate the disease. With adult vaccination coverage maintained at 50%, canine rabies was maintained at around 65–100 cases/year.

For Region IV-B, curve fitting was difficult to achieve due to the low number of confirmed cases annually. Simulating the current situation in Region IV-B, vaccine coverage was set at 53% for adults and 0% for puppies. The modelled number of annual infected adults closely paralleled that of the confirmed cases. There was minimal difference between that and vaccinating 50% of adults. However, setting vaccine coverage at 50% for adults and 50% for puppies resulted in a very low number of infected – a 52.6% total decrease in cases after 10 years – implying the possibility of eventual elimination of the disease at some point thereafter, unlike in other scenarios wherein 70% of either puppies or adults were vaccinated. Vaccinating only puppies reduced rabies incidence by 33.7% after 6 years. Vaccinating both puppies and adults at 30% was insufficient for disease elimination, but decreased overall prevalence by 40.2% after 6 years. The model's results can be found in Fig. 9.4.

Since the vaccine's duration of immunity is presumed to be 1 year, it is implied that in any given year, multiple mass vaccinations with gaps of at least 3 months in between are needed to sustain the target percentage for vaccinated puppies in particular. This is easily incorporated into current rabies elimination practices, as seasonal dog mass vaccinations are already common practice.

In the absence of demographic data regarding dog populations, the proportion of puppies to adults was presumed. Provided that puppies were not vaccinated, the model showed that at least 57.7% of rabies cases were puppies regardless of the proportion of puppies to adults in the initial population, which is somewhat higher than the results of a previous study, wherein puppies made up 40% of rabies cases (Domingo and Mananggit, 2014).

9.3.5 Sensitivity analysis

Several sensitivity analyses were performed: initially, the birth rate and mortality rate were

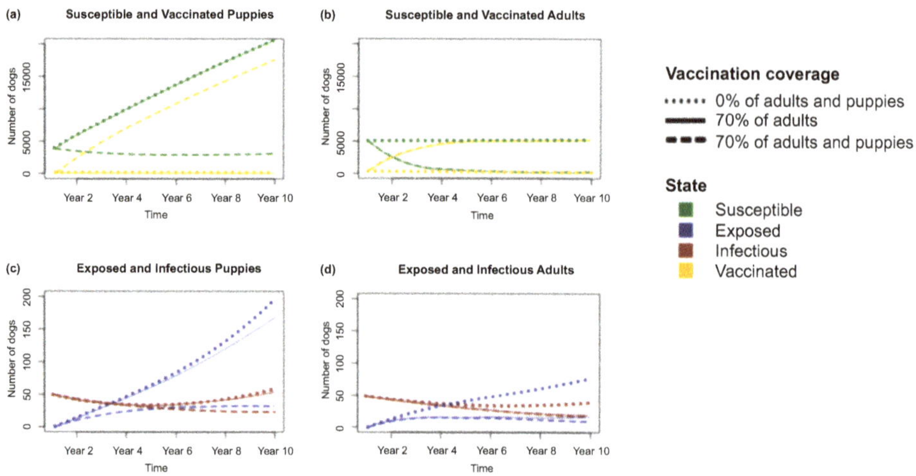

Fig. 9.4. Modelling plots depicting mass dog vaccination strategies in the Philippines. The current dog rabies vaccination coverage in the Philippines is reported to be 53%, but the current goal is to vaccinate at least 70% of the dog population (barring puppies) in aiming to eliminate canine rabies. (a) If 70% of puppies are vaccinated, the number of susceptible puppies is maintained below a certain level but does not decrease considerably even after 10 years. If puppies are not vaccinated, the number of susceptible puppies steadily increases over time. (b) Without vaccination, the number of susceptible adults will remain stable, but is expected to rapidly decrease to zero within 5 years if a 70% vaccination rate is maintained yearly. (c) The number of exposed puppies increases yearly without any vaccination whatsoever, but the rise is less sharp if 70% of adults are vaccinated. Meanwhile, the number of infectious puppies decreases over a period of 5 years before steadily rebounding, but this decrease becomes continuous if 70% of puppies are vaccinated. (d) For exposed adults, their numbers will gradually increase over a 10-year period without vaccination, but the number of infectious adults will be maintained. With a 70% adult vaccination rate (and regardless of whether puppies are vaccinated or not), the number of infectious adults will decrease but not reach zero even after 10 years. Figure created by Mirava Yuson.

adjusted to determine their effect on rabies incidence. Increased birth rates caused a large increase in the epidemic peak. To analyse the effect of mortality rate, life expectancy was adjusted to 5 years to reflect the lifespan of owned but occasionally free-roaming dogs, and 7 years to reflect that of confined dogs (Leung and Davis, 2017) – however, little change in canine rabies incidence was evident. An increased infectious period was associated with a higher number of cases, whereas an increased incubation period resulted in higher epidemic peaks and longer time lags between them. Given that the incubation period of rabies may range from weeks to months, the model suggests that a longer incubation period allows rabies to circulate in a greater number of dogs, with transmission occurring at intermittent times of exposure and eventually resulting in more infected animals.

R_0 was adjusted from 1.1 up to 2.3, echoing the R_0's calculated in similar countries wherein rabies is endemic (Kitala *et al.*, 2002); although these changes had a minimal effect on the model aside from increasing the number of resulting cases. As for carrying capacity, any value below 322.5 dogs/km could not sustain the disease long term. When the carrying capacity was increased, fewer oscillations were observed, and an endemic state was reached within 2 years.

Vaccine immunity was first tested at 3 years (the expected duration of the vaccine), followed by 5 years, then finally, 1 year with presumed fading immunity as annual vaccination is performed in the Philippines based on that assumption. There was minimal difference in the number of cases when the duration of vaccine immunity was factored in.

The three parameters with the greatest impact on the model were vaccination rate, birth rate and carrying capacity.

9.4 Discussion

9.4.1 Main findings

Assuming that the lack of rabies cases between 2014 and 2016 in Region IV-B is an accurate figure, the data compiled seemingly portrays a disease that has recurrent epidemic cycles, potentially with increasing peaks separated by time lags of 3–4 years. There were some disparities between the data and the model, which predicted a mean of 97 rabies cases annually (88% of which were puppies) vs the two or three cases recorded yearly in Region IV-B prior to 2016, implying the vast under-reporting of canine rabies cases.

It is difficult to estimate the true prevalence of canine rabies without knowledge of the reporting rate, which can range between 1 and 30% in rabies-endemic countries (Tohma *et al.*, 2016). A previous study that took place in Region VI notes that up to 1% of dogs/year become rabid in an endemic situation (Ferguson *et al.*, 2015) – thus, in a population of 1,000,000 unvaccinated dogs (estimated with the model), 10,000 cases are expected, a far cry from the reported 12 or 13 cases annually since 2017.

According to the confirmed cases in the compiled Region IV-B data, in some years the ratio of infected puppies to infected adults closely followed the demographics shown in the Bicol study (Domingo and Mananggit, 2014), wherein puppies made up a significant portion (40%) of rabid cases. Conversely, puppies were greatly under-represented in the Greater Manila Area animal rabies data.

The overall data show that information regarding the number of rabid dogs is often incomplete: calculating the basic reproduction number based on the cases was not possible as each one was isolated and could not be linked with genomic sequencing, and no dog was observed as having bitten another dog. The age of the dog was not known for 65% of the cases reported in Region IV-B from 2017 to 2019. It was, however, known for 75.5% of dogs recorded in the Greater Manila Area.

Every scenario tested in the model showed that the majority of rabies cases were found in puppies, even when annual vaccination was concentrated exclusively on their age group. The model's parameters were imitative of other countries' but simulating the number of infected dogs as presented in the sparse data of Region IV-B (and similarly, the Greater Manila Area's data) required a much smaller total dog population than estimated, showing further evidence of under-reporting canine rabies cases.

The model showed that maintaining the current vaccination coverage but exempting puppies causes an overestimation of the total percentage vaccinated. Since 70% of dogs should be vaccinated in order to eliminate the disease (WHO *et al.*, 2019), that entails vaccinating 70% of the total canine population (including puppies), as covering only adults would not have the same encompassing effect. In one scenario, the vaccination rate in the model was set so the vaccination coverage achieved in adult dogs was 100%, but dog rabies incidence still did not decline to zero in the following years, and therefore was maintained entirely in the puppy population. When adjusting the vaccination rate, it was shown to be more effective to vaccinate one-third of puppies and adults than it was to vaccinate half of adults. Vaccinating half of all adults and puppies was also shown to be more effective than vaccinating 70% of adults, which is the current target percentage for mass vaccinations.

It must be noted that the model did not take into account vaccine efficacy, which may differ slightly depending on the manufacturer. In one study, a 90% effectiveness of up to 3 years has been declared for unspecified rabies vaccines used in China (Zhang *et al.*, 2011), though the rabies vaccines readily supplied to the Philippines by the World Organization for Animal Health (OIE) are thought to be 100% effective.

There is also some contention regarding the duration of rabies vaccine immunity. Three out of ten canine rabies cases in 2019 were found to have been vaccinated within 3 years of their rabies diagnosis. Ten out of the 38 vaccinated rabid dogs overall died within 10 days of their last vaccination, and perhaps may have been vaccinated as a preventive measure following a biting incident with another animal, or upon showing signs and symptoms of rabies. The age group with the highest percentage of previous vaccinations was the senior dog group, hinting at waning immunity due to a lack of revaccination efforts. Because Region IV-B's data pre-2017 lacks details on vaccination and ownership status, a comparison cannot be made with the Greater Manila Area regarding the aforementioned variables. Vaccination details are lacking in both datasets, and rabies vaccine efficacy in animals continues to be hotly debated, leading many government officials to adopt a golden rule wherein a pet is not considered truly vaccinated for rabies unless it has undergone at least two consecutive vaccinations.

Alternate vaccination strategies (such as pulse vaccination) could not be incorporated into the model, which only assumed a once-yearly homogenously vaccination of one or both age groups. What is commonly practised in the Philippines is the hosting of free mass vaccination events that rotate through various communities at different times of the year. Residents are invited to bring their pets, though are unable to if they are too busy to attend, cannot restrain their dog, or live too far from the event location. House-to-house vaccination is also performed, depending on the choice of the local government unit. Participation is not mandatory, making it challenging to calculate the vaccination coverage for areas such as Mimaropa. However, data from the Department of Agriculture report that the proportion of dogs vaccinated in Mimaropa steadily increased from 25.42% in 2015 to 38.63% in 2018 (Bureau of Animal Industry, 2019a).

Stray dogs are not covered by mass vaccination events. Should puppies be included in the vaccination coverage, the process comes with its own challenges: such as newborn puppies belonging to stray dogs being hard to locate as they would usually be hidden out of sight.

Carrying capacity was shown to have a large impact on the model, substantiating how rabies transmission increases with dog density. This has been disputed by other studies claiming that rabies in dogs is not density dependent but persists due to frequency-dependent contact (Leung and Davis, 2017). The true canine carrying capacity of the Philippines is difficult to estimate and most likely varies according to the province. The birth rate also had a notable impact on the total number of infected, but further study is required to ascertain that value for dogs in the Philippines in order for the model to produce more accurate results.

9.4.2 Strengths and limitations of the study

One of the primary strengths of the study is the extensive canine rabies data used, which spans 10 years and multiple provinces throughout the Philippines. Age in canine rabies is not an often-discussed topic, and the data provided a framework of rabies incidence from two different regions through which various model scenarios could be compared and contrasted. The hundreds of samples from the Greater Manila Area, where very few rabid puppies were recorded, were juxtaposed against that of Region IV-B, wherein few rabid dogs, in general, were confirmed, but a substantial number of them were puppies.

A particular drawback in conducting an age-oriented study is that extricating the age of a dog can be difficult if it is not already known by the owner. Gauging dog ages is highly dependent on many factors, such as breed, and there is no specific physical marker that would differentiate a 3-month-old puppy from a 4-month-old puppy, as size can be a misleading factor.

Undeniably, there was more than one owner that gave an 'impossible' dog age (such as 20 years or more), indicating that memory recall in terms of pet ages may be questionable; owners are also prone to giving rough estimates, as it is uncommon for them to know the precise day of birth of their pet. For strays, knowledge of their exact age is even less likely. This was expected as it is admittedly difficult to estimate the age of an unfamiliar animal, although comments included in rabies reports would sometimes specify whether a dog was a puppy. Since there is no way to truly ascertain a dog's age (although estimates can be made based on size and teeth), discerning which data resulted from wrong answers by the owner may be an insuperable challenge.

The vaccination status of a stray dog is also difficult to ascertain: none of the stray dogs were listed as vaccinated, although some may have been runaways, previously owned or owned but free roaming, having died without the owner's knowledge.

Because the model used was deterministic, unrealistically low numbers of dog rabies persist. Non-spatial models struggle to capture the dynamic of transmission, which happens on a much more local level. Hence the overestimation of rabies incidence, and the assumption that rabies control happens a lot quicker than in practice as control measures are presumed to have been implemented perfectly. A 70% vaccination coverage means it is rare that this is achieved. Pulse vaccination rather than continuous would cause mass dog vaccination results to be more heterogeneous.

Data regarding dog demography in the Philippines is sparse, and in certain regions, completely absent. In the Greater Manila Area, puppies appear to make up only a small fraction of canine rabies, while records in Bicol and Region IV-B depict the opposite. The Greater Manila Area consists of several highly urbanized cities – it is possible this may have an effect on the dog demographics and dog-to-dog rabies transmission in general, especially if dogs residing there happen to live longer, drastically skewing the adult to puppy ratio. There may also be other confounding factors resulting in the differentiation in the Greater Manila Area rabies laboratory data from that of other Philippine regions.

Nevertheless, the fact that the model emphasizes the puppy's role as the main vector in the transmission of rabies shows that it is a topic that merits further investigation.

9.4.3 Broader context

The increase in canine rabies cases in recent years is probably due to improved surveillance methods, seeing as the Department of Agriculture introduced an initiative in 2012 to maintain the rabies-free status of Marinduque and eliminate rabies in the other regions. In 2015, four suspect human rabies cases reported within a span of 1 month in Oriental Mindoro made national headlines (*The Manila Times*, 2015), yet only one rabid dog was submitted to RITM the very next year. RADDL's opening in Oriental Mindoro in 2017 resulted in a marginal increase in rabid specimens, with at least ten being confirmed per year – barring 2020, when the number of submitted specimens drastically reduced nationwide, a phenomenon largely attributed to COVID-19 pandemic restrictions implemented from March onwards.

A growing number of ABTCs has contributed greatly to the sharp increase of reported animal bite incidents (Amparo *et al.*, 2018), as they are established in locations that are considered more accessible to bite victims, who would otherwise seek treatment at hospital ABTCs in neighboring municipalities. More ABTCs also result in a bigger supply of vaccine stocks (within the province), which are often at risk of running out at public hospitals. However, the main hurdle in confirming canine rabies cases continues to be contact tracing of rabid dogs. Despite mandatory reporting of suspicious dog behaviour or bite incidents, only 181 dogs were reported and tested in Oriental Mindoro from 2016 to mid-2019. Ultimately, the continuing underestimation of canine rabies incidence stems more from the lack of reporting and testing of suspicious dogs regardless of the number of biting incidents reported to ABTCs. Adhering to the One Health approach by maintaining a steady communication line between ABTC health workers and their animal health counterparts is one such strategy that could rectify the issue.

Philippine dog population data is limited and the proportion of puppies to adults in the population is unknown, though a recent study on free-roaming dog populations in Bangladesh revealed that 18% of the surveyed dog population were puppies (Tenzin *et al.*, 2015). Mass spaying and neutering programmes are held regularly nationwide and may also be skewing the demographics over time. While sterilization has been a go-to method for reducing the dog population, with decreasing rabies incidence as one of the aims in mind (Taylor *et al.*, 2017), mass vaccinations remain the far more cost-effective option – an important factor in developing countries such as the Philippines – in working towards eventual disease elimination and achieving the Zero by 30 goal.

Whereas the model's curve was consistent with the trends in the data collected for the Greater Manila Area, for Region IV-B, precise curve fitting was difficult to achieve due to the extremely low number of confirmed cases annually.

The severe under-reporting could very well stem from geographical variations: the five provinces of Region IV-B – Oriental Mindoro, Occidental Mindoro, Marinduque, Romblon and Palawan – are distinct from each other in terms of land area and demographics. Mindoro and Palawan, the provinces with higher population densities, have human and canine rabies cases that are detected annually, whereas Marinduque and Romblon have been devoid of canine rabies cases; a single human rabies case originating from Marinduque in 2018, and two suspect human rabies cases in Romblon in 2020 notwithstanding. The two island provinces share notably lower human (and consequently, dog) populations. Any dog samples would have to be transported to the diagnostic laboratory in Mindoro for confirmation, a factor which could affect the number of rabies cases detected in other islands of Region IV-B. This is in stark contrast with the Greater Manila Area, composed of provinces that are all contiguous with each other and accessible by land transport. Despite Cavite's 30 km distance from Metro Manila – where RITM is located – the number of samples from that area has matched, and in certain years even exceeded, the numbers from Metro Manila.

In terms of establishing recommended vaccination rates, is difficult to say whether the 53% national vaccine coverage is accurate especially at the time of writing, considering that Philippine dog populations are based on rough estimations only. Dog censuses have not been performed in Oriental Mindoro and the majority of mass vaccination events nationwide were cancelled in 2020, with similar issues befalling most other provinces in both Region IV-B and the Greater Manila Area.

9.5 Conclusion

In the Philippines, not many human rabies cases go undetected and human rabies surveillance has been reliable thus far since cases are diagnosed on a clinical basis. This is a practice done in most other countries; though human rabies cases can be misdiagnosed as malaria cases (and vice versa) in malaria-endemic countries, this has not been an issue locally. However, canine rabies surveillance still needs a lot of improvement due to the need for laboratory diagnosis before a dog can be declared rabid. If the human rabies incidence exceeds the number of animal rabies cases in a given year, that is an irrefutable signal that animal surveillance is not picking up on as many cases as it should.

A number of studies have attempted to address the question of whether vaccinating puppies under 3 months of age would sufficiently protect them from rabies. While some research reports that vaccinated puppies suffer from declining post-vaccine antibody titre by the time they are 1 year old (Hodgins and Shewen, 2012), field studies have otherwise noted that puppy vaccination does induce a sufficient and prolonged protective antibody response (Morters *et al.*, 2015). To account for potential failure of vaccinated puppies to gain protection from rabies, the authors suggest that further studies be done to review the viral neutralizing antibody titres in vaccinated puppies every time, before developing a proper vaccination regime. Repeated rabies vaccinations may be recommended – such as vaccinating puppies at least twice, with the second shot occurring once they hit 3 months of age, then annually thereafter – in order to maintain post-vaccine antibody levels.

According to the model, following WHO's guidelines and vaccinating puppies that are below 3 months of age would serve to help bring the overall vaccination coverage to 70%, thus improving the possibility of achieving a Zero by 30 scenario by eliminating rabies in problem areas. The majority of modelled infected dogs were puppies, but targeting puppies specifically was not shown to eliminate the disease. Doing so would also be difficult to replicate in real life as contact tracing of puppies (especially strays) requires more resources, since they make up a smaller portion of the canine population. However, if puppies were no longer exempted from vaccination, the total number of infected dogs yearly would significantly decrease.

Authors' Declaration

All authors declare that they have no conflict of interest.

All authors have approved this manuscript, agree with its submission, and share collective responsibility and accountability.

This manuscript has not been published or is not under review elsewhere.

References

Amparo, A.C.B., Jayme, S.I., Roces, M.C.R., Quizon, M.C.L., Mercado, M.L.L. *et al.* (2018) The evaluation of Animal Bite Treatment Centers in the Philippines from a patient perspective. *PLoS ONE* 13(7), e0200873. DOI: 10.1371/journal.pone.0200873.

Barroga, T.R.M., Basitan, I.S., Lobete, T.M., Bernales, R.P., Gordoncillo, M.J.N. *et al.* (2018) Community awareness on rabies prevention and control in Bicol, Philippines: pre- and post-project implementation. *Tropical Medicine and Infectious Disease* 3(1), 16. DOI: 10.3390/tropicalmed3010016.

Bureau of Animal Industry (2019a) Adoption of the NRPCP: Manual of Procedures (MOP) 2019. Available at: https://doh.gov.ph/sites/default/files/publications/Rabies%20Manual_MOP_2019%20nov28.pdf (accessed 14 June 2022).

Bureau of Animal Industry (2019b) Adoption of the NRPCP: Strategic plan 2020 to 2025. Available at: https://rr-asia.woah.org/wp-content/uploads/2020/03/final-mtp-rabies_philippines.pdf (accessed 14 June 2022).

Department of Health (2012) National rabies prevention and control program manual of operations (2012). Available at: https://doh.gov.ph/sites/default/files/publications/FINALMOP6.4.13WORDRADMay30.pdf (accessed 3 June 2022).

Department of Health (2018) The 2018 Philippine health statistics. Available at: https://doh.gov.ph/sites/default/files/publications/2018%20Philippine%20Health%20Statistics.pdf (accessed 7 February 2023).

Domingo, R.D. and Mananggit, M.R. (2014) Animal rabies patterns in Central Luzon, Philippines and implications for disease control. *Philippine Journal of Veterinary Medicine* 51(2), 117–124.

Ferguson, E.A., Hampson, K., Cleaveland, S., Consunji, R., Deray, R. *et al.* (2015) Heterogeneity in the spread and control of infectious disease: consequences for the elimination of canine rabies. *Scientific Reports* 5, 18232. DOI: 10.1038/srep18232.

Hodgins, D.C. and Shewen, P.E. (2012) Vaccination of neonates: problem and issues. *Vaccine* 30(9), 1541–1559. DOI: 10.1016/j.vaccine.2011.12.047.

Kitala, P.M., McDermott, J.J., Coleman, P.G. and Dye, C. (2002) Comparison of vaccination strategies for the control of dog rabies in Machakos District, Kenya. *Epidemiology and Infection* 129(1), 215–222. DOI: 10.1017/s0950268802006957.

Lapiz, S.M.D., Miranda, M.E.G., Garcia, R.G., Daguro, L.I., Paman, M.D. *et al.* (2012) Implementation of an intersectoral program to eliminate human and canine rabies: the Bohol Rabies Prevention and Elimination Project. *PLoS Neglected Tropical Diseases* 6(12), e1891. DOI: 10.1371/journal.pntd.0001891.

Leung, T. and Davis, S.A. (2017) Rabies vaccination targets for stray dog populations. *Frontiers in Veterinary Science* 4, 52. DOI: 10.3389/fvets.2017.00052.

Morters, M.K., McNabb, S., Horton, D.L., Fooks, A.R., Schoeman, J.P. *et al.* (2015) Effective vaccination against rabies in puppies in rabies endemic regions. *The Veterinary Record* 177(6), 150. DOI: 10.1136/vr.102975.

Philippine Statistics Authority (2016) Highlights on household population, number of households, and average household size of the Philippines (2015 Census of Population). Available at: https://psa.gov.ph/content/highlights-household-population-number-households-and-average-household-size-philippines (accessed 9 August 2021).

RStudio Team (2020) RStudio: integrated development for R. Available at: www.rstudio.com/ (accessed 15 August 2019).

Soetaert, K., Petzoldt, T. and Setzer, R.W. (2010) Solving differential equations in R: package deSolve. *Journal of Statistical Software* 33(9), 1–25. DOI: 10.18637/jss.v033.i09.
Taylor, L.H., Wallace, R.M., Balaram, D., Lindenmayer, J.M., Eckery, D.C. *et al.* (2017) The role of dog population management in rabies elimination – A review of current approaches and future opportunities. *Frontiers in Veterinary Science* 4(10), 109. DOI: 10.3389/fvets.2017.00109.
Tenzin, T., Ahmed, R., Debnath, N.C., Ahmed, G. and Yamage, M. (2015) Free-roaming dog population estimation and status of the dog population management and rabies control program in Dhaka City, Bangladesh. *PLoS Neglected Tropical Diseases* 9(5), e0003784. DOI: 10.1371/journal.pntd.0003784.
The Manila Times (2015) DOH raises alarm on rabies. *The Manila Times*. Available at: www.manilatimes.net/2015/12/26/news/regions/doh-raises-alarm-on-rabies/236408 (accessed 14 June 2022).
Tohma, K., Saito, M., Demetria, C.S., Manalo, D.L., Quiambao, B.P. *et al.* (2016) Molecular and mathematical modeling analyses of inter-island transmission of rabies into a previously rabies-free island in the Philippines. *Infection, Genetics and Evolution* 38, 22–28. DOI: 10.1016/j.meegid.2015.12.001.
WHO (World Health Organization) (2018) WHO expert consultation on rabies: WHO TRS No.1012. Available at: www.who.int/publications/i/item/WHO-TRS-1012 (accessed 30 October 2021).
WHO, FAO and OIE (World Health Organization, Food and Agriculture Organization of the United Nations andWorld Organisation for Animal Health) (2019) *Zero by 30: The Global Strategic Plan to End Human Deathsfrom Dog-Mediated Rabies by 2030*. WHO, Geneva, Switzerland.
Zhang, J., Jin, Z., Sun, G., Zhou, T. and Ruan, S. (2011) Analysis of rabies in China: transmission dynamics and control. *PLoS ONE* 6(7), e20891. DOI: 10.1371/%20journal.pone.0020891.

10 Oral Vaccination of Dogs as a Complementary Tool for Canine Rabies Control: The Thai Protocol

Kansuda Leelahapongsathon[1], Vilaiporn Wongphruksasoong[2], Ad Vos[3], Suwicha Kasemsuwan[1], Tipsarp Kittisiam[1], Onpawee Sagarasaeranee[2] and Karoon Chanachai[4*]

[1]Kasetsart University, Kamphaeng Saen, Nakhon Pathom, Thailand; [2]Ministry of Agriculture, Bangkok, Thailand; [3]CEVA Innovation Center GmbH, Dessau-Rosslau, Germany; [4]USAID Regional Development Mission Asia, Bangkok, Thailand

Abstract

In Thailand, human and animal rabies incidence has decreased dramatically recently. Most rabies cases are among unvaccinated and free-roaming (owned or ownerless) dogs, underscoring that free-roaming dog vaccination is a key factor in controlling the disease. Oral rabies vaccination (ORV) has been intensively studied in dogs in Thailand. The concept of ORV is based on the following three components: (i) vaccines; (ii) baits; and (iii) distribution systems. A suitable, safe and efficacious vaccine candidate was identified and tested in local dogs. Furthermore, a bait readily accepted by the dogs was selected during field studies and further optimized. Finally, the handout-and-retrieve model for bait distribution was adapted to local conditions to reach a significant proportion of free-roaming dogs inaccessible to parenteral vaccination. Large-scale field studies in different settings have indicated that ORV may not only be feasible in Thailand but can also contribute to eliminating dog-mediated human rabies in other endemic countries.

10.1 Introduction

Rabies is an endemic disease in Thailand and many other Asian countries. Thailand has made enormous efforts to control and prevent animal rabies through vaccination, surveillance, animal population control, public awareness campaigns, and multisectoral collaboration (OIE, n.d.). Nevertheless, rabies elimination is yet to be achieved because of various obstacles. This chapter describes the situation of animal rabies in Thailand by analysing national animal rabies surveillance data from 2013 to 2020. Based on this analysis, free-roaming dogs are unlikely to be vaccinated because of difficulties in being restrained. We further discuss the issue of free-roaming dogs as a significant obstacle to rabies control in Thailand. Finally, we address oral rabies vaccination (ORV) of dogs and relevant experimental and field studies conducted in Thailand. ORV has been successfully applied to control rabies in wildlife in many parts of the world (Müller *et al.*, 2015; Maki *et al.*, 2017), and it could be a game changer for controlling dog rabies. Innovative strategies and tools are required to enhance dog-mediated human rabies elimination by 2030.

*Corresponding author: kchanachai@usaid.gov

DOI: 10.1079/9781800622975.0010

10.2 Animal Rabies in Thailand

The national animal rabies surveillance system in Thailand is operated by the Department of Livestock Development (DLD), Ministry of Agriculture and Cooperatives. The rabies surveillance system in Thailand includes both active and passive surveillance. In passive surveillance, animal health officers, animal owners or villagers send the animal samples, including the animal's brain, head or body, to one of ten laboratories located in different regions of Thailand whenever they find suspected rabies cases. Following the DLD guideline of animal rabies surveillance, the symptoms experienced by a suspected animal rabies case include the sudden death of an unknown cause, neurological symptoms, behavioural changes, aggressiveness, and biting without provocation. Local DLD staff coordinate with the animal owner or the community for safe sample collection. They are advised to send the whole animal body or the animal's head to the laboratory.

Active clinical surveillance has been implemented to strengthen the surveillance system as part of rabies-free zone projects applied in specific areas. The targeted samples for this active surveillance are animals that die in accidents and other unknown causes. Local DLD officers collect these animal samples as supportive information to declare rabies-free areas. Data of all submitted samples, laboratory and investigation results were collected from the online national database Thai Rabies Net, or TRN (available at: http://www.thairabies.net/trn/ (accessed 24 June 2022). The TRN has been fully operational since 2013.

From 2013 to 2020, 58,798 samples were submitted to the ten laboratories in Thailand. Our analysis excluded 147 records of unidentified laboratory results (n=103) and untested samples owing to poor sample conditions (n=44). Of the remaining 58,651 samples, most were obtained from active surveillance (82.44%). The samples were mainly from dogs (66.22%), cats (22.63%), rats (12.16%), cattle (1.24%) and squirrels (1.18%), respectively. The total number of positive samples was 4239, accounting for 7.23% of the total sample. Interestingly, despite the higher number of tests, only 17.39% (737/4239) of the positive samples were recorded as active surveillance samples. Positive percentages of passive surveillance samples were significantly higher than those of active surveillance samples in all years ($P \leq 0.001$) (Table 10.1). Despite national efforts to control and prevent rabies, animal rabies cases have been reported nationwide throughout the study period (Fig. 10.1). Positive cases were reported during the entire period from 2013 to 2020. The highest number of animal cases was reported in 2018 (1476 cases) (Table 10.2). This peak was attributed to a nationwide vaccine supply shortage during 2016–2018 (Department of Disease Control, Ministry of Public Health, Thailand, n.d.).

Most of the positive samples were from dogs (3715/4239, 87.64%), demonstrating their

Table 10.1. Animal rabies sample submission and number of positive rabies cases in Thailand, 2013–2020. Table created by Karoon Chanachai.

	Number of submitted samples (%)			Number of positive samples (%)		
Year	Passive	Active	Total	Passive	Active	Total
2013	834 (20.41)	3,252 (79.59)	4,086	99 (98.02)	2 (1.98)	101
2014	912 (20.82)	3,469 (79.18)	4,381	244 (97.60)	6 (2.40)	250
2015	1,221 (13.99)	7,507 (86.01)	8,728	324 (98.18)	6 (1.82)	330
2016	1,369 (15.42)	7,510 (84.58)	8,879	471 (76.34)	146 (23.66)	617
2017	1,439 (16.76)	7,149 (83.24)	8,588	584 (68.87)	264 (31.13)	848
2018	3,105 (32.24)	6,527 (67.76)	9,632	1,228 (83.20)	248 (16.80)	1,476
2019	909 (12.45)	6,391 (87.55)	7,300	343 (90.98)	34 (9.02)	377
2020	509 (7.21)	6,548 (92.79)	7,057	209 (87.08)	31 (12.92)	240
Total	10,298 (17.56)	48,353 (82.44)	58,651	3,502 (82.61)	737 (17.39)	4,239

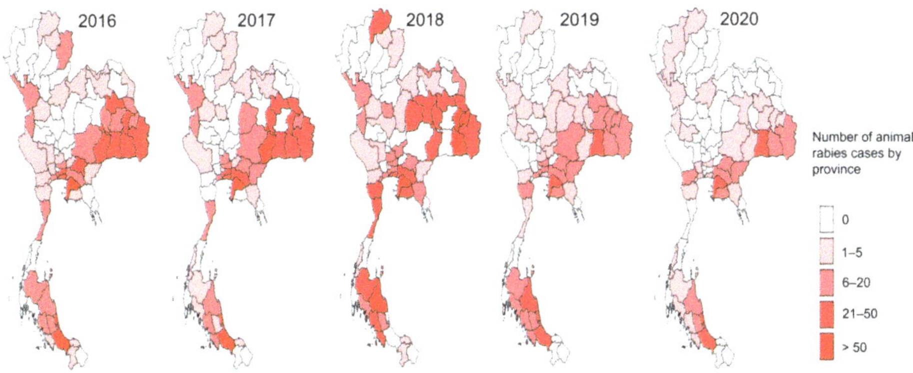

Fig. 10.1. Animal rabies distribution map in Thailand, 2016–2020. Figure created by Vilaiporn Wongphruksasoong.

Table 10.2. Number of rabies cases by animal species in Thailand, 2013–2020. Table created by Karoon Chanachai.

	Year								
Animal	2013	2014	2015	2016	2017	2018	2019	2020	Total
Dog	90	231	308	539	742	1287	301	217	3715
Cattle	5	12	19	56	64	133	51	11	351
Cat	6	6	3	22	41	51	21	6	156
Other livestock (pig, horse and goat)	0	1	0	0	1	4	4	1	11
Wildlife (deer and jackal)	0	0	0	0	0	1	0	5	6
Total	101	250	330	617	848	1476	377	240	4239

Table 10.3. Number of dog rabies cases according to ownership status. Table created by Karoon Chanachai.

	Year								
Ownership status	2013	2014	2015	2016	2017	2018	2019	2020	Total
Owned	60	116	160	263	353	668	116	65	1801
Ownerless	22	94	124	228	291	488	127	88	1462
Unknown	8	21	24	48	98	131	58	64	452
Total	90	231	308	539	742	1287	301	217	3715

importance as reservoir species. Other animal species contributing substantially to rabies were cattle (351 samples, 8.28%) and cats (156 samples, 3.68%) (Table 10.2). When we focused on dog rabies, the surveillance data further highlighted that 51.52% (1914/3715) of rabid dogs were either ownerless or had unclear ownership status (Table 10.3). Unsurprisingly, only 2.82% (54/1914) of these dogs were reported as vaccinated (data not presented). Additionally, 26.93% (485/1801) of the owned-rabid dogs were claimed to have been vaccinated against rabies in the past. Many owned-rabid dogs may be inaccessible to parenteral vaccination because only 12% of owned-rabid dogs were kept in a confined area (Table 10.4). Fifty-four per cent of the vaccinated owned rabid dogs (261/485) were reported receiving vaccination during the

Table 10.4. History of rabies vaccination in owned rabid dogs by level of dog's confinement in Thailand, 2013–2020. Table created by Karoon Chanachai.

	Vaccination history (no. of samples)						
	Vaccinated						
Level of confinement	1–30 days[a]	31–365 days[a]	> 365 days[a]	Data unavailable	Unvaccinated	Data unavailable	Total
Confined	16	14	15	8	110	42	205
Partially free roaming	47	42	60	20	296	77	542
Always free roaming	22	29	20	19	174	49	313
Data unavailable	46	45	36	46	344	224	741
Total	131	130	131	93	924	392	1801

[a]Days that have passed since the last vaccination until the sample submission date.

recommended period (once a year). Fifty per cent of them (131/261) were recently vaccinated (within 30 days). These dogs may have been exposed to the virus before vaccination. If a dog has already been infected, the onset of protective immunity after vaccination may come too late, and the animal succumbs to the disease. This issue may be linked to owners' attitudes towards having their pets vaccinated after an incidence of rabies in the area and not during the annual vaccination campaign, negatively impacting the rabies elimination strategy. Further studies are needed to identify the cause of this problem.

Our analysis showed that the inability to vaccinate a large segment of the owned and ownerless dog populations was apparent. Achieving the elimination of dog-mediated human rabies in Thailand by 2030 is a significant challenge. This challenge has also been reported in many developing countries, such as Latin America (Castillo-Neyra *et al.*, 2017, pp. 7–14), and South, South-east and East Asian countries (Tenzin and Ward, 2012, pp. 452–461).

10.3 Free-Roaming Dogs in Thailand

According to World Animal Protection, free-roaming dogs are defined as dogs that live in public areas and are not currently under direct control (World Society for Protection of Animals Companion & Working Animals Unit, 2005). This term includes both owned and ownerless dogs and is often used interchangeably with 'free-ranging' or 'stray' dogs. This population may comprise owned dogs that roam freely, abandoned owned dogs including their puppies born from uncontrolled breeding, and ownerless dogs that reproduce successfully (OIE, 2021, p. 3).

Most dogs in Thailand are free roaming, regardless of their ownership status. An extensive survey of the dog population in Thailand between May 2018 and September 2019 found that the majority of dogs (75%) were free roaming, and only 25% of the dogs were confined. In Thailand, most people in rural areas keep dogs unconfined, and the dogs roam freely in villages. As such, most free-roaming dogs were owned; 44% of the dogs were owned and free roaming, 18% were owned and semi-confined, and 13% were ownerless and thus also free roaming (Thanapongtharm *et al.*, 2021, p. 5). Free-roaming dogs in Thailand generally comprise street dogs, feral dogs, village or community dogs, temple dogs and owned dogs.

Different stakeholders within the community may disagree on the status of a dog, especially when ownership status is unclear or when a single owner cannot be identified. Often, these dogs are indistinguishable from true ownerless dogs and are thus referred to as free-roaming dogs irrespective of their ownership status. Accurately defining dogs' ownership status, namely differentiating between true ownerless dogs and others, can reflect the socialization and level of familiarity between humans and dogs in the community, which can sometimes affect the outcome of a vaccination campaign (Boitani *et al.*, 2016, pp. 364–365). Promoting responsible ownership of an individual dog is challenging if the ownership status is unclear. Therefore, these free-roaming dogs might not have been included in the mass dog vaccination (MDV) campaigns.

Free-roaming dogs in Thailand share similar characteristics with those previously described in Latin America (Castillo-Neyra *et al.*, 2019, pp. 8–9; Capellà Miternique and Gaunet, 2020, pp. 6–16), Africa (Ortolani *et al.*, 2009, pp. 213–215), and other countries in Asia (Corrieri *et al.*, 2018, pp. 4–8). Most free-roaming dogs in Thailand are the result of irresponsible dog ownership. Buddhist temples are one of the most common areas for people to abandon their dogs. Most temples keep their gates open, allowing dogs to roam independently.

Sociocultural and religious beliefs motivate Thai people to feed free-roaming dogs, regardless of their ownership status (or lack thereof). Buddhist temples often provide a habitat for free-roaming dogs. Food sources vary depending on the resources available to the temple. Commonly, there are leftovers from the food offered to monks. Almost every temple in Thailand houses at least one abandoned animal. It is common to see a mixture of dogs, cats and even chickens in and around Buddhist temples in Thailand. Most free-roaming dogs in Thailand appear to be in good body condition, and very few are described as thin or emaciated. This reflects abundant food sources, including those directly provided by humans, known as their caretakers (Sirasoonthorn *et al.*, 2020). Dog caretakers commonly offer food in the community either inside or outside the temple. Dogs are also likely to associate human settlements with food. Therefore, dogs frequently develop a preference for living or gathering near areas where humans feed them (Sen Majumder *et al.*, 2016, p. 4).

A particular dynamic between free-roaming dogs and their caretakers has been observed in Thailand. Caretakers volunteer their time and financial resources to provide ownerless dogs with food and, sometimes, medicine. The caretaker's self-appointed duties include ensuring that the dogs in their care are fed. They are commonly depicted as older women (over 40 years old) with lower incomes (Sirasoonthorn *et al.*, 2020). Caretakers typically operate individually. However, there may be multiple caretakers in the neighbourhood, causing an overlap. In some areas, such as the Cha-um Municipality, where previous field study trials with ORV of dogs were conducted, there were enough volunteers to establish a systematic distribution of territories among the group of caretakers (Chanachai *et al.*, 2021). When dogs associated with caretakers are reported to cause disturbances or injuries, caretakers are often blamed. Therefore, it is apparent that they often have conflicts with the local administrative organization (LAO) since the community may perceive them as an enabler of dog-related nuisances. However, in many places, the caretakers cooperate with the LAO to reduce cases of dog rabies.

10.4 Oral Rabies Vaccination Protocol in Thailand

10.4.1 ORV: vaccine bait and distribution system

In European and North American countries, wildlife species, such as striped skunks, red foxes, raccoons and raccoon dogs, are the most prominent reservoir species for rabies (Müller *et al.*, 2015; Maki *et al.*, 2017). In the past, ORV programmes have successfully eliminated fox-mediated rabies in Western and Central Europe (Freuling *et al.*, 2013; Müller *et al.* 2015, pp. 4–7; Robardet *et al.*, 2019, pp. 5–7; Vega *et al.*, 2021, p. 5). It has also been successfully applied to raccoon-, fox- and coyote-mediated rabies in some other parts of Europe, Canada and the USA (Maki *et al.*, 2017, pp. 13–22; Fehlner-Gardiner, 2018). Although in the past, Canada and Western Europe were able to successfully eliminate rabies in red foxes through large-scale ORV programmes with live rabies virus vaccines, many safety concerns and limitations are associated with these first-generation vaccines, including residual virulence and vaccine-induced rabies (Vos *et al.*, 1999, pp. 170–172; Fehlner-Gardiner *et al.*, 2008, pp. 78–79; Müller *et al.*, 2015, p. 12). Since then, alternatives, such as recombinant vaccines with higher safety profiles, have been developed (Rupprecht *et al.*, 2005, pp. 99–101; Maki *et al.*, 2017, pp. 3–13).

The use of ORV of dogs, in addition to (parenteral) MDV, has been widely suggested (Cliquet *et al.*, 2018, pp. 3–8; Undurraga *et al.*, 2020, pp. 6167–6169; Wallace *et al.*, 2020, pp. 2–7). Before the adoption of field use of ORV, the World Health Organization (WHO) suggests that the safety and efficacy of the vaccine and bait in the target species should be determined. The vaccines and baits should also be tested under field conditions to establish their benefits and constraints. The candidate vaccine should not cause any adverse effects on the target or non-target species, including humans (WHO, 2007, pp. 12–15). Hence, attenuated vaccines with safety margins higher than those of first-generation oral rabies vaccines have been developed. A second-generation vaccine, a low-virulence monoclonal antibody selected variant – SAG2 – did not elicit a pathogenic response in rodents or any of the studied mammalian species (Mähl *et al.*, 2014, pp. 5–7). Third-generation recombinant vaccines, including SPBN GASGAS used in Thailand's ORV studies, could further reduce the risk of reversion to virulence. A controlled study of SPBN GASGAS in Thai dogs revealed no vaccine-induced adverse effects over the entire 1-year observation period (Leelahapongsathon *et al.*, 2020, p. 5). Subsequent field studies in Thailand have corroborated the safety of the vaccine. No adverse effects have been reported in dogs or unintentionally exposed humans during vaccination campaigns (Chanachai *et al.*, 2021).

A limitation of ORV is its inability to determine whether the dog has received adequate immunization, in contrast to parenteral vaccines. A particular challenge is the ability to assess the amount of vaccine released and taken up by the tonsils or mucus membrane in the oral cavity and whether it was sufficient to induce an immune response to protect against rabies. If the vaccine is rapidly swallowed, which does not allow sufficient time to target the mucus membranes and tonsils, it will lose its immunogenicity during its passage through the gastrointestinal tract; thus, it is unable to provide immunization. Alternatively, the vaccine could spill on the ground if the sachet is perforated.

Visible markers, which stain the oral cavity, have proven to be useful in identifying the most suitable bait system for vaccine release. Staining intensity (categorized as weak or strong) can be correlated with the amount of virus released in the dog's oral cavity (Langguth *et al.*, 2021, p. 11). However, the results of a previous serological study suggest that intensive dog chewing on the bait is a good indicator that the vaccine is released from the sachet and subsequently elicits an adequate immune response (Leelahapongsathon *et al.*, 2020, p. 5). The immune response induced by the oral vaccine SPBN GASGAS was not different from that induced by a parenteral vaccine commercially available in Thailand (Leelahapongsathon *et al.*, 2020, p. 9). Furthermore, underscoring its efficacy, no rabies cases have been reported following the latest MDV campaign with adjunct ORV in selected areas within 1 year, even though there were rabies reports in nearby areas before the ORV field trials (according to local stakeholders and cross-checking with the national surveillance database TRN, December 2021).

Oral rabies vaccines are administered using a bait matrix system, whereby a sachet filled with the liquid vaccine is incorporated into an attractive bait. The bait matrix mainly acts as an attractant for vaccine delivery. Several types of bait materials have been tested and studied; however, currently, there are no universally applicable baits. Baits must successfully deliver the vaccine to the oral cavity and have a high acceptance rate in local dogs to consume the baits. Multiple studies have found that dogs from different regions appear to have different bait preferences (Berentsen *et al.*, 2016, p. 22; Kasemsuwan *et al.*, 2018, pp. 4–8; Gibson *et al.*, 2019, p. 6). Hence, it is strongly recommended that bait acceptance tests determine the most effective type of bait in both dog preference and vaccine uptake efficacy.

Thai free-roaming dogs tended to have a higher preference for handmade pig (swine) intestinal bait (Fig. 10.2c) than other bait types tested. However, this intestine bait had a lower success rate in releasing the vaccine into the oral cavity. Dogs tended to swallow vaccine sachets in the intestinal casing. Conversely, the slightly less attractive egg bait (Fig. 10.2b) proved to be more successful in delivering and releasing the vaccine in the oral cavity than the intestinal

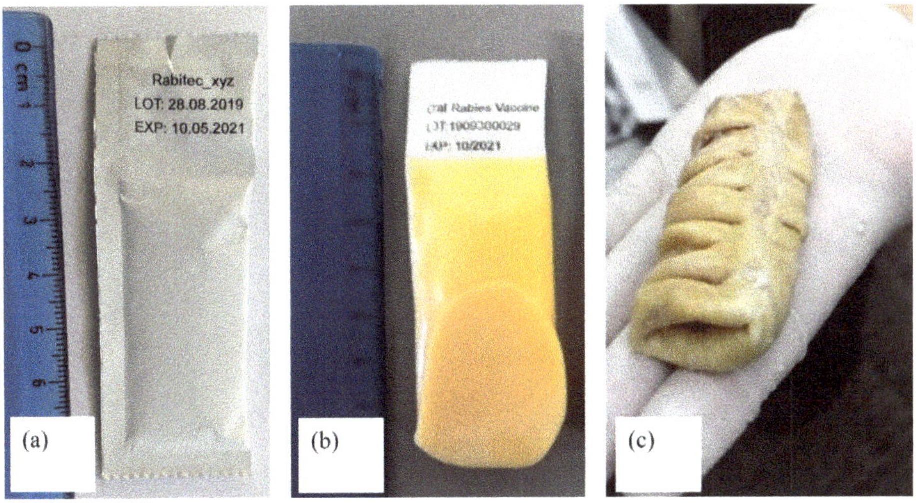

Fig. 10.2. Oral rabies vaccine sachet and baits used in Thailand's field studies. (a) Vaccine sachet without bait; (b) egg-flavoured vaccine bait; and (c) pig-intestine vaccine bait. Photos courtesy of Kansuda Leelahapongsathon.

bait. Dogs are more likely to chew egg baits for an extended period, providing greater vaccine exposure to the oral cavity. Furthermore, local field personnel in Thailand expressed a preference for manufactured egg bait because of its ease of use (Chanachai *et al.*, 2021). Local personnel are required to perform the laborious task of incorporating a vaccine-filled sachet into the intestinal bait. Egg baits can be manufactured with the vaccine sachet contained inside on an industrial scale, thus eliminating the need for additional labour. Egg bait can be further optimized to induce local dog preference, such as pasting liquid cat snacks on to the egg bait to improve bait acceptance by dogs (Chanachai *et al.*, 2021).

There are two possible methods of bait distribution that target free-roaming dogs (Vos, 2019). The first, called the wildlife model, involves the mass distribution of baits at selected sites where the animals must locate the distributed baits themselves. The second is the handout-and-retrieve model, which involves directly handing out the vaccine to the free-roaming dogs encountered and retrieving the vaccine sachet discarded by the dogs. The handout-and-retrieve model is highly advantageous for reducing unintentional human contact with vaccine viruses. Compared to the wildlife model, it limits the number of unaccepted baits and reduces the risk of non-target animals consuming the bait.

Vaccine-bait distributors can be recruited from government staff, volunteers, dog caretakers and owners. In studies conducted in Thailand, the government and LAO staff, volunteers and caretakers were recruited and trained to assist in the handout-and-retrieve model. The caretaker's knowledge of the dogs' estimated population size, feeding habits and area preference was crucial in conducting the vaccination. Moreover, local personnel's familiarity with dogs also positively influenced the success of ORV (Chanachai *et al.*, 2021). After multiple vaccination campaigns, the accumulating experience of vaccination teams undoubtedly leads to a higher success rate in ORV, increasing vaccination coverage (Güzel *et al.*, 1998, p. 129).

Thailand has successfully reduced the number of human rabies cases in the past few decades: 185, 50, 15 and three human cases were reported in 1990, 2000, 2010 and 2020, respectively (Bureau of Epidemiology, Department of Disease Control, MoPH, Thailand, n.d.). However, as described previously, suboptimal vaccination coverage of evasive free-roaming dogs has always been an essential limitation for success (Kasempimolporn *et al.*, 2007, p. 29). ORV has shown potential in reaching a population that was previously not accessible to parenteral vaccination, increasing the overall vaccination coverage, especially in one of the essential rabies reservoir groups – free-roaming dogs. In a previous field study in Thailand, 65.6% of free-roaming dogs in targeted areas were successfully vaccinated through ORV (Chanachai *et al.*, 2021). Therefore, the risk of free-roaming dogs transmitting rabies to animals and humans can be reduced.

10.4.2 The Thai protocol

The satisfactory outcome of these ORV studies in Thailand has led to a larger-scale distribution of oral rabies vaccine bait in local municipalities across Thailand. Five thousand doses of baits containing the SPBN GASGAS vaccine construct were allocated among all nine regional jurisdictions of the DLD in 2021. The annual MDV campaigns consist of the usual parenteral vaccinations and the addition of oral vaccines as an adjunct. Local knowledge of the dogs' location preferences and activities will help develop plans for vaccination campaigns, such as the ideal vaccination time and staffing requirements. Successful countrywide application of oral vaccines in the field should promote ORV usage as an adjunct to MDV in order to control and eliminate dog-mediated rabies. From studies conducted in Thailand, we suggest the following protocol for implementing the ORV campaign in this country.

Identify hot spots

Hot spots, identified by the uninterrupted presence of dog rabies with high numbers of rabid free-roaming dogs, are areas that require oral vaccines. In Thailand, unvaccinated, free-roaming dogs are the leading cause of recurring rabies. ORV campaigns can specifically target these areas to vaccinate this subpopulation of dogs and conserve time and resources. Organizers should identify high-risk areas with free-roaming dog populations and rabies hotspots. The campaign organizer can identify these targeted areas by analysing rabies surveillance data and discussing it with local staff and the community. It is also feasible to identify locations with free-roaming dogs by conducting a quick field survey while implementing the MDV campaigns.

Identify vaccine construct, bait and distribution system

In this case, the handout-and-retrieve model is the most suitable distribution system. The vaccine construct SPBN GASGAS has been used in Thailand because of its high safety profile. Targeted dogs sometimes live near human habitation, which could lead to unintentional human exposure to the vaccine. Another factor considered was the onset and duration of immunity induced by this vaccine candidate: preferably the vaccine should induce a rapid onset of immunity and also protect the animals for a prolonged period of time (Leelahapongsathon *et al.*, 2020). The cold-chain requirements for the selected vaccine must also be considered. The storage of SPBN GASGAS requires a deep freeze (−20°C) to keep the vaccine viable throughout its shelf life. Dry ice has been used to transport vaccines from central to local levels. A household refrigerator that can store ice cream (−18°C) is used for local vaccine storage. After thawing, the vaccine can be maintained in a standard cool box or refrigerator for at least several days. The egg-flavoured bait was highly successful in delivering the vaccine. The simple storage and ease of use also make the egg-flavoured bait preferable to locally made baits for the field staff. The slightly lower attractiveness of the egg bait than that of the intestine bait can be mitigated by local products (e.g. cat paste).

Campaign organization

Vaccination teams must identify possible methods to reach the free-roaming dog population. Non-traditional stakeholders (such as villagers, dog caretakers, monks and dog care societies) at the local level should be approached and engaged as soon as possible. In Thailand, most free-roaming dogs have caretakers and are accustomed to being fed by humans. Caretakers sometimes form informal networks, and thus it may be possible to reach others after reaching out to individual ones. The involved stakeholders must be identified in the planned area. Responsibilities and tasks also need to be assigned to participating organizers because various components are involved in conducting a successful vaccination campaign. Significant considerations include: (i) the appropriate handling of the materials (bait and vaccine); (ii) ensuring a proper cold chain for vaccines; (iii) personnel training; and (iv) planning advocacy and educational campaigns. Multisector collaboration between stakeholders provides a holistic approach to the campaign, thus increasing the potential for success.

Implementation of the campaign

The timing of the ORV campaign is crucial to ensure that the target dog population is reached. We recommend conducting an ORV campaign shortly after the MDV campaign because the information on free-roaming dogs received during the MDV campaign should not significantly change during this short period. The sharing of local expertise between the stakeholders involved can result in a practical and efficient strategy for campaign implementation. Thus, caretakers can also provide valuable insights. For example, they can help define the time of day that would be most appropriate to conduct the vaccination campaign and estimate the number of dogs in each area. Many dogs tend to gather during their feeding time (early morning or late afternoon), which may not be the standard working hours for staff. Flexible and adaptive management of campaigns leads to higher vaccination coverage. Moreover, knowledge of the number of dogs in an area can govern the number of vaccines and baits required. Local personnel training and experience in the area can also be advantageous when planning campaigns. The organizer must prepare for appropriate personal protective equipment and vaccine roll-out logistic supplies such as gloves, infective disposal bags, cool boxes and recording equipment. Vaccine handlers receive complete pre-exposure prophylaxis for rabies vaccination.

Follow up after the campaign

ORV of dogs has recently been introduced in Thailand; therefore, continuous research and development are crucial. Technical cooperation among stakeholders, including manufacturers, educational institutes, authorities and local communities, should be formalized. After conducting the vaccination campaign, it is essential to include a post-campaign monitoring plan to receive reports on vaccination success and adverse events in humans and animals. Follow-up monitoring plans should be actively conducted. The contact information of volunteers and dog caretakers should be collected during the first field visit. The organizer should contact them to inquire about any adverse events, complaints or appreciations during post-campaign monitoring. The organizer may conduct serological monitoring following ORV to ensure that the campaign reaches acceptable seroconversion rates. In such cases, a comprehensive serological monitoring protocol must be developed. Multiple-time capturing of free-roaming dogs for drawing blood is probably impossible. The organizer may consider accessible free-roaming dogs living in similar environments to replace inaccessible free-roaming dogs for serological monitoring.

Possible alternatives, adaptations and improvements

If ORV campaigns are conducted simultaneously with MDV, visually marking parenterally vaccinated dogs is recommended to be able to identify 'unmarked' dogs that need to be offered oral baits. In Thailand, attaching a vaccination tag to a dog's collar is recommended. However, owners are not likely to follow this recommendation. The ORV organizer should emphasize this guideline to dog owners to facilitate ORV campaigns. They may consider conducting an ORV campaign as soon as possible after the MDV campaign to reduce the problem of recalling the vaccination history of owned free-roaming dogs that may have been vaccinated in the recent MDV campaign.

The campaign organizer should systematically plan to assess vaccination coverage whenever possible. The number of owned and ownerless dog populations is crucial for assessing the overall vaccination coverage. Practical and good coverage of owned-dog registrations serves as the baseline for the overall vaccination coverage. A periodic systematic survey of ownerless dog populations helps estimate the ORV target and its performance. This information facilitates the adaptation and adjustment of strategies to achieve the elimination of dog-mediated rabies.

10.4.3 One Health

MDV has been the single most cost-effective tool to reduce the number of dog rabies cases, and thus human rabies, as more than 95% of all human cases are caused by dogs infected with the rabies virus. In the past, Thailand spent a large amount of the budget for rabies prevention on providing rabies post-exposure treatment for people bitten by a potentially rabid animal (Wilde *et al.*, 1999, p. 241). However, a more economical approach would be eliminating the disease from its principal reservoir in the country, the domestic dog. Unfortunately, due to the high proportion of free-roaming dogs in Thailand that are inaccessible for parenteral vaccination, this approach is unlikely to succeed.

Including ORV as a component of the MDV campaigns could therefore increase the vaccination coverage above the critical threshold to interrupt rabies transmission among dogs. Without the circulation of the rabies virus in the dog population, the risk of human rabies is significantly reduced.

10.5 Conclusion

The population of free-roaming dogs inaccessible to parenteral vaccination has historically been a primary limitation to the elimination of dog-mediated rabies. ORV has proven to be successful in immunizing dogs in controlled settings as well as in field conditions. The familiarity of free-roaming dogs with their caretakers facilitates the acceptance of oral vaccines given by the handout-and-retrieve model. The high safety profile of the vaccine makes it suitable for use in the handout-and-retrieve model.

In our previous studies, no vaccine-induced adverse effects were reported in humans or animals. Integration of manufactured bait is recommended as it is highly successful in vaccine delivery and scalability. The simplicity of storage and handling of the bait-vaccine matrix is also advantageous compared to locally sourced materials, such as pig intestines. Proper communication, education and training of local personnel are also crucial for successful oral vaccination campaigns. The involvement of cooperative local communities can increase the likelihood of success by leveraging familiarity and knowledge of local free-roaming dogs. Finally, the addition of ORV as an adjunct to parenteral vaccination campaigns increases the coverage of free-roaming dogs and can contribute to eliminating dog-mediated human rabies in Thailand by 2030.

Hopefully, the Thai initiative to use ORV as a complementary tool in dog rabies control will be followed by other countries in the region confronted with a significant number of free-roaming dogs that are not accessible to parenteral vaccination.

Acknowledgements

We express our sincere gratitude to staff from different organizations for contributing to a series of studies in Thailand. Special thanks to Daniel Schar, Regional Emerging Infectious Diseases Advisor, USAID Regional Development Mission for Asia, for editing the English in this chapter.

Authors' Declaration

All of the authors, except one, declare that they have no conflict of interest; Ad Vos is a full-time employee of a pharmaceutical company that manufactures oral rabies vaccine baits.

All authors have approved this manuscript, agree with its submission and share collective responsibility and accountability.

This manuscript has not been published or is not under review elsewhere.

References

Berentsen, A.R., Bender, S., Bender, P., Bergman, D., Gilbert, A.T. *et al.* (2016) Bait flavor preference and immunogenicity of ONRAB baits in domestic dogs on the Navajo Nation, Arizona. *Journal of Veterinary Behavior* 15, 20–24. DOI: 10.1016/j.jveb.2016.08.007.

Boitani, L., Francisci, F., Ciucci, P. and Andreoli, G. (2016) The ecology and behavior of feral dogs: a case study from central Italy. In: Serpell, J. (ed.) *The Domestic Dog*. Cambridge University Press, Cambridge, pp. 342–368. DOI: 10.1017/9781139161800.017.

Bureau of Epidemiology, Department of Disease Control, MoPH, Thailand (n.d.) *Human Rabies Situation in Thailand*. Ministry of Public Health (MoPH), Thailand. Available at: http://doe.moph.go.th/surdata/disease.php?dcontent=old&ds=42 (accessed 27 March 2022).

Capellà Miternique, H. and Gaunet, F. (2020) Coexistence of diversified dog socialities and territorialities in the city of Concepción, Chile. *Animals* 10(2), 298. DOI: 10.3390/ani10020298.

Castillo-Neyra, R., Brown, J., Borrini, K., Arevalo, C., Levy, M.Z. *et al.* (2017) Barriers to dog rabies vaccination during an urban rabies outbreak: qualitative findings from Arequipa, Peru. *PLoS Neglected Tropical Diseases* 11(3), e0005460. DOI: 10.1371/journal.pntd.0005460.

Castillo-Neyra, R., Toledo, A.M., Arevalo-Nieto, C., MacDonald, H., De la Puente-León, M. *et al.* (2019) Socio-spatial heterogeneity in participation in mass dog rabies vaccination campaigns, Arequipa, Peru. *PLoS Neglected Tropical Diseases* 13(8), e0007600. DOI: 10.1371/journal.pntd.0007600.

Chanachai, K., Wongphruksasoong, V., Vos, A., Leelahapongsathon, K., Tangwangvivat, R. *et al.* (2021) Feasibility and effectiveness studies with oral vaccination of free-roaming dogs against rabies in Thailand. *Viruses* 13(4), 571. DOI: 10.3390/v13040571.

Cliquet, F., Guiot, A.-L., Aubert, M., Robardet, E., Rupprecht, C.E. *et al.* (2018) Oral vaccination of dogs: a well-studied and undervalued tool for achieving human and dog rabies elimination. *Veterinary Research* 49(1), 61. DOI: 10.1186/s13567-018-0554-6.

Corrieri, L., Adda, M., Miklósi, Á. and Kubinyi, E. (2018) Companion and free-ranging Bali dogs: environmental links with personality traits in an endemic dog population of South East Asia. *PloS ONE* 13(6), e0197354. DOI: 10.1371/journal.pone.0197354.

Department of Disease Control, Ministry of Public Health, Thailand (n.d.) *Report on Disease Control Program Review: The Rabies Elimination Program (in Human)*. Available at: https://ddc.moph.go.th/uploads/ckeditor2//9.%20การกำจัดโรคพิษสุนัขบ้า%20(ในคน).pdf (accessed 27 March 2022).

Fehlner-Gardiner, C. (2018) Rabies control in North America – past, present and future. *OIE Revue Scientifique et Technique* 37(2), 421–437. DOI: 10.20506/rst.37.2.2812.

Fehlner-Gardiner, C., Nadin-Davis, S., Armstrong, J., Muldoon, F., Bachmann, P. *et al.* (2008) Era vaccine-derived cases of rabies in wildlife and domestic animals in Ontario, Canada, 1989-2004. *Journal of Wildlife Diseases* 44(1), 71–85. DOI: 10.7589/0090-3558-44.1.71.

Freuling, C.M., Hampson, K., Selhorst, T., Schröder, R., Meslin, F.X. *et al.* (2013) The elimination of fox rabies from Europe: determinants of success and lessons for the future. *Philosophical Transactions of the Royal Society B: Biological Sciences* 368(1623), 20120142. DOI: 10.1098/rstb.2012.0142.

Gibson, A.D., Yale, G., Vos, A., Corfmat, J., Airikkala-Otter, I. *et al.* (2019) Oral bait handout as a method to access roaming dogs for rabies vaccination in Goa, India: a proof of principle study. *Vaccine* X(1), 100015. DOI: 10.1016/j.jvacx.2019.100015.

Güzel, N., Leloglu, N. and Vos, A. (1998) Evaluation of a vaccination campaign including oral vaccination in Kusadasi, Turkey. *Journal of Etlik Veterinary Microbiology* 9, 121–134.

Kasempimolporn, S., Sichanasai, B., Saengseesom, W., Puempumpanich, S., Chatraporn, S. *et al.* (2007) Prevalence of rabies virus infection and rabies antibody in stray dogs: a survey in Bangkok, Thailand. *Preventive Veterinary Medicine* 78(3–4), 325–332. DOI: 10.1016/j.prevetmed.2006.11.003.

Kasemsuwan, S., Chanachai, K., Pinyopummintr, T., Leelalapongsathon, K., Sujit, K. *et al.* (2018) Field studies evaluating bait acceptance and handling by free-roaming dogs in Thailand. *Veterinary Sciences* 5(2), 47. DOI: 10.3390/vetsci5020047.

Langguth, A., Leelahapongsathon, K., Wannapong, N., Kasemsuwan, S., Ortmann, S. *et al.* (2021) Comparative study of optical markers to assess bait system efficiency concerning vaccine release in the oral cavity of dogs. *Viruses* 13(7), 1382. DOI: 10.3390/v13071382.

Leelahapongsathon, K., Kasemsuwan, S., Pinyopummintr, T., Boodde, O., Phawaphutayanchai, P. *et al.* (2020) Humoral immune response of Thai dogs after oral vaccination against rabies with the SPBN GASGAS vaccine strain. *Vaccines* 8(4), 573. DOI: 10.3390/vaccines8040573.

Mähl, P., Cliquet, F., Guiot, A.-L., Niin, E., Fournials, E. *et al.* (2014) Twenty year experience of the oral rabies vaccine SAG2 in wildlife: a global review. *Veterinary Research* 45(1), 77. DOI: 10.1186/s13567-014-0077-8.

Maki, J., Guiot, A.-L., Aubert, M., Brochier, B., Cliquet, F. *et al.* (2017) Oral vaccination of wildlife using a vaccinia-rabies-glycoprotein recombinant virus vaccine (RABORAL V-RG®): a global review. *Veterinary Research* 48(1), 57. DOI: 10.1186/s13567-017-0459-9.

Müller, T.F., Schröder, R., Wysocki, P., Mettenleiter, T.C. and Freuling, C.M. (2015) Spatio-temporal use of oral rabies vaccines in fox rabies elimination programmes in Europe. *PLoS Neglected Tropical Diseases* 9(8), e0003953. DOI: 10.1371/journal.pntd.0003953.

OIE (World Organisation for Animal Health) (2021) Dog population management (Chapter 7.7). In: *Terrestrial Animal Health Code*. Available at: https://www.woah.org/en/what-we-do/standards/codes-and-manuals/terrestrial-code-online-access/ (accessed 1 April 2022).

OIE (World Organisation for Animal Health) (n.d.) Rabies updates in Thailand. Available at: https://rr-asia.oie.int/wp-content/uploads/2020/03/thailand_final_30nov18.pdf (accessed 27 March 2022).

Ortolani, A., Vernooij, H. and Coppinger, R. (2009) Ethiopian village dogs: behavioural responses to a stranger's approach. *Applied Animal Behaviour Science* 119(3–4), 210–218. DOI: 10.1016/j.applanim.2009.03.011.

Robardet, E., Bosnjak, D., Englund, L., Demetriou, P., Martín, P.R. *et al.* (2019) Zero endemic cases of wildlife rabies (Classical Rabies Virus, RABV) in the European Union by 2020: an achievable goal. *Tropical Medicine and Infectious Disease* 4(4), 124. DOI: 10.3390/tropicalmed4040124.

Rupprecht, C.E., Charlton, K.M., Artois, M., Casey, G.A., Webster, W.A. *et al.* (1990) Ineffectiveness and comparative pathogenicity of attenuated rabies virus vaccines for the striped skunk (*Mephitis mephitis*). *Journal of Wildlife Diseases* 26(1), 99–102. DOI: 10.7589/0090-3558-26.1.99.

Rupprecht, C.E., Hanlon, C.A., Blanton, J., Manangan, J., Morrill, P. *et al.* (2005) Oral vaccination of dogs with recombinant rabies virus vaccines. *Virus Research* 111(1), 101–105. DOI: 10.1016/j.virusres.2005.03.017.

Sen Majumder, S., Paul, M., Sau, S. and Bhadra, A. (2016) Denning habits of free-ranging dogs reveal preference for human proximity. *Scientific Reports* 6, 32014. DOI: 10.1038/srep32014.

Tenzin and Ward, M.P. (2012) Review of rabies epidemiology and control in South, South East and East Asia: past, present and prospects for elimination. *Zoonoses and Public Health* 59(7), 451–467. DOI: 10.1111/j.1863-2378.2012.01489.x.

Sirasoonthorn, P., Suesat, W., Thongyuan, S. and Pinyopummintr, T. (2020) Social interaction between labors and stray dogs, and information of rabies in urban temporary construction sites, Phitsanulok Province. In: *Proceedings of the 14th Ubon Ratchathani University National Research Conference,* Ubon Ratchathani, Thailand, 3–4 September 2020, Ubon Ratchathani University, Ubon Ratchathani, Thailand, pp. 270–283.

Thanapongtharm, W., Kasemsuwan, S., Wongphruksasoong, V., Boonyo, K., Pinyopummintr, T. *et al.* (2021) Spatial distribution and population estimation of dogs in Thailand: implications for rabies prevention and control. *Frontiers in Veterinary Science* 8, 790701. DOI: 10.3389/fvets.2021.790701.

Undurraga, E.A., Millien, M.F., Allel, K., Etheart, M.D., Cleaton, J. *et al.* (2020) Costs and effectiveness of alternative dog vaccination strategies to improve dog population coverage in rural and urban settings during a rabies outbreak. *Vaccine* 38(39), 6162–6173. DOI: 10.1016/j.vaccine.2020.06.006.

Vega, S., Lorenzo-Rebenaque, L., Marin, C., Domingo, R. and Fariñas, F. (2021) Tackling the threat of rabies reintroduction in Europe. *Frontiers in Veterinary Science* 7, 613712. DOI: 10.3389/fvets.2020.613712.

Vos, A. (2019) Oral vaccination of dogs against rabies. *International Animal Health Journal* 6(1), 25–29.
Vos, A., Neubert, A., Aylan, O., Schuster, P., Pommerening, E. *et al.* (1999) An update on safety studies of SAD B19 rabies virus vaccine in target and non-target species. *Epidemiology and Infection* 123(1), 165–175. DOI: 10.1017/s0950268899002666.
Wallace, R.M., Cliquet, F., Fehlner-Gardiner, C., Fooks, A.R., Sabeta, C.T. *et al.* (2020) Role of oral rabies vaccines in the elimination of dog-mediated human rabies deaths. *Emerging Infectious Diseases* 26(12), 1–9. DOI: 10.3201/eid2612.201266.
WHO (World Health Organization) (2007) Oral vaccination of dogs against rabies: guidance for research on oral rabies vaccines and field application of oral vaccination of dogs against rabies. Available at: https://apps.who.int/iris/handle/10665/331036 (accessed 25 March 2022).
Wilde, H., Tipkong, P. and Khawplod, P. (1999) Economic issues in postexposure rabies treatment. *Journal of Travel Medicine* 6(4), 238–242. DOI: 10.1111/j.1708-8305.1999.tb00525.x.
World Society for Protection of Animals Companion & Working Animals Unit (2005) *Surveying Roaming Dog Population: Guidelines on Methodology*. Available at: https://www.worldanimalprotection.org.cn/sites/default/files/media/cn_files/cn_attachment/surveying_roaming_dog_populations_guidelines_on_methodology.pdf (accessed 1 April 2022).

11 Immunogenicity Following Dog Rabies Vaccination: A Sri Lankan Experience

Mangala Gunatilake[1]*, Ruwini M.S. Pimburage[2], Omala Wimalaratne[3] and K.A.D.N. Perera[3]
[1]University of Colombo, Colombo, Sri Lanka; [2]Public Health Veterinary Services, Ministry of Health, Colombo, Sri Lanka; [3]Medical Research Institute, Colombo, Sri Lanka

Abstract
Proper canine vaccination and the efficacy of vaccines play a major role in controlling rabies in rabies-endemic countries. This chapter focuses on two studies that demonstrated humoral immunity following dog rabies vaccination (RV) of puppies, juveniles and adult dogs. The results of these studies indicated the necessity of a primary and a booster vaccination when given for the first time to a puppy, juvenile or adult dog, followed by an annual booster. Quantitative analysis of antibody responses in these studies, and the recommendations made, were considered for the revision of the dog RV protocol in Sri Lanka to maintain a good immune response in the dog population. This chapter also discusses two ongoing studies that were initiated to determine the effectiveness of this revised protocol. The ultimate aim is to control and prevent dog rabies in Sri Lanka and to achieve zero dog-mediated human rabies deaths by the year 2030.

11.1 Introduction

Sri Lanka is a rabies-endemic country; therefore, rabies is a significant health issue, having an impact on the country's health budget. Public Health Veterinary Services, which is under the purview of the Ministry of Health, has initiated several measures to control its transmission and prevention. The Department of Animal Production and Health (DAPH), Ministry of Local Government, Faculty of Veterinary Medicine and Animal Science, private veterinary practitioners, and several non-governmental organizations (NGOs) also extend their support in this endeavour.

Animal rabies vaccination (RV), human post-exposure prophylaxis (PEP), health education of the general public, and dog population control (sterilization of dogs) are the measures adopted in the country indicating the multi-stakeholder involvement in the process of rabies control (Harischandra *et al.*, 2016; Kanda *et al.*, 2021). Furthermore, indiscriminate killing or elimination of free-roaming dogs for the control of rabies was discontinued in 2006, after the introduction of the no-kill policy in the country (Madies, 2020). Taking measures to prevent or minimize dog-mediated human rabies deaths is one of the best examples of adoption of the One Health concept in the country.

*Corresponding author: mangalagunatilake@hotmail.com

DOI: 10.1079/9781800622975.0011

It is an accepted fact that over 99% of human rabies deaths are due to domestic dogs (Hampson *et al.*, 2015). In Sri Lanka, too, the urban cycle is more prominent than the sylvatic cycle of rabies transmission as the dog is the principal reservoir and transmitter of rabies. Hence, dog RV plays a crucial role in controlling this zoonotic disease in the country (Gunatilake *et al.*, 2003; Gunatilake, 2015).

There had been instances where puppies less than 3 months of age died of rabies in addition to reporting RV failures which were encountered during our clinical practice. In the mid-1990s, 3 months was the recommended age to commence vaccination against rabies in puppies (Gunatilake *et al.*, 2003). Furthermore, studies conducted in several countries indicated a failure in producing long-lasting protective levels of antibodies in a significant group of dogs that received only one dose of rabies vaccine (Tepsumethanon *et al.*, 1991, pp. 627–630; Sage *et al.*, 1993, pp. 593–595).

These results led us to initiate the first study in the country to investigate whether the puppies born to vaccinated dams have protective antibody levels (based on the recommended minimum threshold level of 0.5 IU/ml) until they receive the first RV and the duration of protection in dogs following vaccination (Gunatilake *et al.*, 2003). At the time of planning the second study in 2006, the vaccine protocol in Sri Lanka was the same as in the mid-1990s – that is, puppies could receive the first RV at 3 months of age, followed by an annual booster. This protocol remained unchanged until 2013. This chapter explores two canine immunogenicity studies conducted in Sri Lanka and the impact these studies had on the revision of dog RV protocol in the country. Finally, this chapter discusses two ongoing studies, and the ways Sri Lanka is working towards ending human deaths by 2030.

11.2 Dog RV and Immunity

The impact of dog RV and human post-exposure treatment (in the form of RV) leading to a progressive reduction in the number of human deaths is shown in Fig. 11.1. These measures have reduced the number of human deaths in Sri Lanka from 377 in 1973 to ≤31 since 2013 (Harischandra *et al.*, 2016; Kanda *et al.*, 2021).

Vaccination is aimed at the development of immunity against rabies infection. As per the World Health Organization (WHO) and the World Organization for Animal Health (OIE), 70% of vaccination coverage should be the target to establish and maintain herd immunity in the dog population (Hampson *et al.*, 2009; Hampson *et al.*, 2015). Achieving the recommended level of vaccination coverage and thereby herd immunity prevents transmission of infection from a rabid dog to a non-infected dog (dog-to-dog transmission) and from a rabid dog to a human (dog-to-human transmission).

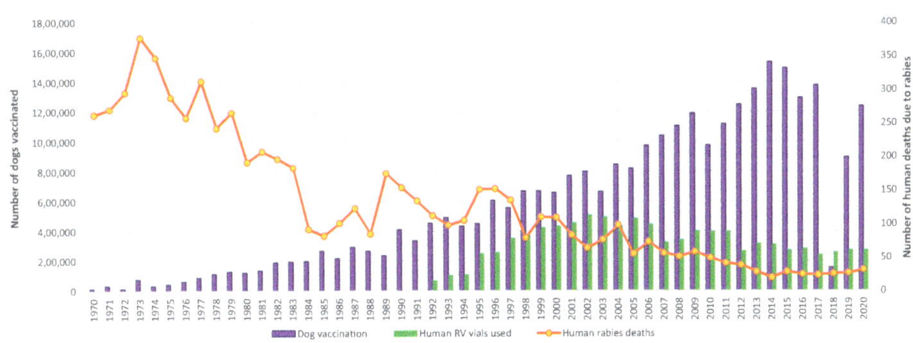

Fig. 11.1. Impact of human and dog rabies vaccination (RV) leading to reduction of human rabies deaths in Sri Lanka, 1970–2020. Human post-exposure treatment comprises RV. From updated information provided by Ruwini M.S. Pimburage, Department of Public Health Veterinary Services, Ministry of Health who was responsible for maintaining this information.

In this way, the urban cycle of rabies transmission could be interrupted.

Immediately after vaccination, an animal is not protected against rabies as it takes a short time period to develop immunity. As per the Centers for Disease Control and Prevention (n.d.) and *Today's Veterinary Practice* (2020), an animal could be considered immunized against rabies when a peak antibody titre is developed within 28 days following the initial vaccination.

Different brands of rabies vaccines intended for animals marketed in the country contain killed (inactivated) rabies virus to initiate an immune response following vaccination; antibody production (humoral immunity) by plasma cells and development of T lymphocytes mediates cellular immunity. Hence, the response to RV, and thus protection from exposure to the rabies virus could be determined by measuring humoral immunity and cellular immunity.

WHO and OIE recommend methods for the determination of humoral immunity including the rapid fluorescent focus inhibition test (RFFIT) and the fluorescent antibody virus neutralization (FAVN) test. These two tests use live virus to challenge the cells in the BHK21 cell lines and measure the virus-neutralizing antibody titres (humoral immune response) in serum samples (Smith *et al.*, 1973, pp. 535–541; Cliquet *et al.*, 1998; Cliquet and Wasniewski, 2015, pp. 217–223; Rupprecht *et al.*, 2019, p. 141). RFFIT is a well-established method in the Department of Rabies Research and Vaccine Quality Control of the Medical Research Institute (MRI) of Sri Lanka. In addition, there are enzyme-linked immunosorbent assay (ELISA) kits for the measurement of humoral immunity semi-quantitatively (Bahloul *et al.*, 2005, pp. 749–757; Wasniewski and Cliquet, 2012, pp. 166–175; Wasniewski *et al.*, 2014, pp. 211–220). These ELISA-based semi-quantitative assays do not require cell culture laboratories and also do not require handling of live rabies virus; therefore, in undersupplied facilities, this method could be undertaken to estimate the immune response following vaccination (i.e. whether the immune response is at or above 0.5 IU/ml, which is considered an acceptable level for a protective immune response if exposed to the rabies virus).

There are several different ELISA kits meant for semi-quantitative assay of the humoral immune response following RV. ELISA kits measure not only neutralizing antibodies (antibodies directed to surface glycoproteins of the rabies virus) but other binding antibodies (i.e. antibodies binding to nucleoproteins of the viral antigens), which is a disadvantage of this semi-quantitative assay method (Moore and Hanlon, 2010; Realegeno *et al.*, 2018, p. e0207009). However, this assay as a screening test is much easier to perform, and is rapid and convenient in a resource-limited setting. Hence, this method was established in 2019 at the Department of Physiology, Faculty of Medicine, University of Colombo for a study conducted to measure humoral immune response using BioPro Rabies ELISA Ab kit (O.K. Servis BioPro, n.d.). The FAVN test and facilities for measurement of cellular immunity following RV are not available in the country at present.

11.3 First Study: Immunogenicity in Domestic Dogs After RV

The first study conducted during 1999–2002 included only domestic canines (puppies and adult dogs with owners) of various breeds (ranging from crossbreeds to pedigree breeds) brought from Colombo and the suburbs (Borella, Rajagiriya, Battaramulla, Malambe, Kotte and Nugegoda) for RV to a veterinary clinic in Kotte, Sri Jayewardenepura (Gunatilake *et al.*, 2003).

The vaccine, Nobivac® Rabies (Merck & Co, n.d.), used in this study was an inactivated vaccine, containing Pasteur RIV (rabies inactivated virus) strain rabies virus grown in cell cultures. The antigenic content in this vaccine, ≥ 2.00 IU per dose, was higher than the antigenic content per dose recommended, as per the information provided by the vaccine manufacturer. According to the product leaflet included in the vaccine boxes, this vaccine was known to induce higher antibody titres with better cell-mediated immunity, protecting dogs and cats for up to 3 years against rabies. The selection of this vaccine in this study was based on approval for its use in the national rabies control programme. Two studies which were conducted in the recent past show the stability of this vaccine at different temperatures and its immunogenic capacity which provides evidence to support the manufacturer's claims (Lankester *et al.*, 2016, pp. 5504–5511; Lugelo *et al.*, 2021, p. 728271).

11.3.1 Methods

Adult dogs and puppies were recruited and categorized into the following groups based on the age of the animal, past vaccination history, or the vaccination status of the dam (past vaccination history and the vaccination status of the dam were based on the availability of the vaccination record issued at the time of vaccination). Only domestic dogs and puppies with owners were considered for this study. When recruiting puppies, age was determined following examination of the teeth growth at the time of presentation, which was cross-checked with the information provided by the animal owners. The groups were:

- **Group A:** adult dogs aged 1–5 years with a past vaccination history against rabies;
- **Group B:** adult dogs aged 1–5 years without a past vaccination history;
- **Group C:** 3-month-old puppies of dams with the evidence of RV; and
- **Group D:** 3-month-old puppies of unvaccinated dams.

A decision was made to recruit a minimum of 15 animals per group at the inception of the study. With the confirmation of the health status following a general clinical examination, these recruited puppies and adult dogs were vaccinated intramuscularly with a single dose of selected rabies vaccine, which was Nobivac® Rabies single-dose vials, after collecting the day 0 (pre-vaccination; D_0) blood sample. Recruited animals were followed up for a period of 1 year with post-vaccination blood sample collection at identified time points: (i) day 30 (D_{30}); (ii) day 180 (D_{180}); and (iii) day 360 (D_{360}). Serum samples were analysed by RFFIT at the Department of Rabies Research and Vaccine Quality Control of the MRI.

The usual practice is that when dogs are brought to a veterinary clinic, RV is given if the animal is clinically healthy in order to prevent dog-to-dog and dog-to-human transmission of rabies. During the planning and recruitment of dogs and puppies for this study, certain variables were not considered due to the complexity that could be faced during data analysis. These variables included: (i) the interval between RVs in adult dogs of Group A; (ii) nutritional status between crossbreed and pedigree dogs and puppies; (iii) sex of recruited animals; (iv) age variation of adult dogs (1–5 years) in Groups A and B; (v) the presence or absence of maternal immunity at the time of vaccinating puppies; and (vi) the frequency of worm treatment – especially in crossbreed dogs.

11.3.2 Results and discussion

Pre-vaccination and post-vaccination serum samples collected from recruited adult dogs and puppies at the identified time points were analysed by RFFIT. Results of the serum sample analysis are given in Table 11.1.

The results of this study showed a wide variation in the humoral immune response following RV in all the groups (within and between groups). The exact reasons for the wide variation observed in this study cannot be worked out as we considered only the age and previous vaccination status during recruitment. There may be other factors, such as the interval between vaccinations, nutritional status, breed, sex, maternal immunity and/or parasitic infections, affecting the immunity development following vaccination. In addition, the type of vaccine being administered is another factor which could potentially affect the development of immunity following vaccination (Kennedy *et al.*, 2007, pp. 5–17; Hampson *et al.*, 2015; Wera *et al.*, 2022). However, this factor was not considered in our study since a single brand of rabies vaccine was used for all subjects.

Our results show that the vaccine used in this study was immunogenic because almost all dogs (except four puppies in Group D) responded with ≥0.5 IU/ml immune response on D_{30} following inoculation. Results indicated that the immunity developed following RV in some adult dogs in Group B and almost all the puppies in Groups C and D were not maintained at or above the indicated threshold level, which should not be considered a vaccine failure, but rather a waning immune response.

Table 11.1. Distribution of dogs and puppies and resulting antibody titres for each group on day 0, pre-vaccination (D_0) and days 30, 180 and 360 (D_{30}, D_{180}, D_{360}, respectively). Table created by the authors.

Group	Number of dogs recruited	Number of dogs considered for analysis	Pre- and post-vaccination mean antibody titres (IU/ml)				Number (percentage) of samples which had antibody titres below threshold level < 0.5 IU/ml			
			D_0	D_{30}	D_{180}	D_{360}	D_0	D_{30}	D_{180}	D_{360}
A	21	17[a]	5.73	26.69	10.4	4.95	2 (11.8%)	0 (0%)	0 (0%)	0 (0%)
B	16	16	0.23[c]	9.37	2.78	1.38	16 (100%)	0 (0%)	2 (12.5%)	8 (50%)
C	4[b]	4	0.20[c]	0.84	0.40	0.12	4 (100%)	0 (0%)	3 (75%)	4 (100%)
D	15	15	0.15[c]	0.74	0.21	0.15	15 (100%)	4 (26.7%)	14 (93.3%)	15 (100%)

[a]All four samples of two dogs were contaminated. A sample obtained from one dog on D_0 was haemolysed and one dog was not available for sample collection on D_{360}. Findings of these four recruited dogs were not included in the analysis.

[b]The targeted number of dogs were recruited as planned to Groups A, B and D. Owners of puppies of vaccinated dams (Group C) brought for rabies vaccination (RV) were reluctant to give their consent for drawing four blood samples on identified time points from their puppies, as they were pedigree breeds. Therefore, it was not possible to obtain the target number of puppies for that group. The four puppies recruited to Group C were from three different dams. The dam of the first puppy in Group C was given a rabies vaccine for the first time during the pregnancy period (4–5 weeks pregnant). She was 2 years old at the time of her first RV.

[c]Adult dogs and puppies in Groups B, C and D were recruited based on the information provided by the owners of these animals. There can be instances where the vaccination record books were misplaced by the owners indicating that their animals were not vaccinated; specifically, the adult dogs in Group B. Also, there can be a situation where owners were unaware of the vaccination status of the dams of puppies in Group D. Without any evidence there can be animals who had received the RV in Groups B and D. In these instances, pre-vaccination mean antibody titres, even in one or two animals, could affect the mean antibody titre in the group. A similar situation may have been encountered in puppies of Group C, leading to the calculated mean antibody titre. The antibody titres of vaccinated dams were not measured to see whether they had the protective threshold or not before recruiting puppies to Group C. However, the mean antibody titres in animals in Groups B, C and D are below the recommended threshold level.

The period of maintenance of immunity following primary (or initial) vaccination is of a shorter duration compared to a secondary (booster) vaccination, which may also be dependent on the health and nutritional status of the animal. This shows the necessity of a booster vaccination for dogs vaccinated for the first time, prior to the annual vaccine booster, in order to maintain the immune response at or above the threshold level until the annual revaccination. During a primary vaccination, memory B and T lymphocytes are developed in the immune system against a pathogen, and the memory B cells (differentiated into plasma cells) produce a higher immune response of longer duration compared to a primary vaccination when a booster vaccination is inoculated against a specific pathogen (Clem, 2011, pp. 73–78).

Adult dogs with past vaccination history (Group A) showed a good humoral immune response following RV, which was maintained ≥0.5 IU/ml until the annual revaccination. Only eight (50%) adult unvaccinated dogs in Group B maintained antibody titres ≥ 0.5 IU/ml on D_{360}. Although the number of puppies in Group C was four, none of them showed the presence of maternal antibodies (antibodies transferred from the dam) at the age of the first RV. The presence of low levels of antibodies (well below the protective threshold) in puppies of group C does not indicate that they have an adequate level of maternal antibodies on D_0 to protect them until the first vaccination against rabies. Additional information is included under footnote c in Table 11.1.

Based on the results of the humoral immune response, the importance of the following was highlighted: (i) annual booster vaccination for adult dogs with a previous vaccination history (per the results, an annual booster is required to maintain the immune response at or above the threshold level in dogs in Sri Lanka); (ii) a primary booster vaccination for the adult dogs without a previous vaccination history needs to be given at a suitable interval, with annual boosters to maintain immunity at least at threshold level in this rabies-endemic country; and (iii) the need for puppies to have the first RV before 3 months of age (irrespective of the dam's vaccination status against rabies) with a primary booster at a suitable interval before going for annual revaccination in the country (Gunatilake *et al.*, 2003).

The time points of sample analysis were limited to four (D_0, D_{30}, D_{180} and D_{360}) due to the high cost of the analysis of samples. This did not allow us to determine the suitable time for the primary booster vaccination to be completed for adult dogs in Group B and puppies of Groups C and D before immunity could wane below the threshold level for adequate protection against rabies.

In the first study, free-roaming (stray) animals were not included, and the number of adult dogs and puppies recruited was at a minimal level which was not adequate for statistical comparison of the results. Therefore, a second, more comprehensive study was planned and conducted with the inclusion of both free-roaming and domestic dogs in adequate numbers for statistical comparisons (Kongkaew *et al.*, 2004, pp. 105–115; Pimburage *et al.*, 2017a, pp. 133–140).

11.4 Second Study: Immunogenicity in Free-Roaming and Domestic Dogs After RV

A representative dog population was included in this comprehensive study conducted from 2009 to 2013. In order to assess the effectiveness of vaccination (post-vaccination response), the humoral immune response was determined by RFFIT at the MRI. The objectives of this study were to determine the pattern of immunogenicity, the suitable time for primary booster vaccination among dogs from different age groups of free-roaming (stray) and domestic dogs (animals with owners), and to make recommendations regarding dog vaccination for rabies control in Sri Lanka.

11.4.1 Methods

The geographical area selected for the second study was the Kalutara district in the western province. There were 14 Divisional Secretariat Divisions and 11 Medical Officer of Health (MOH) areas at the time of conducting the study.

The average dog population was estimated at 143,000, considering the human population as per the National Census published in 2012 and human-to-dog ratio (Matter *et al.*, 2000, pp. 95–108; Department of Census and Statistics, 2012; Pimburage *et al.*, 2017b). Out of the total dog head samples submitted to the MRI from Kalutara district in 2011, 55% had been positive for rabies.

The dog RV campaign in Kalutara district is conducted under the authority of the Regional Director of Health Services and nearly 2000 temporary vaccination centres were created in 11 MOH areas in the district. Out of this, 37 temporary vaccination centres were randomly selected for the study. Furthermore, some areas in Kalutara district which were not covered by the government rabies control programme were selected for the recruitment of unvaccinated free-roaming (stray) dogs. These dogs were caught and restrained by specially trained dog catchers. A special tattooing number and a photograph were used for easy identification of free-roaming dogs during the follow-up period. Free-roaming dogs from the areas covered by the RV programme were identified by the presence of a tattooing mark and/or a red dog collar (Pimburage *et al.*, 2017a, pp. 133–140). Previous vaccination status against rabies among domestic juvenile and adult dogs (animals with owners) was determined based on vaccination records presented by the owners at the time of vaccination of animals. The different groups under which animals were recruited for this second study and the criteria used for grouping are as follows:

- **Group A:** adult free-roaming dogs > 1 year of age with a past RV history;
- **Group B:** adult free-roaming dogs > 1 year of age without a past RV history;
- **Group C:** domestic juveniles, 3 months–1 year of age with a previous vaccination;
- **Group D:** domestic juveniles, 3 months–1 year of age without a previous vaccination;
- **Group E:** domestic adult dogs, 1–6 years of age with regular (annual) vaccination;
- **Group F:** domestic adult dogs, 1–6 years of age without regular (annual) vaccination;
- **Group G:** domestic puppies (> 6 weeks–≤ 3 months of age) of vaccinated dams; and
- **Group H:** domestic puppies (> 6 weeks–≤ 3 months of age) of unvaccinated dams.

In order to have high statistical power, the sample size required for this comprehensive study was calculated to be 384. The number recruited was 510, considering about 33% contingency margin for the loss of dogs during the follow-up period. Dogs and puppies were recruited to eight different groups based on age and vaccination status as indicated in Table 11.2.

The Nobivac® Rabies single-dose vaccine was injected intramuscularly into recruited animals in this comprehensive study. Pre- and post-vaccination blood samples were collected on day 0 (D_0), day 30 (D_{30}), day 180 (D_{180}) and day 360 (D_{360}) and the serum was analysed by RFFIT to determine antibody titre levels.

11.4.2 Results and discussion

Not all of the recruited animals were available for sample collection through to the end of the study and some samples collected were haemolysed. These animals were not included in the analysis and therefore, the number of dogs having all four samples considered for analysis was 380. Multivariate analysis was applied using a computer software package (SPSS version-10) to compare antibody titres. The results of serum sample analysis are shown in Table 11.2.

The results of the second study also showed a wide variation in humoral immune response, similar to the first study. Vaccination has produced higher mean antibody titres on D_{30} in regularly vaccinated adult dogs and juveniles compared to the antibody development in the animals without regular vaccination or no vaccination previously. Their mean antibody titres are also at a higher level on D_0 compared to unvaccinated or irregularly vaccinated animals. The presence of memory B cells (secondary immune response) should be the reason for this boost in the immune response following revaccination.

It is a well-accepted scientific fact that when an antigen is introduced for the first time, the body's immune system of humans and animals gets sensitized and activates T

Table 11.2. Distribution of dogs and puppies, and resulting antibody titres for each group, in the second study. From Pimburage *et al.* (2017a).

Group, (sample size) and age of dogs considered for analysis	Sample collection time points[a]	Mean antibody titre (IU/ml) (95% CI)[b]	Median antibody titre (IU/ml) (95% CI)[b]	Interquartile range	Dogs with ≥ 0.5 IU/ml titres (%)
A (*n*=47)	D_0	6.66(4.05–9.92)	2.0(1.32–8.74)	9.4	70.21
Previously vaccinated adult	D_{30}	51.85(35.90–71.27)	36.44(13.74–49.05)	41.1	100
free-roaming dogs	D_{180}	22.89(15.11–31.20)	10.53(8.43–14.32)	31.17	100
(> 1 year)	D_{360}	7.17(4.25–10.88)	2.0(1.62–4.01)	8.08	82.98
B (*n*=47)	D_0	0.13(0.08–0.19)	0.65(0.2–0.08)	0.08	6.4
Unvaccinated adult free-roaming	D_{30}	12.61(7.23–19.89)	4.0(2.10–9.80)	11.68	87.24
dogs	D_{180}	9.16(4.74–15.29)	2.15(1.78–5.48)	7.75	87.24
(> 1 year)	D_{360}	3.89(1.93–5.45)	0.67(0.47–1.70)	1.85	59.82
C (*n*=47)	D_0	15.99(8.06–26.10)	2.7(1.63–6.78)	8.53	78.72
Previously vaccinated owned	D_{30}	34.77(18.07–56.20)	10.74(5.40–11.59)	25.11	95.74
juveniles	D_{180}	27.09(13.40–45.05)	8.34(2.90–13.84)	17.76	93.62
(3 months–1 year)	D_{360}	21.59(9.00–37.74)	2.90(1.98–7.78)	8.31	78.72
D (*n*=63)	D_0	0.11(0.07–0.14)	0.06(0.02–0.08)	0.07	1.59
Unvaccinated owned juveniles	D_{30}	17.39(11.34–24.13)	9.40(4.07–10.74)	17.44	93.65
(3 months–1 year)	D_{180}	11.21(6.05–17.26)	2.50(2.12–5.20)	6.34	93.65
	D_{360}	3.04(1.53–4.71)	0.44(0.28–0.52)	1.7	42.86
E (*n*=51)	D_0	13.62(6.01–24.24)	29(0.56–2.96)	7.28	76.47
Owned adult dogs with regular	D_{30}	29.81(14.55–46.80)	10.02(2.50–11.04)	18.87	96
annual vaccination	D_{180}	39.47(20.48–60.20)	10.50(2.35–12.46)	18.03	88
(> 1–6 years)	D_{360}	24.23(11.11–40.90)	2.56(2.19–8.93)	10.21	78
F (*n*=48)	D_0	6.84(3.14–11.06)	0.93(0.55–2.28)	5.6	77.08
Owned adult dogs without regular	D_{30}	25.73(15.20–40.34)	10.74(4.30–19.54)	17.54	100
vaccination	D_{180}	16.95(11.12–23.80)	9.41(2.35–15.56)	17.85	95.84
(> 1–6 years)	D_{360}	9.69(5.91–13.97)	2.95(2.07–8.93)	9.15	83.33
G (*n*=40)	D_0	0.10(0.07–0.14)	0.08(0.08–0.08)	0.03	0
Domestic puppies (> 6 weeks–≤3	D_{30}	10.66(7.93–13.74)	9.80(9.70–10.20)	6.3	97.5
months) of	D_{180}	4.63(3.16–6.46)	3.31(2.00–5.35)	5.64	82.5
previously vaccinated dams	D_{360}	0.23(0.15–0.35)	0.09(0.08–0.19)	0.32	7.5
H (*n*=37)	D_0	0.07(0.06–0.09)	0.08(0.04–0.08)	0.04	0
Domestic puppies (> 6 weeks–≤	D_{30}	12.56(8.32–17.15)	8.50(4.70–10.45)	9.75	94.59
3 months) of	D_{180}	4.76(3.00–6.80)	2.09(0.59–5.40)	7.31	78.38
previously unvaccinated dams	D_{360}	0.32(0.21–0.44)	0.19(0.09–0.40)	0.36	10.81

[a]D_0, day 0, pre-vaccination; D_{30}, day 30; D_{180}, day 180; D_{360}, day 360.
[b]CI, confidence interval.

and B lymphocytes in order to produce cellular and humoral immune responses. During this process, memory T cells and memory B cells are produced and maintained. When the same or similar antigen is introduced again, these memory cells work to produce their specified effects (secondary immune response). During RV, inactivated rabies virus is introduced with the vaccine to sensitize and initiate an immune response in the body, while revaccination, or booster vaccination, is done to maintain the immune response through memory cells.

It is interesting to note the immune response is greater than the threshold level on D_0 in adult dogs of Group F, irrespective of irregular vaccination. When compared to the immune response in adult dogs of Group E with regular vaccination, the mean immune response in Group F dogs is at a lower level at each and every time point of sample collection. However, the percentage of animals having the protective threshold is higher at each time point in Group F than in Group E. Based on this observation between the two groups, one might argue that after several years of successive vaccinations there is no necessity for regular revaccination, as dogs in Group F are apparently protected based on the immune response. Variations such as sex, nutritional status and/or previous vaccine brand used for vaccination may have contributed to the change observed in this study.

It is not possible to detect which dogs are adequately protected against the rabies virus, due to the non-availability of free antibody titre assessment facilities in the country. Dogs and juveniles possessing antibody titres higher than the minimum threshold cannot be challenged with the rabies virus for experimental purposes in the country for ethical reasons. Therefore, it is not possible to determine whether the recruited animals are actually protected against rabies or not. The vaccine could be considered efficacious based on the immune response of all categories of animals on D_{30} compared to the pre-vaccination titre.

Mean antibody titres on D_{360} of juvenile and adult dog groups were maintained at ≥0.5 IU/ml threshold level, although there is a decline compared to D_{30} titres until the annual revaccination (antibody titres are not maintained at the same level following a primary vaccination or booster vaccination and there is a natural decline in the immune response over time). This study also shows the absence of antibodies in puppies, transferred from the vaccinated dams, supporting the requirement of vaccination of puppies at an early age. Furthermore, this study demonstrates that these puppies should receive a booster vaccine before reaching 1 year of age since most puppies do not sustain adequate titre levels by D_{360}.

Considering the results of the second study and the rabies-endemic status in Sri Lanka, recommendations made were to: (i) revaccinate dogs on an annual basis; (ii) provide a primary booster vaccination at a suitable interval when RV is given to adult dogs and juveniles for the first time; and (iii) to vaccinate puppies at a suitable age and provide a primary booster before the annual revaccination. Other studies should be considered to determine an appropriate suitable interval for primary boosters in adult dogs or what age is suitable to vaccinate puppies. Organizations should consider the challenges these recommendations may bring – such as logistical challenges to vaccinating free-roaming dogs or locating puppies at the desired age for subsequent booster vaccinations.

11.5 Revision of Dog RV Protocol in Sri Lanka

Due to the non-availability of a common protocol for dog RV, government institutions, private veterinary practitioners and NGOs who care for free-roaming animals were following different protocols in Sri Lanka. At the same time, if a puppy or an adult dog has not received two consecutive doses of RV, it was considered unprotected against rabies by the Ministry of Health, especially when a decision is to be made regarding post-exposure treatment for a dog-bite victim.

In order to address these issues, the Veterinary Teaching Hospital, Faculty of Veterinary Medicine and Animal Science, University of Peradeniya organized two workshop sessions in August and September 2013. Research-based information was considered in this endeavour to design and revise the dog RV protocol in the country. The results of our two studies were the only available evidence, based on the WHO- and OIE-recommended RFFIT method for analysis of serum samples to determine humoral immune response following the RV.

There was another study conducted on puppies, juveniles and adult dogs kept at an animal home in Peradeniya, Kandy (Gunawardena *et al.*, 2011, pp. 13–18). The sample analysis had been conducted using a different ELISA kit and this was not presented or considered. The decision taken at the workshop sessions was publicized by the DAPH in order to emphasize the need for practising veterinarians to follow the revised schedule (DAPH, Sri Lanka, 2013, pp. 1–4; Gunatilake, 2015). This was adopted as a step forward to control dog-mediated rabies in Sri Lanka. Furthermore, it follows the recommendation of the WHO and OIE to adopt a suitable vaccination protocol in rabies-endemic countries to combat this deadly, vaccine-preventable, zoonotic disease.

The revised RV protocol for dogs in Sri Lanka indicates primary vaccination at the age of 6 weeks regardless of the vaccination status of the dam, the primary booster vaccination at the age of 14 weeks, and thereafter with annual revaccinations. In addition to the above-mentioned schedule for domestic puppies of vaccinated dams, the recommendation for domestic puppies of unvaccinated dams was to give the first rabies vaccine at the time of presentation (after opening eyes) and the first booster in the 14th week of age with annual revaccinations. For stray animals, the recommendation made was to inoculate first at the time of presentation (after opening eyes) and the first booster after 10 weeks with annual revaccinations (DAPH, Sri Lanka, 2013, pp. 1–4).

Adequate research findings were not available to make the guidelines for the exact timing of booster vaccination for puppies of vaccinated dams, although it was decided to vaccinate at the age of 14 weeks. Considering the danger faced specifically by small children, it was decided to vaccinate puppies of unvaccinated dams and stray puppies at the time of presentation after opening eyes. Information for the primary booster vaccination for juveniles was also not available. Therefore, it was crucial to find out the starting point of declining antibody response after primary vaccination against rabies in puppies and juveniles. Also, it was worthwhile to find out how the immune responses vary after the first booster vaccination against rabies in puppies. Therefore, the third study which is ongoing was planned.

One important aspect to consider is whether there was a reduction in the number of rabies-positive dog cases and any reduction in human deaths by dog-mediated rabies following the revision of the dog vaccination protocol in 2013. As per the data given on the website of the Public Health Veterinary Services, Ministry of Health, there is a reduction in the number of human deaths due to rabies from 38 in 2012 to equal or less than 28 in subsequent years until 2019 (Public Health Veterinary Services, n.d.). Similarly, there had been a gradual reduction in rabies-positive dog cases from 664 in 2013 to 383 in 2020. Published data do not clearly indicate whether this reduction could be attributed to the revision of the vaccination protocol. The reduction in human deaths could be due to improved post-exposure treatment protocols.

11.6 Ongoing Studies

11.6.1 Third study: Testing the effectiveness of the revised dog RV protocol

The effectiveness of the revised RV protocol was the next target in the third study. This study was also initiated in collaboration with the MRI using a different brand of rabies vaccine (Rabisin) recommended by the WHO and OIE to use in rabies-endemic countries. This vaccine is the recommended vaccine to be used in the selected area of study by the authorities of Dehiwela Municipality in the western province.

This study included several different groups containing domestic puppies and juveniles of mixed breeds. Animals recruited were given a booster vaccination, 2 months after the primary vaccination to puppies and 1 month after the primary vaccination to juveniles, and sample collection was done at appropriate intervals (more sample collection time points were included). Sample analysis will need to be completed to find out if this protocol is effective. The results of this study would help the Sri Lankan authorities engaged in rabies control to establish a new protocol, if required, for puppies and juveniles.

11.6.2 Fourth study: Comparison of immunogenicity of different brands of animal rabies vaccines

Nine different brands of rabies vaccines imported from different countries are currently marketed in the country. Some of these brands are available in single-dose or multi-dose preparations. The efficacy and potency of these vaccines are not subjected to any testing before their registration at the Veterinary Drug Control Authority (VDCA) under the DAPH, and the registration is based on the dossier submitted by the vaccine-importing companies.

Previously there was a vaccine marketed in the country which had zero potency when tested using the mouse inoculation test at a WHO reference laboratory for rabies in Europe. Subsequently, the local company which was involved in the marketing of this vaccine withdrew it from VDCA registration. For ethical reasons, the brand of the vaccine is not disclosed here. With this background, the fourth study was initiated and is ongoing.

The results of this study will enable us to select the animal rabies vaccines with high immunogenic capacity and thereby make recommendations to the VDCA of DAPH. When vaccines of high immunogenic capacity are used, the immune response in the animals may be enough to maintain a higher level of herd immunity. This would promote control of dog-mediated rabies in the country and thereby achieve zero human rabies deaths by 2030.

11.7 Conclusion

Many factors play a role in immunity development following RV and maintenance of titres at the recommended minimum protective level. The results of two completed studies described above showed a positive relationship between previous vaccinations and the duration of maintenance of humoral immunity until the annual booster vaccination. On the other hand, the dam's vaccination status did not have any impact on the protection of puppies against rabies as the humoral immune response after vaccination seems to be closely similar in puppies of vaccinated and unvaccinated dams. The results of this study emphasized the need to vaccinate puppies as soon as possible to commence an immune response.

Humoral immune response against rabies follows the norm of the gradual waning of immune response over time when it reaches the end point of D_{360} in these two studies. This emphasizes the need for annual revaccination in the country, as rabies is endemic. Also, when animals were given only one vaccine in the first year of life, data indicates that it is not sufficient for the maintenance of antibody titres until the time of annual booster vaccination. Therefore, puppies and juveniles should be given two rabies vaccines in the first year of life at suitable time intervals with annual revaccinations. Revision of the dog RV protocol in Sri Lanka was timely in order to control this deadly zoonotic disease.

Although the first two studies generated valuable information, there were several limitations including: (i) small sample size; (ii) unreliable vaccine history (especially for dogs believed to be unvaccinated, which could account for a titre response when there should be none); (iii) inability to separate dogs with a history of one vaccine from those with two or more; and (iv) inability to consider other factors that could play a role in the immune response. In this context, more studies need to be conducted to validate the results.

The third study, which is ongoing, will generate new data in order to revise the primary booster vaccination time point indicated in the revised protocol in 2013 for puppies, if required, and also to establish a protocol for juveniles. The fourth study which focuses on the immunogenic capacity of the rabies vaccines marketed in the country would enable us to select the vaccines with high immunogenic capacity. The results of all these studies will help Sri Lanka to achieve the zero dog-mediated human rabies deaths by 2030 target with the support of human post-exposure prophylaxis operated by the Ministry of Health.

Acknowledgements

The two principal studies described in the chapter were funded by the National Science Foundation of Sri Lanka (grant numbers RG/99 /V/01 and RG/2007/HS/03).

Ethics approval was granted by the Ethics Review Committees of the Faculty of Medicine, University of Colombo, and the Medical Research Institute (MRI).

Authors' Declaration

All authors declare they have no conflict of interest.

All authors have approved this manuscript, agree with its submission, and share collective responsibility and accountability.

This manuscript has not been published or is not under review elsewhere.

References

Bahloul, C., Taieb, D., Kaabi, B., Diouani, M.F., Ben Hadjahmed, S. *et al.* (2005) Comparative evaluation of specific ELISA and RFFIT antibody assays in the assessment of dog immunity against rabies. *Epidemiology and Infection* 133(4), 749–757. DOI: 10.1017/s095026880500381x.

Centers for Disease Control and Prevention (n.d.) Rabies vaccination. Available at: www.cdc.gov/rabies/specific_groups/veterinarians/vaccination.html (accessed 6 October 2022).

Clem, A.S. (2011) Fundamentals of vaccine immunology. *Journal of Global Infectious Diseases* 3(1), 73–78. DOI: 10.4103/0974-777X.77299.

Cliquet, F. and Wasniewski, M. (2015) Direct fluorescent antibody test for rabies diagnosis. In: Rupprecht, C. and Nagarajan, T. (eds) *Current Laboratory Techniques in Rabies Diagnosis*. Elsevier, Amsterdam, pp. 217–223.

Cliquet, F., Aubert, M. and Sagné, L. (1998) Development of a fluorescent antibody virus neutralisation test (FAVN test) for the quantitation of rabies-neutralising antibody. *Journal of Immunological Methods* 212(1), 79–87. DOI: 10.1016/s0022-1759(97)00212-3.

DAPH (Department of Animal Production and Health), Sri Lanka (2013) Rabies: a major zoonosis. *Veterinary Epidemiological Bulletin, Sri Lanka* 6(2), 1–4.

Department of Census and Statistics (2012) Census of population and housing 2012. Department of census and statistics, Ministry of Finance and Planning, Colombo. Available at: www.statistics.gov.lk/Population/StaticalInformation/CPH2011/Census2012ResultsPopulationHousingBased5Sample (accessed 10 February 2023).

Gunatilake, M. (2015) New challenges to anti-rabies vaccination schedule in dogs, based on research done in Sri Lanka. *Sri Lanka Medical Association News* November, 18.

Gunatilake, M., Wimalaratne, O. and Perera, K.A.D.N. (2003) Persistence of antibody titres in adult dogs and puppies following anti-rabies immunization. *The Ceylon Journal of Medical Science* 46(2), 67. DOI: 10.4038/cjms.v46i2.4849.

Gunawardena, G.S.P. de S., Randunu, R.S., Harischandra, P.A.L. and de Silva, D.D.N. (2011) Serological detection of anti-rabies virus antibodies in dogs and its implications to the vaccination schedules. *Sri Lanka Veterinary Journal* 58(A), 13–18.

Hampson, K., Dushoff, J., Cleaveland, S., Haydon, D.T., Kaare, M. *et al.* (2009) Transmission dynamics and prospects for the elimination of canine rabies. *PLoS Biology* 7(3), e53. DOI: 10.1371/journal.pbio.1000053.

Hampson, K., Coudeville, L., Lembo, T., Sambo, M., Kieffer, A. *et al.* (2015) Estimating the global burden of endemic canine rabies. *PLoS Neglected Tropical Diseases* 9(4), 1041–1058. DOI: 10.1371/journal.pntd.0003709.

Harischandra, P.A.L., Gunesekera, A., Janakan, N., Gongal, G. and Abela-Ridder, B. (2016) Sri Lanka takes action towards a target of zero rabies death by 2020. *WHO South-East Asia Journal of Public Health* 5(2), 113–116. DOI: 10.4103/2224-3151.206247.

Kanda, K., Jayasinghe, A., Jayasinghe, C. and Yoshida, T. (2021) Public health implication towards rabies elimination in Sri Lanka: a systematic review. *Journal of Acta Tropica* 223, 106080. DOI: 10.1016/j.actatropica.2021.106080.

Kennedy, L.J., Lunt, M., Barnes, A., McElhinney, L., Fooks, A.R. *et al.* (2007) Factors influencing the antibody response of dogs vaccinated against rabies. *Vaccine* 25(51), 8500–8507. DOI: 10.1016/j.vaccine.2007.10.015.

Kongkaew, W., Coleman, P., Pfeiffer, D.U., Antarasena, C. and Thiptara, A. (2004) Vaccination coverage and epidemiological parameters of the owned-dog population in Thungsong district, Thailand. *Preventive Veterinary Medicine* 65(1–2), 105–115. DOI: 10.1016/j.prevetmed.2004.05.009.

Lankester, F.J., Wouters, P.A.W.M., Czupryna, A., Palmer, G.H., Mzimbiri, I. *et al.* (2016) Thermotolerance of an inactivated rabies vaccine for dogs. *Vaccine* 34(46), 5504–5511. DOI: 10.1016/j.vaccine.2016.10.015.

Lugelo, A., Hampson, K., Czupryna, A., Bigambo, M., McElhinney, L.M. *et al.* (2021) Investigating the efficacy of a canine rabies vaccine following storage outside of the cold-chain in a passive cooling device. *Frontiers in Veterinary Science* 8, 728271. DOI: 10.3389/fvets.2021.728271.

Madies, T. (2020) Animal Welfare Organizations taking over government's role in dealing with stray dogs. Available at: https://economynext.com/animal-welfare-organisations-taking-over-governments-role-in-dealing-with-stray-dogs-73406/ (accessed 5 June 2022).

Matter, H.C., Wandeler, A.I., Neuenschwander, B.E., Harischandra, L.P. and Meslin, F.X. (2000) Study of the dog population and the rabies control activities in the Mirigama area of Sri Lanka. *Acta Tropica* 75(1), 95–108. DOI: 10.1016/s0001-706x(99)00085-6.

Merck & Co (n.d.) Nobivac® rabies. Available at: www.merck-animal-health-usa.com/nobivac/nobivac-1-rabies-vaccine (accessed 5 June 2022).

Moore, S.M. and Hanlon, C.A. (2010) Rabies-specific antibodies: measuring surrogates of protection against a fatal disease. *PLoS Neglected Tropical Diseases* 4(3), e595. DOI: 10.1371/journal.pntd.0000595.

O.K. Servis BioPro (n.d.) BioPro rabies ELISA Ab Kit. O.K. Servis BioPro, Prague. Available at: www.rabieselisa.com (accessed 5 June 2022).

Pimburage, R.M.S., Gunatilake, M., Wimalaratne, O., Balasuriya, A. and Perera, K.A.D.N. (2017a) Sero-prevalence of virus neutralizing antibodies for rabies in different groups of dogs following vaccination. *BMC Veterinary Research* 13, 133–141. DOI: 10.1186/s12917-017-1038-z.

Pimburage, R.M.S., Harischandra, P.A.L., Gunatilake, M., Jayasinhe, D.N., Balasuriya, A. *et al.* (2017b) A cross-sectional survey on dog ecology and dog anti-rabies vaccination coverage in selected areas in Sri Lanka. *Sri Lanka Veterinary Journal* 64(1), 1–9. DOI: 10.4038/slvj.v64i1A.18.

Public Health Veterinary Services (n.d.) History of rabies control activities in Sri Lanka. Public Health Veterinary Services, Ministry of Healthcare and Indigenous Medical Services, Sri Lanka. Available at: www.rabies.gov.lk/english/history.php (accessed 5 June 2022).

Realegeno, S., Niezgoda, M., Yager, P.A., Kumar, A., Hoque, L. *et al.* (2018) An ELISA-based method for detection of rabies virus nucleoprotein-specific antibodies in human antemortem samples. *PLoS ONE* 13(11), e0207009. DOI: 10.1371/journal.pone.0207009.

Rupprecht, C.E., Fooks, A.R. and Abela-Ridder, B. (eds) (2019) Generation of anti-rabies single domain antibodies by display technologies. In: *Laboratory Techniques in Rabies*. World Health Organization, Geneva, Switzerland, p. 141.

Sage, G., Khawplod, P., Wilde, H., Lobaugh, C., Hemachudha, T. *et al.* (1993) Immune response to rabies vaccine in alaskan dogs: failure to achieve a consistently protective antibody response. *Transactions of the Royal Society of Tropical Medicine and Hygiene* 87, 593–595. DOI: 10.1016/0035-9203(93)90099-C.

Smith, J.S., Yager, P.A. and Baer, G.M. (1973) A rapid reproducible test for determining rabies neutralizing antibody. *Bulletin of the World Health Organization* 48, 535–541.

Tepsumethanon, W., Polsuwan, C., Lumlertdaecha, B., Khawplod, P., Hemachudha, T. *et al.* (1991) Immune response to rabies vaccine in Thai dogs: a preliminary report. *Vaccine* 9(9), 627–630. DOI: 10.1016/0264-410x(91)90186-a.

Today's Veterinary Practice (2020) Rabies vaccination in dogs. Available at: https://todaysveterinarypractice.com/preventive-medicine/rabies-vaccination-in-dogs/ (accessed 6 October 2022).

Wasniewski, M. and Cliquet, F. (2012) Evaluation of ELISA for detection of rabies antibodies in domestic carnivores. *Journal of Virological Methods* 179(1), 166–175. DOI: 10.1016/j.jviromet.2011.10.019.

Wasniewski, M., Labbe, A., Tribout, L., Rieder, J., Labadie, A. *et al.* (2014) Evaluation of a rabies ELISA as an alternative method to seroneutralisation tests in the context of international trade of domestic carnivores. *Journal of Virological Methods* 195, 211–220. DOI: 10.1016/j.jviromet.2013.10.021.

Wera, E., Warembourg, C., Bulu, P.M., Siko, M.M. and Dürr, S. (2022) Immune response after rabies vaccination in owned free-roaming domestic dogs in Flores Island, Indonesia. *Frontiers in Veterinary Science* 9, 868380. DOI: 10.3389/fvets.2022.868380.

12 The Role of Dog Ecology in Canine Rabies Prevention and Control in Asia: Lessons from Indonesia and the Oceanic Region

Salome Dürr[1]*, Ewaldus Wera[2], Victoria J. Brookes[3], Charlotte Warembourg[1], Silja Griss[1] and Anna S. Fahrion[4]

[1]Veterinary Public Health Institute, University of Bern, Bern, Switzerland; [2]Kupang State Agricultural Polytechnic (Politeknik Pertanian Negeri Kupang), West Timor, Indonesia; [3]The University of Sydney, Camperdown, New South Wales, Australia; [4]Institute for International Animal Health and One Health, Friedrich-Loeffler-Institut (Bindestriche), Riems, Germany

Abstract

Constraints to canine rabies elimination include low political priority, poor implementation of a One Health approach, low accessibility of rabies post-exposure treatment, and often insufficient information on local dog populations. Studies on dog ecology can provide essential insights to guide the implementation of effective dog vaccination campaigns, but they are rare in Asia. While population size estimations and demographics are relatively well-researched, information about contact networks, population dynamics, and human-mediated dog movements is limited. In this chapter, dog ecology research in Asia is reviewed, and two dog ecology case studies from Indonesia and Oceania using GPS (Global Positioning System) collars, contact sensors, camera trapping, serology and surveys are presented. Differences within and between dog populations reveal the heterogeneity of dog behaviour that is bound to the societal context in which dogs are kept. Such findings can be used for scenario simulations in disease spread models, and can inform rabies elimination strategies by dog vaccination, supported by dog confinement and population control.

12.1 Introduction

12.1.1 Dogs as a reservoir of rabies

Domestic dogs (*Canis familiaris*) are by far the most important reservoir species of rabies, causing more than 99% of human rabies cases worldwide (WHO, 2013). In Asia, over 35,000 people are estimated to die annually from dog-mediated rabies and the disease causes an economic burden of over US$6 billion/year, substantially more than on other continents (Hampson *et al.*, 2015, p. 9). After being exposed to a rabid dog, typically through a bite, effective post-exposure prophylaxis (PEP) can prevent clinical rabies and death in people; however, PEP availability, access and affordability are often limiting factors in rabies-endemic countries.

*Corresponding author: salome.duerr@unibe.ch

DOI: 10.1079/9781800622975.0012

Also, preventing human rabies by PEP alone is not sustainable because the virus still circulates in the reservoir population (i.e. the dogs). Investment in controlling rabies in the reservoir hosts is thus key to sustainably eliminating the disease.

It has been demonstrated that in countries with high rabies incidence in humans, investment in dog vaccination is low, and vice versa (Hampson *et al.*, 2015), and several studies have shown that dog mass-vaccination campaigns can dramatically reduce canine rabies incidence. For example, in 1983, a regional control programme for rabies, including canine mass vaccination, was introduced in Latin America and the Caribbean (Vigilato *et al.*, 2013). The programme achieved an overall reduction of almost 90% of cases in humans and dogs with the elimination of dog-mediated rabies in many countries of the continent (Del Rio Vilas *et al.*, 2017). Surveillance strategies are still required despite the elimination of dog-mediated rabies, because rabies in wildlife remains endemic in this region, with potential risk of rabies transmission to domestic animals and people. An example from Asia is the successful national strategy for dog-mediated rabies elimination that was implemented in Sri Lanka. This strategy includes dog mass vaccination and led to a significant reduction in the number of human deaths caused by rabies (Harischandra *et al.*, 2016). Applying a One Health approach by combining improved PEP access as a public health intervention, investing in dog vaccination, and strengthening communication between the two sectors (Mindekem *et al.*, 2017), is the most efficient route to reach the target of dog-mediated human rabies elimination globally by 2030 – a goal, also named 'Zero by 30', that has been set by the World Organization for Animal Health (WOAH), the World Health Organization (WHO) and the Food and Agriculture Organization of the United Nations (FAO) (Ren *et al.*, 2018). A core contribution from the veterinary sector is the investment in effective and sustainable vaccination of dogs, which not only requires technical expertise, but also knowledge of the respective dog population.

12.1.2 Definition of free-roaming domestic dogs (FRDDs)

WOAH categorizes dogs according to the level of dependency on humans and restrictions in their roaming behaviour (OIE, 2009). Dogs that are not fully restricted are also called stray dogs and are the majority of the worldwide dog population (Gompper, 2013). Based on the level of dependency on humans, stray dogs are divided into three categories: (i) feral dogs that have reverted to the wild state and do not depend on humans for feeding or breeding; (ii) village (or community) dogs that do not have a designated owner but rely on local communities for feeding; and (iii) free-roaming owned dogs that roam without supervision at least part of their time but depend on their owners for shelter and feeding. Village and free-roaming owned dogs are often collectively termed free-roaming domestic dogs (FRDDs) (Warembourg *et al.*, 2021a).

In Africa and Latin America, FRDDs are predominantly owned, with estimations of ownerless dogs ranging from 0 to 20% (Dürr *et al.*, 2009; ; ; Muthiani *et al.*, 2015; Warembourg *et al.*, 2020). In Asia, the proportion of ownerless dogs varies and ranged from 3.3% in Bali, Indonesia (Hiby *et al.*, 2018), and up to 40 and 62%, in Bangladesh and India, respectively (Hossain *et al.*, 2013). Among owned FRDDs, the dog's role (e.g. companionship, hunting) and husbandry practices, such as feeding, access to veterinary care, dog confinement or human-mediated dog movements, vary between and within countries (Villatoro *et al.*, 2016; Warembourg *et al.*, 2021b). These factors greatly influence dog behaviour – including roaming and contact with dogs and other species such as humans, livestock and wildlife – which is relevant to rabies transmission and control. It is therefore essential to generate knowledge on FRDD ecology and the factors that influence this for evidence-based, sustainable rabies control. This aligns with the WHO's recommendations to adapt vaccination strategies to local dog population characteristics (WHO, 2018).

12.1.3 Definition and relevance of dog ecology

Ecology can be defined as the science of 'the distribution and abundance of organisms and the interactions that determine distribution and abundance' (Sagoff, 2017). Applying an ecological lens to FRDDs highlights that they are a species with characteristics of companion animals, livestock (dogs are raised as a meat source in some countries) and wildlife. Relevant ecological studies include research on: (i) population size, density, demography and dynamics; (ii) roaming behaviour and interactions with the (often anthropogenic) environment; (iii) contacts to conspecifics and other species; and (iv) human-, dog- and environment-related factors influencing these parameters.

Methods to estimate dog population size range from simple counts and surveys to advanced models (Tiwari *et al.*, 2019b). Studies investigating roaming behaviour and contacts within and between species have often been based on observations (e.g. Hutabarat *et al.*, 2003; ; Dias *et al.*, 2013). More recently, Global Positioning System (GPS) devices and contact sensors attached to FRDDs via collars have been used and complement previous observation studies (e.g. Vaniscotte *et al.*, 2011; Dürr and Ward, 2014; Laager *et al.*, 2018; Raynor *et al.*, 2020; Warembourg *et al.*, 2021a).

In this chapter, the role of dog ecology in rabies prevention and control is reviewed and discussed. Following an overview of constraints to rabies elimination and of dog ecology in Asia regarding rabies control, two case studies are presented: one from rabies-endemic Indonesia and one from rabies-free Oceania. These case studies are intended to demonstrate and show how dog ecology studies in specific regions can generate in-depth knowledge on dog populations and how such knowledge can be employed to improve rabies prevention and control. Each case provides a different perspective on the activities required to eliminate or prevent rabies at a regional level.

12.2 Constraints to Rabies Elimination

12.2.1 Rabies as a neglected disease

Vaccinating dogs is recognized as the most sustainable measure to control and eventually eliminate dog rabies (Cleaveland and Hampson, 2017), but many hurdles on the way to 'Zero by 30' are linked to dog vaccination campaigns. There are also other challenges to the goal of elimination that are not directly related to dog vaccination.

An important challenge is the poor visibility of rabies. Only if perceived as the danger and burden that it actually presents, will there be sufficient societal pressure, advocacy and political insistence to act upon rabies with the necessary will and the resources required. Since rabies is a disease of poverty, and like other neglected tropical diseases often occur in remote places and under-resourced populations, under-reporting of cases in people, as well as dogs, is likely, and estimates show that 100-fold under-reporting might be usual (Scott *et al.*, 2017, p. 2). The absence of solid evidence induces a 'cycle of neglect': because data is weak or non-existent, rabies is falsely perceived as an insignificant health issue, leading to further neglect (Taylor *et al.*, 2017). Secondly, rabies might not be a disease of priority because of limited awareness in the population, spanning from poor knowledge of the disease and its terrible outcome and post-bite prophylactic measures, to lack of aggression-avoidance behaviour towards dogs. To increase public awareness, efforts mainly consist of promoting rabies information, but whether this translates into the desired behavioural changes remains unclear (Fahrion *et al.*, 2017).

Thirdly, the political will of decision makers to change the situation might be absent due to competing public health priorities, the knowledge that rabies elimination is a lengthy process requiring multiple costly rounds of dog vaccination campaigns, and possibly the perception of rabies as a mainly veterinary issue. As such, the weakness of the veterinary sector in many countries compounds the relatively low priority of interventions involving domestic dogs (Müller *et al.*, 2015). Overall, these factors predominantly put at risk and cause impacts in an under-represented and therefore vulnerable sector of society, the rural poor. Without collective societal pressure, strong advocacy, and local and national champions, rabies continues to be ignored.

12.2.2 Constraints related to human PEP

From a human health perspective, a common constraint to rabies prevention following a potentially rabid dog bite is the inaccessibility or unaffordability of rabies biologicals (Sreenivasan *et al.*, 2019). A course of rabies PEP, combined with repeated, potentially long travel routes to get the vaccine, can exceed a household's resources, and is a bitter truth for many bite victims. Newer developments to fill this gap, including shortened PEP regimens, requiring fewer health facility visits, changed recommendations towards vaccine-saving intradermal administration, and new technologies such as thermostable vaccines and monoclonal antibodies, could contribute to alleviating the problem. However, a change towards better supply and distribution of these biologicals require systematic improvement of universal health coverage. A global push towards universal healthcare and a facilitating shift for many countries may follow the recent uptake of human rabies vaccine on the portfolio of Gavi, the Vaccine Alliance (WHO, 2018).

12.2.3 Constraints related to dog vaccines

From the veterinary perspective, constraints related to dog vaccination are manifold. Studies assessing rabies vaccination coverage often detect vaccination coverage below the WHO recommended threshold of 70% (e.g. Kitala *et al.*, 2001; Atuman *et al.*, 2014; Tenzin *et al.*, 2015b). Besides the hurdles discussed above, well-planned vaccination campaigns require knowledge of the underlying dog populations, which is often not available. Unknown dog population size and demographics, lack of information on the typically rapid population turnover leading to many susceptible puppies, as well as the natural death of vaccinated dogs, or unknown immigration of unvaccinated dogs, challenge the planning of dog vaccination campaigns (Gamble *et al.*, 2018). Also, dogs that are inaccessible can limit vaccination coverage, and the proportion of these varies considerably from country to country even within Asia (Morters *et al.*, 2014b; Hiby *et al.*, 2018).

To reduce the proportion of non-accessible dogs, methods such as trapping with nets are needed to vaccinate parenterally. In addition, oral vaccination of domestic dogs is increasingly discussed as an option to overcome this issue (Cliquet *et al.*, 2018; Wallace *et al.*, 2020). Recently, researchers were able to successfully vaccinate 83% of dogs using oral rabies vaccination in Thailand, which otherwise would have been only partly accessible for parenteral vaccination (Chanachai *et al.*, 2021). An adequate understanding of dog ecology (i.e. population size, dynamics, and accessibility for vaccination in the respective setting) is therefore a key determinant to planning rabies vaccination strategies in endemic countries.

12.2.4 Constraints related to a One Health approach

Addressing rabies solely either from a veterinary or from a public health perspective misses a 'One Health' integration of the sectors and can present a roadblock to the elimination of dog-mediated human rabies. The key justification for a One Health approach lies in the realization that the main strategy to avoid human deaths is to interrupt transmission in the reservoir by dog vaccination. This means that the public health benefit is achieved through a veterinary intervention (Cleaveland and Hampson, 2017). If this is not recognized, and the public health sector only assumes responsibility for providing PEP to bite victims, dog vaccination risks remaining underfunded by the typically lower-resourced veterinary sector. Responding to rabies through PEP provision alone is expensive and symptomatic, lacking the perspective of sustainable reduction in rabies incidence in humans, because the disease is not addressed at its source. Economic models support One Health solutions. Public and veterinary health sectors have to collaborate and ideally, co-fund solutions for dog vaccination (Lavan *et al.*, 2017). This is challenging because existing systems need to be adjusted to work together, requiring harmonizing administrative and management structures and budget lines across sectors; however, the benefits of a One Health approach for rabies control are increasingly being recognized (Seifmann and Kaplan, 2021). Ultimately, alleviating the burden of rabies is cost-effective and can bring indirect benefits to other disease programmes and promote universal health coverage more generally.

12.3 Dog Ecology in Asia

12.3.1 Overview of studies on dog ecology

Returning to the role of dog ecology and its contribution to canine vaccination on the way to rabies elimination, an overview of the current evidence from research in Asia was conducted. Twenty-four studies, published between 1982 and 2021, were identified in a literature scan when searching for dog ecology studies in Asia, and were clustered in only a few countries. India is most represented, while research in about 85% of rabies-endemic Asian countries was not identified (Fig. 12.1a). Studies either focused on owned and unowned free-roaming dogs, and some of them also included owned, confined dog populations. Seventeen studies estimated dog population size, ten included estimates of dog densities, seven estimated human:dog ratios, and three reported the investigation and application of various dog enumeration techniques (Fig. 12.1b). About half of the studies reported on dog demography, such as the population distribution of age, sex and breed, whereas only two studies addressed dog population turnover. For aspects related to dog management, nine studies were found to report on dog confinement, five on neutering, three on the role and origin of dogs, and one on reported feeding practices. Data on vaccination coverage and accessibility of dogs for vaccination were collected in eight and three studies, respectively. Nine studies addressed dog roaming behaviour, home range size, immigration, and group sizes of dogs in the streets, as well as three studies reporting on the transport of dogs. Only one study investigated contact networks among dogs.

12.3.2 Five key inferences for rabies control

The information generated from the identified dog ecology studies spanned five key areas of knowledge required for rabies prevention and control. First, dog population size provides important baseline data to plan and monitor rabies control programmes for the respective dog population (WHO Veterinary Public Health Unit, 1988; Hossain *et al.*, 2013; Tenzin *et al.*, 2013, Tenzin *et al.*, 2015a; Morters *et al.*, 2014a; Rinzin *et al.*, 2016; Pimburage *et al.*, 2017). Methods to estimate dog population size and density must be time- and cost-effective. Tiwari *et al.* (2018) compared and discussed eight such methods, concluding that those accounted for individual heterogeneity of dogs' catchability led to more accurate and robust estimations compared to those not accounting for heterogeneity.

Secondly, dog demography, such as age and sex distribution, and turnover, helps to understand the population dynamics and estimate the annual replacement of vaccinated by unvaccinated dogs (WHO Veterinary Public Health Unit, 1988). This information is needed to determine the interval between dog vaccination campaigns to maintain appropriate vaccination coverage. A study by Kumarapeli and Awerbuch-Friedlander (2009) showed that more frequent than annual vaccination campaigns are required in populations with a high proportion of free-roaming and ownerless dogs due to their rapid population turnover and shorter lifespan.

Thirdly, the accessibility of dogs in the identified studies was estimated to be >70%, which makes adequate vaccination coverage feasible (WHO Veterinary Public Health Unit, 1988; Morters *et al.*, 2014a; Hiby *et al.*, 2018). Nevertheless, 15–25% of the free-roaming dogs – owned and unowned – were inaccessible for parenteral vaccination, because of difficulties bringing them to vaccination posts (WHO Veterinary Public Health Unit, 1988; Matter *et al.*, 2000; Pimburage *et al.*, 2017), and therefore,

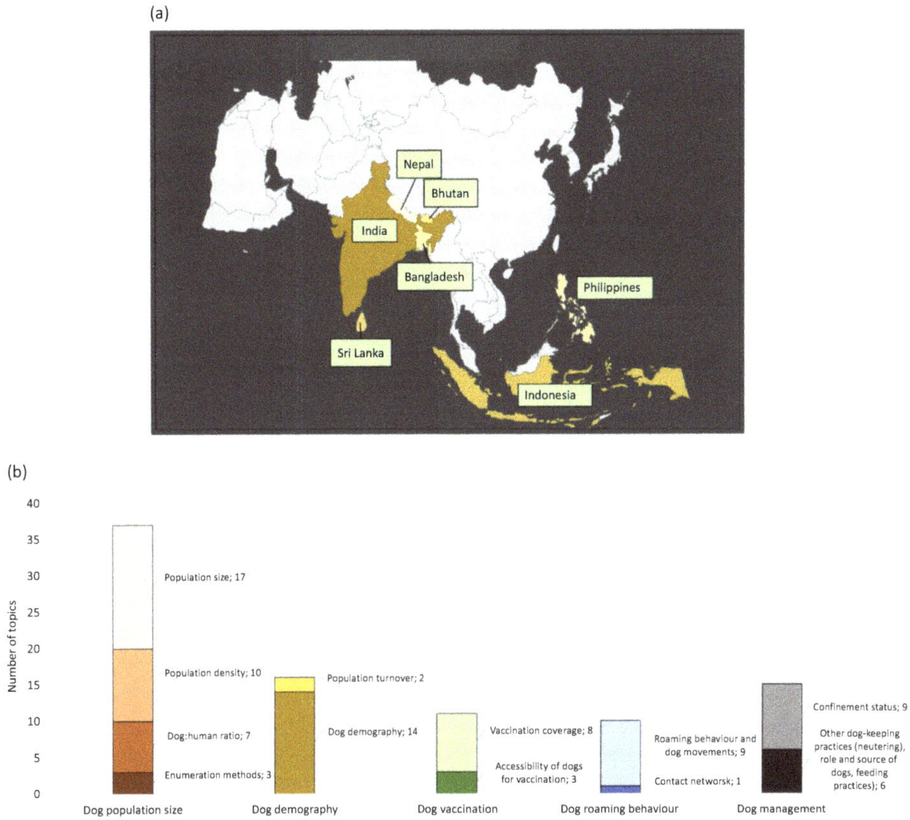

Fig. 12.1. Dog ecology studies in Asia. (a) Map of countries of origin of these studies; and (b) frequency of topics covered by 24 studies identified from PubMed and Google Scholar (published between 1982 and 2021) when searching for dog ecology-related topics of free-roaming dogs in Asian countries. Figure created by Silja Griss.

alternative vaccination strategies, such as oral vaccination, were discussed (Matter *et al.*, 2000).

Fourthly, some studies tried to identify rules for more effective targeted vaccination compared to random vaccination of 70% of all dogs, using outcomes from dog ecology studies. For example, Rinzin *et al.* (2016) and Tenzin *et al.* (2015a) showed that to use resources more effectively, vaccination could be optimized by focusing on areas with high dog density. Another study from India suggested that dogs tended to build large groups with conspecifics, and those dogs being sighted closer to garbage points would have lower accessibility for parenteral vaccination, and should therefore be considered for oral vaccination (Tiwari *et al.*, 2019). Also, by investigating contact networks of dog populations in four countries (Chad, Guatemala, Indonesia and Uganda), it was suggested that targeted vaccination of highly connected dogs (those with more contacts with other dogs than the average) could be more efficient (Warembourg *et al.*, 2021a).

Finally, anthropogenic factors greatly influence dog ecology. A study undertaken in Indonesia and South Africa highlighted that dog population size is regulated by human demands for dogs (Morters *et al.*, 2014b). Also, human-mediated dog movement was observed and discussed as a key factor in rabies spread (Beran, 1982; Morters *et al.*, 2014b; Hutabarat *et al.*, 2003). For example, Beran (1982) reported on rabies outbreaks in mountain settlements and concluded that latently infected dogs could have only reached this remote site with people. Therefore, it is crucial to consider such owner-related factors in the design and implementation of rabies prevention and control programmes and adapt them to the socio-economic characteristics of the communities.

12.4 Dog Ecology Case Study 1: Indonesia

12.4.1 Background

Rabies was first reported in Indonesia in 1889 in Bekasi, West Java and then spread widely to other regions in Indonesia (Ward, 2014). The number of rabies-infected provinces in Indonesia has increased from 20 (concentrated in the four islands Java, Sumatra, Kalimantan and Sulawesi) (Mustiana, 2013) in 1981 to 26 in 2021 (Ministry of Health Republic Indonesia, 2020). The ongoing spread in the eastern islands of Indonesia has included Flores, Ambon, Seram, Buru and Bali (Amaral *et al.*, 2014) and most recently, Sumbawa in 2018. The incursion of rabies into Sumbawa was most likely due to the transportation of latently infected dogs from rabies-infected areas because it is common in Indonesia to transport dogs by ferry (Mustiana, 2013) or via fishing boats (Bingham, 2001). The government of Indonesia now conducts an annual rabies celebration day to increase people's awareness of rabies and annual vaccination of dogs in infected areas, to reduce the number of rabies cases both in humans and in dogs, and to prevent further spread to rabies-free islands.

The number of rabies cases in Indonesia ranks fourth highest in South-east Asia, after India, Bangladesh and Myanmar, with 150–300 human cases reported annually (WHO, 2012). Most victims are children under 19 years old (Wera, 2001). Almost all of these human rabies cases were due to exposure to rabid dogs, who did not receive timely PEP (Wera *et al.*, 2015). Based on data from the Ministry of Health in Indonesia for the period 2015–2019, more than 80,000 humans are bitten by suspected rabid dogs every year (Ministry of Health Republic Indonesia, 2020), and 30% (thus 24,000 victims) do not receive PEP. The high number of bite cases, particularly during outbreaks, could be due to the high number of dogs in rabies-infected areas. The estimated dog population in Indonesia is 16 million dogs, with the human:dog ratio ranging from 4:1 in Bali to 32:1 in Aceh (One Health Roadmap, 2019).

12.4.2 Rabies on Flores Island

Flores Island covers an area of 15,624 km^2, with an estimated human and dog population of 1.8 and 0.2 million, respectively (Wera *et al.*, 2013). Most dogs in Flores are owned and dependent on their owners for food (Wera *et al.*, 2021). Seventy-nine per cent of dog owners in Flores are farmers and keep dogs to guard their house and property (Wera *et al.*, 2015; Warembourg *et al.*, 2021b). Dog owners allow their dogs to roam freely, mostly day and night. Sometimes, they accompany the owners to the field to chase away wildlife, particularly monkeys and wild pigs that destroy the farmers' plants. This is risky behaviour for rabies transmission to wild animals. Unlike in Bali and Lombok (Dalem *et al.*, 2012; Mustiana *et al.*, 2015), most dogs in Flores are female (63%) (Warembourg *et al.*, 2021b) and dog owners tend to prioritize vaccination for female over male dogs (Wera *et al.*, 2015). This may be due to the use of the dogs in Flores for meat consumption, which is typical during traditional ceremonies. People in Flores prefer to slaughter male dogs and keep female dogs for breeding, giving them a higher economic value. Therefore, dog owners benefit more from keeping and vaccinating reproductive female dogs than male dogs.

The rabies incursion to Flores happened in 1998, with currently more than 2000 people being bitten by suspected rabid dogs and 15 deaths every year (Wera, 2017). However, these official numbers do not reflect the complete burden of human rabies in Flores, because the records only capture the number of rabies patients who seek help in hospitals or public health centres for PEP, or when they show symptoms (Wera *et al.*, 2015). To reduce the incidence of rabies, the regencies on Flores Island implemented annual dog vaccination campaigns (Wera, 2017). Although dog vaccination is compulsory, it is difficult to enforce due to the absence of a dog registration system and lack of resources to catch and restrain dogs, resulting in an estimated 55% vaccination coverage (Wera *et al.*, 2015). This coverage level has not yet achieved rabies elimination from Flores Island.

12.4.3 Fieldwork undertaken on Flores Island

Research undertaken in Sikka district, Flores, Indonesia demonstrates how a better understanding of dog ecology can inform rabies control programmes. The first aspect covers the loss of rabies antibodies over 12 months following a vaccination campaign to inform policy makers on vaccination campaign frequency in this region, and identifies dog factors associated with faster loss of antibodies to characterize dogs that could be prioritized for additional booster vaccination (Wera *et al.*, 2021). Secondly, research is described to identify dogs that are more likely to transmit rabies because of their behaviour, such as having large home ranges, high centrality measures in their contact networks, or high activity levels, because such dogs could be prioritized for targeted vaccination (Griss *et al.*, 2021; Warembourg *et al.*, 2021a). Thirdly, research to identify habitat resources preferentially used by FRDDs is described, because such habitats could be targeted for bait distribution in oral vaccination programmes (Cunha Silva *et al.*, 2022, unpublished data).

Free-roaming, owned dogs were studied in two rural and one semi-urban area located close to Maumere on Flores Island (Warembourg *et al.*, 2021a). A total of 256 dogs were included in the study. A questionnaire survey about dog characteristics and dog management practices was conducted with dog owners (Warembourg *et al.*, 2021b). Blood was sampled from each dog included in the study and 217 dogs were collared with georeferenced contact sensors (see Fig. 12.2). These devices combine GPS to track dog movements and a contact sensor to record proximity events between collared dogs (Laager *et al.*, 2018, p. 3; Warembourg *et al.*, 2021a). Of the collared dogs, 37 were randomly selected to be equipped with activity trackers to investigate their physical activity on a continuous timescale (Griss *et al.*, 2021).

Fig. 12.2. Dog wearing a GPS (Global Positioning System) collar in Flores Island. Photo courtesy of C. Warembourg, 2018.

12.4.4 Maintenance of rabies antibody titres

Blood samples were analysed for rabies antibodies 30, 90, 180, 270 and 360 days after vaccination, using a locally produced rabies ELISA kit (Wera *et al.*, 2021). The results showed that of the dogs that have developed antibody levels above 0.5 EU/ml 1 month after vaccination, only 58% (67/115) still had titres above this threshold after 3 months. This percentage dropped to 35% (11/31) 1 year after vaccination. Although the level of antibodies measured through the ELISA does not directly reflect the level of neutralizing antibodies, nor the level of protection against rabies (Aubert, 1992; Dodds *et al.*, 2020), the level of practically measurable immunity against rabies through the titre of antibodies quantified in this study suggests that providing booster vaccination more frequently than once a year would contribute to the maintenance of sufficient rabies vaccination coverage (Wera *et al.*, 2021). A regression model showed that dogs that were never vaccinated, vaccinated more than 1 year ago, dogs younger than 12 months, or with poor body condition scores, were good candidates for booster vaccination (Wera *et al.*, 2021).

12.4.5 Roaming behaviour and predictors

Data collected by the contact recording devices were used to generate contacts in the dog populations and compute network metrics such as degree centrality, which is the number of dogs with which a given dog is in contact (Warembourg *et al.*, 2021a). In addition, GPS data were used to compute the dogs' home range sizes, defined as the area used for normal activities such as breeding or foraging (Burt, 1943; Warembourg *et al.*, 2021b). Finally, activity trackers were used to compute dogs' mean activity levels and the proportion of time spent 'resting', or in 'moderate' and 'high' activity during a circadian cycle. These studies showed that the younger the dog, the more active they were (Griss *et al.*, 2021). Conversely, the home ranges of younger dogs were observed to be significantly smaller (Warembourg *et al.*, 2021b) and age did not significantly affect the dogs' degree centrality (i.e. the number of contacts to other dogs) (Warembourg *et al.*, 2021a). This suggests that young dogs might be more active in small areas but do not encounter a high number of other dogs. In addition, dogs displaying poor body condition had smaller home ranges. It is interesting to link this finding with the results of the rabies antibody study. Promoting feeding of dogs according to their biological need as part of responsible ownership practices will improve body condition, which might lead to an increase in the dog home range size if dogs are not confined. However, it might also induce a better response to rabies vaccination. This example highlights that combining knowledge of several aspects of dog ecology and rabies transmission is crucial when making recommendations on rabies control measures. Also, dogs that were part-time restrained had smaller home ranges and a smaller network degree centrality, highlighting that some sort of confinement should be recommended as responsible dog ownership, in addition to rabies vaccination and appropriate feeding practices.

It is important to highlight that extrapolation of such results should always be made with caution. Comparing these results from Flores with those generated using the same study protocol in other countries, namely Chad, Guatemala and Uganda, showed wide heterogeneity (Warembourg *et al.*, 2021b). For example, being a young dog was significantly negatively associated with the home range size in Indonesia, but not in Uganda. Similarly, male dogs had larger core home ranges in Chad, but not in the three other countries. Dog behaviour is complex and the differences observed between dog populations are probably related to dog husbandry and the general impact of humans on dog behaviour (Warembourg *et al.*, 2021a), and is an area requiring further investigation.

The behaviour of FRDDs is not only influenced by dogs' owners, but also driven by the environment in which they live. A spatial habitat selection model was built to identify the environmental resources – including forests, open fields, roads, tree-covered areas, beaches and buildings – preferred by the study dogs in Sikka regency. The analyses revealed that FRDDs tended to prefer man-made resources such as roads and buildings, suggesting that those locations would be useful locations to deploy oral baits for rabies vaccination. Oral vaccination would be particularly useful for dogs that are inaccessible for parenteral vaccination or are owned but fearful or aggressive. Providing baits to such dogs under observation of consumption would be required if baits are deployed close to buildings, so that they are not consumed by non-target species including children. Such an approach has already been tested in India (Gibson *et al.*, 2019) and successfully applied in Haiti (Wallace *et al.*, 2020).

These studies of FRDD ecology in Flores Island informed the following recommendations about dog vaccination campaigns in this region: (i) young dogs, dogs with no vaccination history or vaccinated more than 1 year ago should be targeted for booster vaccination; (ii) promoting responsible ownership, including rabies vaccination, dog confinement and good-quality dog feeding, would contribute to maintaining sufficient vaccination coverage; and (iii) buildings and roads could be a target location for oral baits to reach and maintain a high rabies vaccination coverage.

12.5 Dog Ecology Case Study 2: The Oceanic Region

12.5.1 Background

The Oceanic region includes Australasia, Polynesia, Micronesia and Melanesia, and is the only continent except for Antarctica that is canine-rabies free. Although the region is geographically large, spanning western and eastern hemispheres, the total human population is relatively small (41 million) in the global context. Countries in the Oceanic region are geographically, culturally, linguistically and economically diverse. While Australia has one of the highest human development indexes (HDIs) in the world (HDI 0.944 in 2019), the HDI of Papua New Guinea (PNG; the second largest population in the region with almost 9 million people) is only 0.555, with approximately 20% of the population living below the national poverty line (Human Development Reports, n.d.).

The spread of rabies in South-east Asia threatens the rabies-free status of Oceania, especially given the accelerating trade and movement of people throughout the South-east Asian and Oceanic regions over recent decades (Testaverde *et al.*, 2017). The ability to detect and respond in a timely manner to a rabies incursion in the Oceanic region is challenging due to limited resources and competing needs in some countries, the sparse population in some areas, and the logistical difficulties of surveillance that needs to span the vast coastline of islands throughout the region. Sparkes *et al.* (2015) highlighted potential incursion scenarios to the Northern Territory and northern Queensland in Australia due to unregulated activities such as those built on centuries of traditional trade and travel for fishing and cultural events, as well as tourism such as itinerant yachting. A large amount of research has been conducted over the last decade to identify priority risk pathways and potential impacts in the region, so that rabies preparedness strategies, such as targeted surveillance, can be implemented (Ward and Brookes, 2021). Qualitative and quantitative risk assessments for routes into PNG and northern Queensland, Australia, have demonstrated that although the risk is low, hotspot regions could include Western Province in PNG, and the Torres Strait Islands and Cape York Peninsula in Queensland due to hunting and fishing activities (Brookes*et al.*, 2017; Hudson *et al.*, 2017b).

Due to the lack of disease data from this currently rabies-free region, understanding the ecology of dogs has underpinned preparedness research. This case study focuses on how collecting detailed information about dog ecology has enabled modelling to predict impacts and identify potential mitigation strategies following a rabies incursion. Models of infectious disease transmission play an increasing role in informing and directing animal health and public health policy (Alahmadi *et al.*, 2020; Kirkeby *et al.*, 2021). In addition, data about the demographics, dynamics and density of dog populations, as well as home range and contact rates between dogs, are critical for accurate predictions from disease spread models. As fieldwork provided increasingly detailed information about these factors, models of rabies spread in dog populations in Australia have become more sophisticated in terms of modelling contacts between dogs, and their specificity to regions and types of dog populations.

Dogs in the Oceanic region include both domestic and wild dogs. Consistent with terminology presented earlier in this chapter, domestic dogs in the Northern Peninsula Area (NPA), Torres Strait and Treaty villages of Western Province can be owned and confined, owned and free-roaming (the most common dog type, and hereafter referred to as FRDDs), stray (previously owned dogs) and feral (living free over generations) (Brookes *et al.*, 2019; Hudson *et al.*, 2017b; Ward and Brookes, 2021). The wild dog population in Australia consists of a mixture of dingoes (*Canis dingo*), feral dogs and dingo-domestic dog hybrids, and is collectively referred to as dingoes hereafter (Stephens *et al.*, 2015).

12.5.2 Simulation of rabies spread in wild dog populations

Initial models of rabies spread in wild dog populations in Australia included a state transition model based on ordinary differential equations by , and a structurally more complex stochastic transition network (percolation) model by Johnstone-Robertson *et al.* (2017). In the latter model, parameters were estimated from the literature and expert opinion, or set to span an order of magnitude in the case of parameters for which empirical information was lacking. This model predicted that canine rabies would spread at least 120km following initiation in 21% of incursions, and in these events, the median speed of spread was 67km/year. The model used percolation to simulate spatial spread, but was not location specific and therefore, did not account for geographic variation in suitable dingo habitats and natural barriers to dingo movement in a defined region. Sensitivity analysis of model parameters demonstrated that parameters associated with dog movement and behaviour were responsible for most of the variation in model outputs. Although much is known about dingo ecology in some parts of Australia, a scoping review confirmed that information about these parameters in the equatorial and tropical climate zones of northern Australia was limited (Gabriele-Rivet *et al.*, 2019). Therefore, to parameterize a location-specific model in a region in which a rabies incursion in Australia is a relatively high risk, camera-trap data were collected in the NPA region of Cape York Peninsula over a 2-year period to determine the density and home range of dingoes, as well as examine some behavioural aspects such as group sizes and breeding activity (Gabriele-Rivet *et al.*, 2020; Ward *et al.*, 2020).

Camera-trap surveys and analysis using spatially explicit capture-recapture and mark–resight methods are frequently used to estimate wildlife population density, with the latter method able to estimate population density when only a subset of the animals can be individually identified (e.g. by observation of natural marks) (McGregor *et al.*, 2015; Carter *et al.*, 2019). Using data from a linear array of 21 camera traps placed approximately 2km apart through community and bushland areas for 1year, mark–resight models estimated home range sizes from 7.95 km^2 to 29.40 km^2 (variations in home range size occurred throughout the year), and dingo density from 0.135 dingoes/km^2 (95% confidence interval (CI) = 0.127–0.144) to 0.147 dingoes/km^2 (95% CI = 0.135–0.159) during the dry and wet seasons, respectively (Gabriele-Rivet *et al.*, 2020). When these estimates were incorporated into a stochastic, agent-based, spatially explicit model of rabies spread in the dingo population of the NPA (Gabriele-Rivet *et al.*, 2021), the model predicted that rabies spread between distinct dingo packs in nearly 60% of simulations. In these events, a median of 22 dingoes (95% predicted range of 2–101 dingoes) were infected within the study area of 1131 km^2. The speed of spread was 0.52 km/week for a median of 191 days, and while this was slower than that predicted by Johnstone-Robertson *et al.* (2017), the predicted R_0 (basic reproductive rate, a metric for an outbreak's transmissibility) was higher, ranging from median $R_0 = 2$ at pack level to $R_0 = 3$ at dingo level, with variation also observed due to season. In addition, more extensive spatial spread was predicted when the simulated incursion occurred close to human settlements, especially during the dry season. Therefore, although it was concluded that the dingo population in the NPA would most likely not support endemic rabies, this spatially explicit approach to modelling rabies spread in the NPA region using region-specific ecological data to parameterize dingo population density and home range enabled a greater understanding of the heterogeneity of risk in the region.

12.5.3 Simulation of rabies spread in domestic dogs

The collection of ecological data from domestic dogs in the NPA region has also enabled increasingly detailed modelling of rabies spread and mitigation strategies. Dürr and Ward (2014) trialled GPS collars on 69 FRDDs in communities of the NPA and on Elcho Island, Northern Territory, and compared three methods to estimate their home range sizes. Analysis using the biased random bridge method (Benhamou, 2011) estimated the core and extended home ranges sizes at 0.3–0.9 ha and 2.7–19.8 ha (95% CI), respectively, and provided a basis for further investigations using telemetry on FRDDs. In an initial stochastic, agent-based, spatially explicit rabies simulation model for the NPA region (Dürr and Ward, 2015) seven of the 37 parameters, all of which parameterized contact rates between dogs for rabies transmission, were informed by these locally collected empirical data. GPS data informed the distance kernel for the contact rates between dogs of different households, and survey data informed the frequency of dog movements between the NPA communities. The model tested pre-emptive and reactive vaccination, reactive targeted (i.e. those dogs that have been contacted by a rabid dog) and non-targeted culling, and movement bans within and between communities. Outputs predicted an estimated median R_0 of approximately 1.7, which is consistent with other reports of the R_0 for rabies in dogs (Sparkes *et al.*, 2015). Model simulations confirmed the important role of dog vaccination, although, in these settings a 70% vaccination coverage was not enough to prevent an outbreak in most cases.

This model was then refined to further investigate vaccination following the collection of longer duration telemetry data from domestic dogs, which demonstrated that dogs could be characterized into three groups based on their free-roaming patterns – 'stay-at-home', 'roamer' and 'explorer' dogs (Hudson *et al.*, 2017a). The home range sizes of explorer dogs were relatively large (median extended home range 9.5 ha) in comparison to the roamer (6.0 ha) and stay-at-home dogs (3.7 ha), and dog-type specific parameterization of contact rates in the model constructed by Dürr and Ward (2015) predicted that vaccination strategies targeted at explorer-type dogs resulted in outbreaks of shorter duration with fewer dogs affected (Hudson *et al.*, 2019). In a further extension of the use of telemetry data to parameterize models of rabies spread in this region, Brookes *et al.* (2020) used GPS collars to describe social networks of domestic dogs in three communities on the Torres Strait Islands. A network-based, stochastic model of rabies spread in similar-sized populations of dogs to Torres Strait Island communities predicted that rabies could persist for 15–275 days (95% CI) at very low incidence even at 70% vaccination coverage due to the way in which rabies altered behaviour and, thus changed the network connections of dogs and effectively increased their contacts (Brookes *et al.*, 2019). This demonstrates that increasing heterogeneity of contacts in rabies spread models, as well as the inclusion of heterogeneity in behaviour, increases the threshold required for herd immunity in dog populations and creates questions around the recommendation that 70% vaccination coverage is sufficient. To what extent such observations depend on the characteristics of dog populations, such as density and mobility within and between communities, should be topics for future research.

12.6 Conclusion

To reach the goal to eliminate dog-mediated human rabies by 2030, several constraints must be overcome, one of them being to understand the ecology of dogs living in and close to human societies. Dog ecology studies generate knowledge on dog populations and dog behaviour, which is required to plan effective dog vaccination campaigns, for both parenteral or oral vaccines, which is the only sustainable way to control and prevent canine and human rabies. Dog ecology in Asia has been relatively well studied in India, Sri Lanka and Indonesia; however, studies are lacking in most Asian countries. In addition, while population size estimations and dog population demographics are relatively well researched, limited research in Asia has been conducted on contact networks, population dynamics and the accessibility of dogs for parenteral vaccination.

Only a few studies were identified that investigated human-mediated dog movements, a parameter that is important for rabies incursion in rabies-free areas and in areas where rabies has been successfully controlled. While information on dog population size, demographics and dynamics, the accessibility of dogs and dogs' competence to develop and maintain immunity can directly inform frequencies and target groups of dogs for vaccination, other insights gained by dog ecology studies are crucial to parameterize rabies transmission models, in particular information on contact rates between dogs. The case studies from Indonesia and the Oceanic region demonstrate that dog behaviour is specific to individual dogs, and heterogeneity of contacts between dogs was found in both regions. Intensive modelling work in the Oceanic region further showed that such heterogeneities influence how rabies spreads, and that 70% vaccination coverage may not be sufficient to control rabies in all populations and outbreaks. Integrating dog ecology knowledge can refine dog vaccination strategies: for example, by targeting far-roaming dogs or those with greater contact with other dogs. However, no single dog characteristic, such as age, sex, activity level or role of the dog could be identified to help classify the dogs into the roaming types. Therefore, these studies demonstrate that dogs need to be studied within their habitat because differences in dog ecology are influenced by coexistence with humans. Investigating the dog–human bond and understanding how dogs use the natural and built environment would help to understand how rabies persists and spreads and can contribute to achieving the elimination of dog-mediated human rabies by 2030.

Authors' Declaration

All authors declare they have no conflict of interest.

No ethical approval was required for the work presented in this book chapter apart from what is already indicated in the published articles.

All authors have approved this manuscript, agree with its submission, and share collective responsibility and accountability.

This manuscript has not been published or is not under review elsewhere.

References

Alahmadi, A., Belet, S., Black, A., Cromer, D., Flegg, J.A. *et al.* (2020) Influencing public health policy with data-informed mathematical models of infectious diseases: recent developments and new challenges. *Epidemics* 32, 1–12. DOI: 10.1016/j.epidem.2020.100393.

Amaral, A.C., Ward, M.P. and da Costa Freitas, J. (2014) Estimation of roaming dog populations in Timor Leste. *Preventive Veterinary Medicine* 113(4), 608–613. DOI: 10.1016/j.prevetmed.2013.11.012.

Atuman, Y.J., Ogunkoya, A.B., Adawa, D.A.Y., Nok, A.J. and Biallah, M.B. (2014) Dog ecology, dog bites and rabies vaccination rates in Bauchi State, Nigeria. *International Journal of Veterinary Science and Medicine* 2(1), 41–45. DOI: 10.1016/j.ijvsm.2014.04.001.

Aubert, M.F. (1992) Practical significance of rabies antibodies in cats and dogs. *Revue Scientifique et Technique (International Office of Epizootics)* 11(3), 735–760. DOI: 10.20506/rst.11.3.622.

Benhamou, S. (2011) Dynamic approach to space and habitat use based on biased random bridges. *PloS ONE* 6(1), e14592. DOI: 10.1371/journal.pone.0014592.

Beran, G.W. (1982) Ecology of dogs in the Central Philippines in relation to rabies control efforts. *Comparative Immunology, Microbiology and Infectious Diseases* 5(1–3), 265–270. DOI: 10.1016/0147-9571(82)90047-9.

Bingham, J. (2001) Rabies on Flores Island, Indonesia: is eradication possible in the near future? In: *Proceedings of the Southern and Eastern African Rabies Group/World Health Organization Meeting,* Lilongwe, Malawi, June 18–22, 2001. Available at: https://www.mediterranee-infection.com/wp-content/uploads/2019/03/SEARG-report-2001.pdf (accessed 2 February 2023).

Brookes, V.J., Degeling, C. and Ward, M.P. (2018) Going viral in PNG - exploring routes and circumstances of entry of a rabies-infected dog into Papua New Guinea. *Social Science & Medicine (1982)* 196, 10–18. DOI: 10.1016/j.socscimed.2017.11.006.

Brookes, V.J., Dürr, S. and Ward, M.P. (2019) Rabies-induced behavioural changes are key to rabies persistence in dog populations: investigation using a network-based model. *PLoS Neglected Tropical Diseases* 13(9), e0007739. DOI: 10.1371/journal.pntd.0007739.

Brookes, V.J., VanderWaal, K. and Ward, M.P. (2020) The social networks of free-roaming domestic dogs in island communities in the Torres Strait, Australia. *Preventive Veterinary Medicine* 181, 104534. DOI: 10.1016/j.prevetmed.2018.09.008.

Brookes, V.J., Keponge-Yombo, A., Thomson, D. and Ward, M.P. (2017) Risk assessment of the entry of canine-rabies into Papua New Guinea via sea and land routes. *Preventive Veterinary Medicine* 145, 49–66. DOI: 10.1016/j.prevetmed.2017.06.011.

Burt, W.H. (1943) Territoriality and home range concepts as applied to mammals. *Journal of Mammalogy* 24(3), 346. DOI: 10.2307/1374834.

Carter, A., Potts, J.M. and Roshier, D.A. (2019) Toward reliable population density estimates of partially marked populations using spatially explicit mark-resight methods. *Ecology and Evolution* 9(4), 2131–2141. DOI: 10.1002/ece3.4907.

Chanachai, K., Wongphruksasoong, V., Vos, A., Leelahapongsathon, K., Tangwangvivat, R. *et al.* (2021) Feasibility and effectiveness studies with oral vaccination of free-roaming dogs against rabies in Thailand. *Viruses* 13(4), 1–12. DOI: 10.3390/v13040571.

Cleaveland, S. and Hampson, K. (2017) Rabies elimination research: juxtaposing optimism, pragmatism and realism. *Proceedings of the Royal Society B: Biological Sciences* 284(1869), 1–9. DOI: 10.1098/rspb.2017.1880.

Cliquet, F., Guiot, A.L., Aubert, M., Robardet, E., Rupprecht, C.E. *et al.* (2018) Oral vaccination of dogs: a well-studied and undervalued tool for achieving human and dog rabies elimination. *Veterinary Research* 49(1), 61. DOI: 10.1186/s13567-018-0554-6.

Cunha Silva, L., Friker, B., Warembourg, C., Kanankege, K., Wera, E, *et al.* (2022) Habitat selection by free-roaming domestic dogs in rabies endemic countries in rural and urban settings. *Scientific Reports* 12(1), 20928. DOI: 10.1038/s41598-022-25038-z.

Dalem, T.I.A.C., Puja, I. and Kardena, I.M. (2012) Ekologi dan demografi anjing di kecamatan Denpasar Timur. *Indonesia Medicus Veterinus* 1(2), 160–172.

Del Rio Vilas, V.J., Freire de Carvalho, M.J., Vigilato, M.A.N., Rocha, F., Vokaty, A. *et al.* (2017) Tribulations of the last mile: sides from a regional program. *Frontiers in Veterinary Science* 4, 4. DOI: 10.3389/fvets.2017.00004.

Dias, R.A., Guilloux, A.G.A., Borba, M.R., Guarnieri, M.C. de L., Prist, R. *et al.* (2013) Size and spatial distribution of stray dog population in the University of São Paulo campus, Brazil. *Preventive Veterinary Medicine* 110(2), 263–273. DOI: 10.1016/j.prevetmed.2012.12.002.

Dodds, W.J., Larson, L.J., Christine, K.L. and Schultz, R.D. (2020) Duration of immunity after rabies vaccination in dogs: the rabies challenge fund research study. *Canadian Journal of Veterinary Research* 84(2), 153–158.

Dürr, S. and Ward, M.P. (2014) Roaming behaviour and home range estimation of domestic dogs in aboriginal and torres strait islander communities in Northern Australia using four different methods. *Preventive Veterinary Medicine* 117(2), 340–357. DOI: 10.1016/j.prevetmed.2014.07.008.

Dürr, S. and Ward, M.P. (2015) Development of a novel rabies simulation model for application in a non-endemic environment. *PLoS Neglected Tropical Diseases* 9(6), 1–22. DOI: 10.1371/journal.pntd.0003876.

Dürr, S., Mindekem, R., Kaninga, Y., Doumagoum Moto, D., Meltzer, M.I. *et al.* (2009) Effectiveness of dog rabies vaccination programmes: comparison of owner-charged and free vaccination campaigns. *Epidemiology and Infection* 137(11), 1558–1567. DOI: 10.1017/S0950268809002386.

Gabriele-Rivet, V., Arsenault, J., Wilhelm, B., Brookes, V.J., Newsome, T.M. *et al.* (2019) A scoping review of dingo and wild-living dog ecology and biology in Australia to inform parameterisation for disease spread modelling. *Frontiers in Veterinary Science* 6, 47. DOI: 10.3389/fvets.2019.00047.

Fahrion, A.S., Taylor, L.H., Torres, G., Müller, T., Dürr, S. *et al.* (2017) The road to dog rabies control and elimination – what keeps us from moving faster? *Frontiers in Public Health* 5, 1–8. DOI: 10.3389/fpubh.2017.00103.

Gabriele-Rivet, V., Arsenault, J., Brookes, V.J., Fleming, P.J.S., Nury, C. *et al.* (2020) Dingo density estimates and movements in equatorial Australia: spatially explicit mark-resight models. *Animals* 10(5), 1–21. DOI: 10.3390/ani10050865.

Gabriele-Rivet, V., Ward, M.P., Arsenault, J., London, D. and Brookes, V.J. (2021) Could a rabies incursion spread in the northern Australian dingo population? Development of a spatial stochastic simulation model. *PLOS Neglected Tropical Diseases* 15(2), e0009124. DOI: 10.1371/journal.pntd.0009124.

Gamble, L., Gibson, A.D., Shervell, K., Lohr, F., Otter, I. *et al.* (2018) The problem of stray dogs. *Revue Scientifique et Technique (International Office of Epizootics)* 37(2), 543–550. DOI: 10.20506/rst.37.2.2822.

Gibson, A.D., Yale, G., Vos, A., Corfmat, J., Airikkala-Otter, I. *et al.* (2019) Oral bait handout as a method to access roaming dogs for rabies vaccination in Goa, India: a proof of principle study. *Vaccine* X1, 100015. DOI: 10.1016/j.jvacx.2019.100015.

Gompper, M.E. (2013) *Free-Ranging Dogs and Wildlife Conservation*. Oxford University Press, Oxford. DOI: 10.1093/acprof:osobl/9780199663217.001.0001.

Griss, S., Riemer, S., Warembourg, C., Sousa, F.M., Wera, E. *et al.* (2021) If they could choose: how would dogs spend their days? Activity patterns in four populations of domestic dogs. *Applied Animal Behaviour Science* 243, 105449. DOI: 10.1016/j.applanim.2021.105449.

Gsell, A.S., Knobel, D.L., Kazwala, R.R., Vounatsou, P. and Zinsstag, J. (2012) Domestic dog demographic structure and dynamics relevant to rabies control planning in urban areas in Africa: the case of Iringa, Tanzania. *BMC Veterinary Research* 8, 236. DOI: 10.1186/1746-6148-8-236.

Hampson, K., Coudeville, L., Lembo, T., Sambo, M., Kieffer, A, *et al.* (2015) Estimating the global burden of endemic canine rabies. *PLoS Neglected Tropical Diseases* 9(4), 1–20. DOI: 10.1371/journal.pntd.0003786.

Harischandra, P.A.L., Gunesekera, A., Janakan, N., Gongal, G. and Abela-Ridder, B. (2016) Sri Lanka takes action towards a target of zero rabies death by 2020. *WHO South-East Asia Journal of Public Health* 5(2), 113–116. DOI: 10.4103/2224-3151.206247.

Hiby, E., Agustina, K.K., Atema, K.N., Bagus, G.N., Girardi, J. *et al.* (2018) Dog ecology and rabies knowledge of owners and non-owners in Sanur, a sub-district of the Indonesian Island Province of Bali. *Animals* 8(7), 1–18. DOI: 10.3390/ani8070112.

Hossain, M., Ahmed, K., Marma, A.S.P., Hossain, S., Ali, M.A. *et al.* (2013) A survey of the dog population in rural Bangladesh. *Preventive Veterinary Medicine* 111(1–2), 134–138. DOI: 10.1016/j.prevetmed.2013.03.008.

Hudson, E.G., Brookes, V.J., Dürr, S. and Ward, M.P. (2017a) Domestic dog roaming patterns in remote northern Australian indigenous communities and implications for disease modelling. *Preventive Veterinary Medicine* 146, 52–60. DOI: 10.1016/j.prevetmed.2017.07.010.

Hudson, E.G., Brookes, V.J. and Ward, M.P. (2017b) Assessing the risk of a canine rabies incursion in Northern Australia. *Frontiers in Veterinary Science* 4, 141. DOI: 10.3389/fvets.2017.00141.

Hudson, E.G., Brookes, V.J., Dürr, S. and Ward, M.P. (2019) Targeted pre-emptive rabies vaccination strategies in a susceptible domestic dog population with heterogeneous roaming patterns. *Preventive Veterinary Medicine* 172, 104774. DOI: 10.1016/j.prevetmed.2019.104774.

Human Development Reports (n.d.) Human development data. Available at: http://hdr.undp.org/en/data (accessed 1 September 2021).

Hutabarat, T., Geong, M., Newsome, A., Ruben, A. and Cutter, S. (2003) Rabies and dog ecology in Flores. In: *Urban Animal Management Conference Proceedings 2003*, pp. 7–14. Available at: https://aiam.org.au/resources/Documents/2003%20UAM/PUB_Pro03_7_14_Cutter.pdf

Johnstone-Robertson, S.P., Fleming, P.J.S., Ward, M.P. and Davis, S.A. (2017) Predicted spatial spread of canine rabies in Australia. *PLoS Neglected Tropical Diseases* 11(1), e0005312. DOI: 10.1371/journal.pntd.0005312.

Kirkeby, C., Brookes, V.J., Ward, M.P., Dürr, S. and Halasa, T. (2021) A practical introduction to mechanistic modeling of disease transmission in veterinary science. *Frontiers in Veterinary Science* 7, 1–13. DOI: 10.3389/fvets.2020.546651.

Kitala, P., McDermott, J., Kyule, M., Gathuma, J., Perry, B. *et al.* (2001) Dog ecology and demography information to support the planning of rabies control in Machakos District, Kenya. *Acta Tropica* 78(3), 217–230. DOI: 10.1016/s0001-706x(01)00082-1.

Kumarapeli, V. and Awerbuch-Friedlander, T. (2009) Human rabies focusing on dog ecology – a challenge to public health in Sri Lanka. *Acta Tropica* 112(1), 33–37. DOI: 10.1016/j.actatropica.2009.06.009.

Laager, M., Mbilo, C., Madaye, E.A., Naminou, A., Léchenne, M. *et al.* (2018) The importance of dog population contact network structures in rabies transmission. *PLoS Neglected Tropical Diseases* 12(8), 1–18. DOI: 10.1371/journal.pntd.0006680.

Lavan, R.P., King, A.I.M., Sutton, D.J. and Tunceli, K. (2017) Rationale and support for a One Health program for canine vaccination as the most cost-effective means of controlling zoonotic rabies in endemic settings. *Vaccine* 35(13), 1668–1674. DOI: 10.1016/j.vaccine.2017.02.014.

Matter, H.C., Wandeler, A.I., Neuenschwander, B.E., Harischandra, L.P.A. and Meslin, F.X. (2000) Study of the dog population and the rabies control activities in the Mirigama area of Sri Lanka. *Acta Tropica* 75(1), 95–108. DOI: 10.1016/s0001-706x(99)00085-6.

McGregor, H.W., Legge, S., Potts, J., Jones, M.E. and Johnson, C.N. (2015) Density and home range of feral cats in north-western Australia. *Wildlife Research* 42(3), 223. DOI: 10.1071/WR14180.

Mindekem, R., Lechenne, M.S., Naissengar, K.S., Oussiguéré, A., Kebkiba, B. *et al.* (2017) Cost description and comparative cost efficiency of post-exposure prophylaxis and canine mass vaccination against rabies in N'Djamena, Chad. *Frontiers in Veterinary Science* 4, 38. DOI: 10.3389/fvets.2017.00038.

Ministry of Health Republic Indonesia (2020) 8 Dari 34 Provinsi di Indonesia bebas rabies [English translation: eight of 34 provinces are rabies-free]. Available at: www.kemkes.go.id/article/view/20092900001/8-dari-34-provinsi-di-indonesia-bebas-rabies.html (accessed 5 December 2022).

Morters, M.K., Bharadwaj, S., Whay, H.R., Cleaveland, S., Damriyasa, I.M. *et al.* (2014a) Participatory methods for the assessment of the ownership status of free-roaming dogs in Bali, Indonesia, for disease control and animal welfare. *Preventive Veterinary Medicine* 116(1–2), 203–208. DOI: 10.1016/j.prevetmed.2014.04.012.

Morters, M.K., McKinley, T.J., Restif, O., Conlan, A.J.K., Cleaveland, S, *et al.* (2014b) The demography of free-roaming dog populations and applications to disease and population control. *The Journal of Applied Ecology* 51(4), 1096–1106. DOI: 10.1111/1365-2664.12279.

Müller, T., Freunling, C.M., Rupprecht, C.E., Both, L., Fooks, A.R, *et al.* (2015) Elimination of rabies – a missed opportunity. In: Sing, A. (ed.) *Zoonoses – Infections Affecting Humans and Animals*. Springer, Dordrecht, pp. 527–571. DOI: 10.1007/978-94-017-9457-2_21.

Mustiana, A. (2013) Assessment of the risk for rabies introduction and establishment in Lombok, Indonesia. MSc thesis, Sydney University, Sydney, Australia.

Mustiana, A., Toribio, J.-A., Abdurrahman, M., Suadnya, I.W., Hernandez-Jover, M. *et al.* (2015) Owned and unowned dog population estimation, dog management and dog bites to inform rabies prevention and response on Lombok Island, Indonesia. *PloS ONE* 10(5), 1–15. DOI: 10.1371/journal.pone.0124092.

Muthiani, Y., Traoré, A., Mauti, S., Zinsstag, J. and Hattendorf, J. (2015) Low coverage of central point vaccination against dog rabies in Bamako, Mali. *Preventive Veterinary Medicine* 120(2), 203–209. DOI: 10.1016/j.prevetmed.2015.04.007.

OIE (World Organisation for Animal Health) (2009) Stray dog population control. In: *OIE Terrestrial Animal Health Standards Commission*. OIE, Paris, pp. 313–621. Available at: www.woah.org/fileadmin/Home/eng/Internationa_Standard_Setting/docs/pdf/A_TAHSC_Sept_2009_Part_A_b_.pdf (accessed 2 February 2023).

One Health Roadmap (2019) One health roadmap eliminasi rabies nasional 2030. Jakarta. Available at: www.slideshare.net/Naipospos/one-health-roadmap-eliminasi-rabies-nasional-2030-jakarta-27-agustus-2019 (accessed 15 February 2023).

Pimburage, R.M.S., Harischandra, P.A.L., Gunatilake, M., Jayasinhe, D.N., Balasuriya, A. *et al.* (2017) A cross-sectional survey on dog ecology and dog anti-rabies vaccination coverage in selected areas in Sri Lanka. *Sri Lanka Veterinary Journal* 64(1), 1–7. DOI: 10.4038/slvj.v64i1A.18.

Raynor, B., De la Puente-León, M., Johnson, A., Díaz, E.W., Levy, M.Z. *et al.* (2020) Movement patterns of free-roaming dogs on heterogeneous urban landscapes: implications for rabies control. *Preventive Veterinary Medicine* 178, 104978. DOI: 10.1016/j.prevetmed.2020.104978.

Ren, M., Stone, M., Semedo, M.H. and Nel, L. (2018) New global strategic plan to eliminate dog-mediated rabies by 2030. *The Lancet Global Health* 6(8), e828–e829. DOI: 10.1016/S2214-109X(18)30302-4.

Rinzin, K., Tenzin, T. and Robertson, I. (2016) Size and demography pattern of the domestic dog population in Bhutan: implications for dog population management and disease control. *Preventive Veterinary Medicine* 126, 39–47. DOI: 10.1016/j.prevetmed.2016.01.030.

Sagoff, M. (2017) On the definition of ecology. *Biological Theory* 12(2), 85–98. DOI: 10.1007/s13752-017-0263-9.

Scott, T.P., Coetzer, A., Fahrion, A.S. and Nel, L.H. (2017) Addressing the disconnect between the estimated,reported, and true rabies data: the development of a regional African rabies bulletin. *Frontiers in Veterinary Science* 4, 4–9. DOI: 10.3389/fvets.2017.00018.

Seifmann, R. and Kaplan, B. (2021) Seismic shift by G7 in recognizing One Health as critical to everyone's health. *Impakter*. Available at: https://impakter.com/g7-recognizing-one-health/ (accessed December 2022).

Sparkes, J., Fleming, P.J.S., Ballard, G., Scott-Orr, H., Durr, S. *et al.* (2015) Canine rabies in Australia: a review of preparedness and research needs. *Zoonoses and Public Health* 62(4), 237–253. DOI: 10.1111/zph.12142.

Sparkes, J., McLeod, S., Ballard, G., Fleming, P.J.S., Körtner, G. *et al.* (2016) Rabies disease dynamics in naïve dog populations in Australia. *Preventive Veterinary Medicine* 131, 127–136. DOI: 10.1016/j.prevetmed.2016.07.015.

Sreenivasan, N., Li, A., Shiferaw, M., Tran, C.H., Wallace, R. *et al.* (2019) Overview of rabies post-exposure prophylaxis access, procurement and distribution in selected countries in Asia and Africa, 2017-2018. *Vaccine* 37 Suppl 1, A6–A13. DOI: 10.1016/j.vaccine.2019.04.024.

Stephens, D., Wilton, A.N., Fleming, P.J.S. and Berry, O. (2015) Death by sex in an Australian icon: a continent-wide survey reveals extensive hybridization between dingoes and domestic dogs. *Molecular Ecology* 24(22), 5643–5656. DOI: 10.1111/mec.13416.

Taylor, L.H., Hampson, K., Fahrion, A., Abela-Ridder, B. and Nel, L.H. (2017) Difficulties in estimating the human burden of canine rabies. *Acta Tropica* 165, 133–140. DOI: 10.1016/j.actatropica.2015.12.007.

Tenzin, T., McKenzie, J.S., Vanderstichel, R., Rai, B.D., Rinzin, K. *et al.* (2015b) Comparison of mark-resight methods to estimate abundance and rabies vaccination coverage of free-roaming dogs in two urban areas of South Bhutan. *Preventive Veterinary Medicine* 118(4), 436–448. DOI: 10.1016/j.prevetmed.2015.01.008.

Tenzin, T., Ahmed, R., Debnath, N.C., Ahmed, G. and Yamage, M. (2015a) Free-roaming dog population estimation and status of the dog population management and rabies control program in Dhaka City, Bangladesh. *PLoS Neglected Tropical Diseases* 9(5), e0003784. DOI: 10.1371/journal.pntd.0003784.

Tenzin, T., Wangchuk, S., Rai, B.D., Rinzin, K., Dahal, N. *et al.* (2013) Enhancing rabies control in Bhutan. *South Asia Regional One Health Symposium*, pp. 42–45. Available at: www.researchgate.net/publication/296050030_Enhancing_Rabies_Control_in_Bhutan (accessed 15 February 2023).

Testaverde, M., Moroz, H., Hollweg, C.H. and Schmillen, A. (2017) *Migrating to Opportunity - Overcoming Barriers to Labor Mobility in Southeast Asia*. World Bank Publications, Washington, DC. DOI: 10.1596/978-1-4648-1106-7.

Tiwari, H.K., Bruce, M., O'Dea, M. and Robertson, I.D. (2019) Utilising group-size and home-range characteristics of free-roaming dogs (FRD) to guide mass vaccination campaigns against rabies in India. *Vaccines* 7(4), 1–15. DOI: 10.3390/vaccines7040136.

Tiwari, H.K., Robertson, I.D., O'Dea, M., Gogoi-Tiwari, J., Panvalkar, P., *et al.* (2019b) Validation of application superduplicates (AS) enumeration tool for free-roaming dogs (FRD) in urban settings of Panchkula Municipal Corporation in North India. *Frontiers in Veterinary Science* 6, 1–7. DOI: 10.3389/fvets.2019.00173.

Tiwari, H.K., Vanak, A.T., O'Dea, M., Gogoi-Tiwari, J. and Robertson, I.D. (2018) A comparative study of enumeration techniques for free-roaming dogs in rural Baramati, District Pune, India. *Frontiers in Veterinary Science* 5, 104. DOI: 10.3389/fvets.2018.00104.

Touihri, L., Zaouia, I., Elhili, K., Dellagi, K. and Bahloul, C. (2011) Evaluation of mass vaccination campaign coverage against rabies in dogs in Tunisia. *Zoonoses and Public Health* 58(2), 110–118. DOI: 10.1111/j.1863-2378.2009.01306.x.

Vaniscotte, A., Raoul, F., Poulle, M.L., Romig, T., Dinkel, A. *et al.* (2011) Role of dog behaviour and environmental fecal contamination in transmission of *Echinococcus multilocularis* in Tibetan communities. *Parasitology* 138(10), 1316–1329. DOI: 10.1017/S0031182011000874.

Vigilato, M.A.N., Clavijo, A., Knobl, T., Silva, H.M.T., Cosivi, O. *et al.* (2013) Progress towards eliminating canine rabies: policies and perspectives from Latin America and the Caribbean. *Philosophical Transactions of the Royal Society B: Biological Sciences* 368(1623), 1–8. DOI: 10.1098/rstb.2012.0143.

Villatoro, F.J., Sepúlveda, M.A., Stowhas, P. and Silva-Rodríguez, E.A. (2016) Urban dogs in rural areas: human-mediated movement defines dog populations in Southern Chile. *Preventive Veterinary Medicine* 135, 59–66. DOI: 10.1016/j.prevetmed.2016.11.004.

Wallace, R.M., Cliquet, F., Fehlner-Gardiner, C., Fooks, A.R., Sabeta, C.T. *et al.* (2020) Role of oral rabies vaccines in the elimination of dog-mediated human rabies deaths. *Emerging Infectious Diseases* 26(12), 1–9. DOI: 10.3201/eid2612.201266.

Ward, M.P. (2014) Rabies in the Dutch East Indies a century ago - a spatio-temporal case study in disease emergence. *Preventive Veterinary Medicine* 114(1), 11–20. DOI: 10.1016/j.prevetmed.2014.01.009.

Ward, M.P. and Brookes, V.J. (2021) Rabies in our neighbourhood: preparedness for an emerging infectious disease. *Pathogens* 10(3), 1–18. DOI: 10.3390/pathogens10030375.

Ward, M.P., Qing-Ling, S.K. and Brookes, V. (2020) The dingo-domestic dog interface: implications for disease spread. *Australian Zoologist* 41(3), 388–407. DOI: 10.7882/AZ.2020.022.

Warembourg, C., Berger-González, M., Alvarez, D., Maximiano Sousa, F., López Hernández, A. *et al.* (2020) Estimation of free-roaming domestic dog population size: investigation of three methods including an unmanned aerial vehicle (UAV) based approach. *PloS ONE* 15(4), e0225022. DOI: 10.1371/journal.pone.0225022.

Warembourg, C., Fournié, G., Abakar, M.F., Alvarez, D., Berger-González, M. *et al.* (2021a) Predictors of free-roaming domestic dogs' contact network centrality and their relevance for rabies control. *Scientific Reports* 11(1), 12898. DOI: 10.1038/s41598-021-92308-7.

Warembourg, C., Wera, E., Odoch, T., Bulu, P.M., Berger-González, M., *et al.* (2021b) Comparative study of free-roaming domestic dog management and roaming behavior across four countries: Chad, Guatemala, Indonesia, and Uganda. *Frontiers in Veterinary Science* 8, 1–14. DOI: 10.3389/fvets.2021.617900.

Wera, E. (2001) Epidemiologi dan usaha pemberantasan penyakit rabies di kabupaten flores timur, provinsi nusa tenggara timur periode November 1997–April 2000 Skripsi. Fakultas Kedokteran Hewan, Institut Pertanian Bogor. Available at: https://repository.ipb.ac.id/handle/123456789/21527 (accessed 15 February 2023).

Wera, E. (2017) Socio-economic modelling of rabies control in Flores Island, Indonesia. PhD thesis, Wageningen University & Research, Wageningen, The Netherlands.

Wera, E., Velthuis, A.G.J., Geong, M. and Hogeveen, H. (2013) Costs of rabies control: an economic calculation method applied to Flores Island. *PLoS ONE* 8(12), 1–15. DOI: 10.1371/journal.pone.0083654.

Wera, E., Mourits, M.C.M. and Hogeveen, H. (2015) Uptake of rabies control measures by dog owners in Flores Island, Indonesia. *PLoS Neglected Tropical Diseases* 9(3), 1–23. DOI: 10.1371/journal.pntd.0003589.

Wera, E., Warembourg, C., Bulu, P.M., Siko, M.M. and Dürr, S. (2021) Loss of binding antibodies against rabies in a vaccinated dog population in Flores Island, Indonesia. *PLOS Neglected Tropical Diseases* 15(9), e0009688. DOI: 10.1371/journal.pntd.0009688.

WHO (World Health Organization) (2012) Strategic framework for elimination of human rabies transmitted by dogs in the South-East Asia region. Available at: www.searo.who.int/entity/emerging_diseases/links/Zoonoses_SFEHRTD-SEAR.pdf (accessed 5 December 2022).

WHO (World Health Organization) (2013) WHO expert consultation on rabies. Second report. *World Health Organization Technical Report Series* 982, 1–138. Available at: https://apps.who.int/iris/handle/10665/85346 (accessed 5 December 2022).

WHO (World Health Organization) (2018) WHO expert consultation on rabies – Third report. *World Health Organization Technical Report Series* 1012, 1–182. Available at: https://apps.who.int/iris/bitstream/handle/10665/272364/9789241210218-eng.pdf (accessed 5 December 2022).

WHO (World Health Organisation) Veterinary Public Health Unit (1988) Report of WHO consultation on dog ecology studies related to rabies control, Geneva, 22–25 February. Available at: http://apps.who.int//iris/handle/10665/58695 (accessed 5 December 2022).

13 Ecology of Dogs in Sri Lanka and Transmission of Rabies Among Dogs and Wildlife

Nihal Pushpakumara[1]*, Ashoka Dangolla[2] and Oswin Perera[2]
[1]Department of Wildlife Conservation, Ampara, Sri Lanka; [2]University of Peradeniya, Peradeniya, Sri Lanka

Abstract
The estimated dog population in Sri Lanka is 2.5 million, and the annual canine vaccination coverage in the country has ranged from 34% to 56%. Rural and suburban areas have a high proportion of free-roaming dogs with unrestricted access to wild animal habitats. Therefore, there is the possibility of rabies transmission from dogs to wildlife and vice versa. To facilitate a successful rabies control programme, habitats where interspecies interactions occur must be identified and targeted for rabies control. Areas that attract animals, such as food-waste dumping sites, should also be targeted, while robust population surveys and rabies surveillance must be enhanced. The objective of this chapter is to describe the dynamics of dog ecology and highlight the status of canine vaccination in this country. This chapter will also discuss collaborative One Health measures being adopted to prevent rabies transmission between dogs and wildlife and end human rabies deaths by 2030.

13.1 Introduction

Sri Lanka has a human population of approximately 22 million and the population of dogs, although no census has been done, is estimated to be around 2.5 million. The human deaths due to rabies have steadily declined from 250 to 300/year in the 1970s to around 20–30/year in 2016 and have remained at that level (Nihal *et al.*, 2019, p. 7808517). The success in reducing human deaths from rabies has been largely attributed to the effective administration of post-exposure prophylaxis (PEP).

The dogs in Sri Lanka can be categorized as: (i) owned or confined dogs (dogs that are restricted to households in which an owner provides food and other welfare facilities); (ii) semi-owned free-roaming dogs (provided with food, but not leashed or caged and allowed to roam free, also called 'community dogs'); and (iii) ownerless free-roaming dogs (an owner does not provide food or care) (Pimburage *et al.*, 2017). The percentages of dogs in each of these categories in different ecological areas of the country are not known.

The annual coverage of rabies vaccination of the estimated dog population has ranged from approximately 34 to 51% during the years 2005–2017. However, this coverage declined to 6% in 2018 and increased again to 34% in 2019 according to the recorded data of the PHVS, Ministry of Health, Sri Lanka (2020). In 2018, rabies control activities were transferred from the Ministry of Health to the Department

*Corresponding author: npkumara29@gmail.com

©CAB International 2023. *One Health for Dog-mediated Rabies Elimination in Asia* (eds V. Slack *et al.*)
DOI: 10.1079/9781800622975.0013

of Animal Production and Health (DAPH), and data on rabies control were not properly recorded by the PHVS during this period. This is the reason for the large drop in recorded dog vaccination coverage for the year 2018.

Although Sri Lanka is progressing towards the goal of eliminating human rabies, animal rabies (in companion animals, livestock and wildlife) is still a problem throughout the country (Nihal *et al.*, 2019). Animal rabies control is largely designed to limit rabies in dogs, which requires a combination of well-planned and sustained vaccination, together with control of the population of ownerless free-roaming dogs. To eliminate dog rabies, vaccination will need to be rapid, geographically even, and scaled up to cover a high percentage of the population.

Apart from dog-mediated rabies, possible intermittent involvement of wildlife in the activation of a sylvatic cycle can also occur (Gongal and Wright, 2011). A new *Lyssavirus* strain was reported from Indian flying foxes (greater Indian fruit bat, *Pteropus medius*) in Sri Lanka (Gunawardena *et al.*, 2016). This highlights the need for studies on the role of wild mammals in the transmission of rabies to dogs and whether the currently used rabies vaccines are effective against this strain of *Lyssavirus*.

Rabies control and ultimate elimination require an interdisciplinary and collaborative action based on the One Health approach because the activities needed do not fit into a single domain in any one administrative department, institution or agency. Sri Lanka is an island located in the Indian Ocean to the south-west of the Bay of Bengal and has the advantage of having no physical connections to neighbouring countries from which entry of rabies could occur. Therefore, once a rabies-free status is achieved, implementing animal import control and quarantine procedures can prevent re-entry of the disease to the country.

Our objective in this chapter is to describe the ecology of dogs, wildlife habitats in rural areas, and the status of dog vaccination and other measures being adopted to eliminate dog rabies, to bring down dog-mediated human deaths due to rabies to zero by 2030 in Sri Lanka. We describe in detail the current strategies, the need for considering wildlife as potential transmitters of the disease to dogs, and the importance of collaboration between the different government departments and institutions that are responsible for implementing those strategies.

13.2 Animal Population and Rabies Control in Sri Lanka

13.2.1 Implementation of dog rabies control

Sri Lankan government policy assigns dog rabies control to the PHVS within the Ministry of Health, and the Municipalities and Pradesiya Sabha within the Ministry of Local Government. The DAPH, which deals with all other aspects of all domestic animals, is not involved in rabies control activities. However, in 2018, control activities such as vaccination and surgical sterilization of dogs were handed over to DAPH, but after a short period of a few months, it was reassigned to the Ministry of Health through a cabinet paper. The information on suspected animal rabies or vaccinations by government veterinary surgeons or by private veterinarians is not communicated to the Ministry of Health. At present, no nationwide dog population survey or laboratory results-based animal rabies surveillance system is in place. All calculations of dog populations and vaccine coverage are done based on estimates made by Public Health Inspectors (PHI) in their respective areas of Medical Officers of Health (MOH).

13.2.2 Dog sterilization

The birth control activities in ownerless free-roaming dogs through surgical sterilization are done by private veterinary surgeons working on contracts funded by the Ministry of Health. Over the last decade, these programmes were expected to reduce the dog population and thereby reduce the number of dog rabies cases and bite incidences. However, according to the available information, the dog populations, the number of dog rabies cases and the number of dog-bite incidents among humans have not been reduced. Furthermore, the data on dog bites among humans recorded in hospitals in a bite registry is not done in every hospital and is not uniformly recorded or reported.

Success in controlling the dog population through surgical sterilization has been disputed and debated in some countries (Hemachudha, 2005; Wilde *et al.*, 2005; Menezes, 2008). In India, the inability to surgically sterilize a high percentage of ownerless free-roaming dogs in a given geographic area within 6 months, before the next reproductive cycle begins, has been identified as a major problem in population control (Menezes, 2008). It is likely that for the same reason, surgical sterilization has not been able to substantially reduce the dog population in Sri Lanka as well.

13.2.3 Dog removal and no-kill policy

Sri Lanka practised the removal of free-roaming dogs (semi-owned and ownerless) by capturing and euthanizing them until 2005, after which a 'no-kill' policy was introduced by the government. Therefore, currently, only surgical sterilization and chemical birth control of free-roaming dogs are used as population control methods. However, the coverage is low and surgical programmes are held unevenly across different districts. The assumption is that having sufficient vaccination coverage in free-roaming dogs would stop the spread of rabies among them and therefore, human deaths would decline. After a confirmed human death from rabies, the district PHI responsible for rabies control conducts a ring vaccination (i.e. vaccinating dogs in an area with a 1 km radius around the place where the rabies positive case was detected), which is possibly too late to have an impact. However, there is no recommendation on acceptable numbers and percentages of vaccination coverage for free-roaming dogs per administrative district in Sri Lanka, and the level of immunity in such vaccinated dogs cannot be easily monitored. This is mainly due to the absence of a laboratory results-based rabies surveillance system in Sri Lanka as yet (Nihal *et al.*, 2020).

With a strict no-kill policy in practice, free-roaming dogs found in public places can be captured, vaccinated and handed over to a dog shelter managed by non-governmental organizations (NGOs). After capturing an animal, potential owners are given a reasonable period to reclaim their dogs. Buddhism is the major religion in Sri Lanka, which further supports a no-kill policy for animals in this country. Therefore, if a free-roaming dog is unowned and not claimed by an owner during this time, the local authorities may transfer the dog to another dog shelter, managed by animal welfare groups, until the dog dies of natural causes (approximately 4–5 years).

13.3 Rabies in Wildlife and Domestic Animals

The impact and transmission of rabies among dogs, livestock and wild animals have not been well studied and input from the DAPH and wildlife sector is minimal in this regard. It is believed that rabies circulates in two transmission cycles in Sri Lanka: (i) an urban cycle involving the maintenance of infection in dog populations in which livestock is also involved; and (ii) a sylvatic cycle involving wild animals (Matsumoto *et al.*, 2011). There is a possibility of a spillover of the rabies virus from dogs to wildlife and vice versa. The species of wildlife in which rabies has been reported in Sri Lanka include four species of mongoose (*Herpestes fuscus*, *Herpestes edwardsi*, *Herpestes smithii* and *Herpestes vitticollis*), golden jackal (*Canis aureus*), giant squirrel (*Ratufa macroura*), palm cat (*Paradoxurus hermaphroditus*), palm squirrel (*Funambulus palmarum*), monkey (*Semnopithecus* spp. and *Macaca* spp.), bandicoot (*Bandicota bengalensis*), civet cat (*Viverricula indica*) and wild cat (*Felis chaus*) (Nihal *et al.*, 2019).

As stated above, Indian flying foxes were found to be infected with a novel *Lyssavirus*, named Gannoruwa bat lyssavirus, which caused rabies-related clinical symptoms and neuro-pathological changes (Gunawardena *et al.*, 2016). The rabies outbreak that occurred in 2020 in 17 jackals from Kaluthara district, which resulted in two human deaths, suggests that a sylvatic cycle could exist, although a spillover of the infection from dogs to jackals is also a possibility. Studies on rabies virus strains from wild animals had previously revealed similarities with dog strains (Nanayakkara *et al.*, 2003).

Recently a novel strain was identified in bats as referred to above (Gunawardena *et al.*, 2016). It is not known whether the rabies outbreak in jackals was due to the dog strain or a different one since no virus strain identification had been done. Such strain identification facilities must be established at the Medical Research Institute (MRI) since it has the other infrastructure needed for rapid identification purposes. However, the public was extremely concerned about the outbreak of jackal rabies, which is assumed to occur almost annually. Public opinion in the affected area in Kaluthara was that it was the fault of the Department of Wildlife Conservation (DWC), the institution responsible for wildlife health and conservation. Wildlife rabies surveillance must be introduced and closely monitored, and the animal health authorities in the DWC must work more closely with PHVS in the Ministry of Health and DAPH, by adopting a One Health approach if the goal of zero rabies human deaths by 2030 is to be achieved.

Domestic cats have also been found to be transmitters of rabies in Sri Lanka while cattle, buffalo, pigs and goats have also been positive for rabies (Nihal *et al.*, 2019). Table 13.1 illustrates the details of companion, domestic, laboratory and wild animal samples submitted for rabies diagnosis to the MRI. During the period 2005–2020, there were 625 human deaths due to rabies and 10,117 positive animal cases. The proportions of positive cases in each animal species are listed in Table 13.1. These results highlight the need to unite all stakeholder institutes involved in the One Health approach, with improved surveillance to control rabies in Sri Lanka. Though the relative risks of infecting people could vary depending on the different categories of animals, it is only through such a One Health approach that zero human rabies deaths by 2030 could be achieved.

Disease surveillance among domestic animals and wildlife, a better understanding of animal ecological patterns, and studying common environments and methods of transmission would all help to prepare a national strategic plan for rabies control. Identification of logistics, infrastructure, financial and human resources in each stakeholder institute, together with genuine and committed leadership is essential for success. The ordinances and laws about dogs, rabies, animal disease control, nuisance, and animal welfare were formulated long ago and need to be amended to suit the current political and administrative structure in Sri Lanka (Kanda *et al.*, 2021). There are also disparities among the several relevant laws dealing with ownerless free-roaming animals, local governments and police ordinances. Basic teaching on zoonoses, including rabies, has now been included in the school curriculum in science and health streams of study and also in teacher training manuals. It is not clear how these curricula or 'teaching the teacher' programmes are conducted, monitored and supervised. Heavy reliance on conventional public education by PHIs must be changed to include new teaching methods using additional suitably skilled people to reach all strata in society; this is an issue that so far has not been considered, let alone addressed.

13.4 Rabies Diagnosis

Rabies diagnosis is performed in three laboratories. The MRI, located in Colombo under the Ministry of Health, is the central point that undertakes the testing of all human and most animal samples. The other two laboratories are at the teaching hospital Karapitiya in the Southern Province and the Faculty of Veterinary Medicine and Animal Science at the University of Peradeniya in the Central Province, which undertakes the testing of animal samples only. It is mandatory for all suspected human rabies samples to be tested at MRI since it is a notifiable disease and the infrastructure and logistics for such sampling are in place. Though rabies in animals is also notifiable, a post-mortem or confirmation is not mandatory, and no logistics or infrastructure is in place for this as yet.

The PHIs who are attached to the Ministry of Health are responsible for sending animal samples for rabies diagnosis while the respective Judicial Medical Officer (JMO) is responsible for sending human brains for post-mortem laboratory confirmation. In addition to PHIs, the general public, wildlife veterinarians, government veterinarians and municipal veterinarians are also submitting suspected animal carcasses for laboratory diagnosis. The MRI usually issues

Table 13.1. The number of samples from different animal species which were received for rabies confirmation at the Medical Research Institute (MRI) in Sri Lanka from 2005 to 2020 and their results. Created by Nihal Pushpakumara using data from the MRI and Public Health Veterinary Services (PHVS), Ministry of Health, Sri Lanka, 2020.

Category of sample	Species	Number of samples received	Positive (%)	Negative (%)	Proportion of positivity
Companion animal	Dog	13,819	8,676 (62.8)	5,143 (37.2)	85.76
	Cat	5228	1,082 (20.6)	4,146(79.3)	10.69
Subtotal		*19,047*	*9,758 (51.2)*	*9,289 (48.7)*	*96.45*
Domestic animal	Cattle	214	160 (74.7)	54 (25)	1.58
	Rabbit	104	3 (2.8)	101 (97)	0.03
	Goat	63	39 (61.9)	24 (38)	0.39
	Pig	12	8 (66.6)	4 (33.3)	0.08
	Buffalo	17	12 (70.5)	5 (29.4)	0.12
	Donkey	1	0 (0.0)	1 (100)	0.00
	Sheep	1	1(100)	0 (0.0)	0.01
	Horse	4	3 (75)	1(25)	0.03
	Domestic elephant	1	0 (0.0)	1(100)	0.00
Subtotal		*417*	*226 (54.2)*	*191 (45.8)*	*2.23*
Wild animal	Palm squirrel	597	23 (3.8)	574 (96.1)	0.23
	Giant squirrel	150	12 (8)	138 (92)	0.00
	Mongoose	193	49 (25.3)	144 (74.6)	0.48
	Rat	107	0 (0.0)	107 (100)	0.00
	Toque monkey and gray langur	59	3 (5.0)	56 (94.9)	0.03
	Palm cat	51	9 (17.6)	42 (82.3)	0.09
	Civet cat	17	2 (11.7)	15 (88.2)	0.02
	Bandicoot	12	2 (16.6)	10 (83.3)	0.02
	Bat	16	1 (6.2)	15 (93.7)	0.01
	Wild cat	6	2 (33.3)	4 (66.6)	0.02
	Loris	1	0 (0.0)	1 (100)	0.00
	Fishing cat	1	0 (0.0)	1 (100)	0.00
	Jackal	30	28 (93.3)	2 (6.6)	0.28
Subtotal		1,240	*131 (10.5)*	*1,109 (89.4)*	*1.29*
Laboratory animal	Guinea pig	8	2 (25)	6 (75)	0.02
	Hamster	6	0 (0.0)	6 (100)	0.00
	Mouse	1	0 (0.0)	1 (100)	0.00
Subtotal		*15*	*2 (13.3)*	*13 (86.6)*	*0.02*
Grand total		**20,719**	**10,117 (48.8)**	**10,602 (51.1)**	**100.00**

a result within 24 hours and returns a report to the submitter. The sample collecting counter at MRI is open 24 hours a day, 7 days a week. The limitations identified by those submitting samples are: (i) the difficulty of dissecting heads of carcasses; (ii) the distance for transport to the laboratory; and (iii) legal issues regarding the transport of wild animal carcasses.

A fluorescent antibody test (FAT) is undertaken for human brain tissue. Animal brain tissues submitted to any of the three laboratories are examined histologically for the presence of Negri bodies (clinically rabid animals show Negri bodies in about 50% of samples), which is a rapid way of establishing if an animal had rabies. The Negri body negative samples are further subjected to FAT for confirmation if there was human contact. For samples (both human and animal) with inconclusive results, mouse inoculation is also performed at MRI, though it is rare.

13.5 Dog Ecology

Ecology is the study of the relationship between living organisms, including humans, and the physical environment where they exist. When the ecology of dogs in Sri Lanka is considered, their population density and distribution depend on living site quality, space and home range, and availability of food, water and shelter.

Living site quality includes climate extremes, precipitation and drought frequencies, etc. In general, sites that produce a high amount of biomass support large dog populations compared to a site with low amounts. Suburban and rural areas in the country have better quality sites and therefore have relatively higher dog populations (Wandeler *et al.*, 1993) compared to urban areas.

Space and home range are needed by dogs to meet their life necessities. Urban areas have less space compared to rural areas. Dogs, especially males, demarcate and compete with each other for their respective home ranges.

Food and water availability and quality determine the size of dog populations in different areas. Even though the urban areas produce more waste, disposal is done close to the suburban and rural areas, accumulating more household and hotel food waste. These open waste dumps attract dogs and help them breed efficiently.

Dogs require shelter for protection against adverse weather, to escape from predators, and for rearing pups. Abundant common and public buildings, temples, schools and community places without predators are available in suburban and rural areas to meet this need when compared with urban areas.

13.5.1 Dog population dynamics

Population dynamics of dog populations include: (i) age structure; (ii) lifespan; (iii) sex ratio; (iv) natality and mortality; (v) interspecific and intraspecific dynamics; (vi) territoriality and home range; (vii) migrations; and (viii) carrying capacity. None of these criteria has been investigated in the dog population in Sri Lanka.

- *Age structure* involves proportions of young and old age groups in a particular area. Except for one NGO currently operating in Sri Lanka, all other state bodies (such as the Ministry of Health) and NGOs decide on population control methods using data estimated by PHIs. Funds are allocated to the Regional Director of the Health Service (RDHS) for outsourcing dog-population control work by private service providers. These are decided in an ad hoc manner and are not consistently held, hence population growth cannot be estimated and is not monitored anyway. Therefore, when annual vaccinations reach the ownerless free-roaming dogs, pups born to unvaccinated and non-sterilized female dogs may also have matured and had pups of their own.
- *Lifespan* – Though the average lifespan of a dog is between 10 and 12 years, the mean age of ownerless free-roaming dogs is considered to be 3.5 years (Wandeler *et al.*, 1993). This relatively shorter lifespan of ownerless free-roaming dogs reduces the proportion of vaccinated dogs fast and keeps reducing herd immunity faster than anticipated. As we do not have data on surveillance and have not monitored antibody titres in vaccinated dogs, herd immunity

is largely estimated at present. This could potentially lead to severe overestimation of herd immunity and a rapid increase in unprotected dogs. This could be one reason why rabies in dogs, using the available meagre data from MRI, has not reduced despite the reduction in human rabies deaths.

- *Sex ratio* – Despite the optimum ratio being 50% (50:50) for both genders, available Sri Lankan data suggest the country's dog population is 65% males (Pimburage *et al.*, 2017). This could be because population control methods are being adopted largely on a proportion of female dogs due to inhibitory budget restrictions at RDHS levels. However, the impact on rabies control by adopting population control methods on male dogs appears to be minimal (Jackman and Rowan, 2007), hence a substantial proportion of female dogs must be sterilized to have a detectable impact.
- *Natality and mortality* – It is apparent that natality and mortality have a reasonable impact on populations. For reasonable estimations of herd immunity and to calculate the level of protection among dogs, this information must be available. No research study in Sri Lanka has attempted to look into this.
- *Interspecific and intraspecific dynamics* – Within the dog population and between dogs and other animal species there is competition for food, water and shelter. This will impact dog survival and disease spread among the dog population. The jackal rabies outbreak is a fitting example of disease transmission among dogs and wildlife during the competition for common food sources such as poultry waste.
- *Territoriality and home range* – As with other species, dogs need space to obtain their life necessities, such as finding shelter, mates and food. Dogs in urban and peri-urban areas often have less space available for these resources compared to dogs in rural areas and they compete for space. Most of the male dogs demarcate their home range.
- *Migration* – If the environment is not suitable for the dogs' survival, they tend to migrate to a neighbouring village or city to find food, water and shelter. This will lead to the introduction of unvaccinated dog populations to immunized populations.
- *Carrying capacity* – This is the maximum number of individuals that live without any harm while having food, water and shelter for their survival. If the carrying capacity is exceeded, mortality factors overcome the natality factors and reduce the population size back to the carrying capacity.

13.5.2 Dog population distribution

Data on dog ecology, their population densities, population structure and characteristics in different districts in Sri Lanka are scarce and therefore rabies vaccination coverage is calculated based on previous estimates in any given area. A dog ecology study conducted in the 1980s in Sri Lanka revealed a dog:human ratio of 1:8 (WHO, 1988). Subsequent studies in identified specific areas have found the ratio to be 1:4.6 in 1997 in the Meerigama Divisional Secretariat area (Matter *et al.*, 2000), 1:5 in the Central Province in 2007 (Perera *et al.*, 2007) and 1:6.7 in 2017 in Kaluthara and Mathara districts (Pimburage *et al.*, 2017). Matter *et al.* (2000) reported that the estimated dog population per square kilometre in the Meerigama area in the Gampaha district was 87 /km^2 for owned dogs and 108 /km^2 for both owned and ownerless dogs. However, this figure does not match with some suburban and rural areas of south-west Sri Lanka, which has an estimated 3000 dogs/km^2 (Wandeler *et al.*, 1993).

Except for one study, all others are relatively outdated, and new research and studies are needed for a better understanding of the present dog ecology pattern. The high density of dogs, including ownerless free-roaming dogs in some areas, has been attributed to three factors: (i) beliefs of certain religious communities such as Buddhists and animal rights groups that oppose the killing of dogs; (ii) free-roaming dogs being fed by communities, as well as dogs having access to discarded and waste food at open dump sites; and (iii) lack of awareness of the danger of rabies from potential exposure to dog bites (Goonaratna, 2004). Nevertheless, the dog population is lower in some areas since certain religious communities, such as Moors, are less likely to rear dogs as pets. None of these studies has determined the numbers and percentages of dogs that are confined dogs (owned and kept confined to the household), semi-owned free-roaming, or ownerless free-roaming dogs. Further, no information is available on their age and sex distribution.

13.6 Interactions Between Dogs and Wildlife

Many rural and suburban areas in Sri Lanka have access to a range of wild animals such as mongooses, jackals, giant squirrels and bandicoots that live in these habitats. If such dogs are unvaccinated and harbour the rabies virus, the disease is easily transmitted to these wild animals, and the latter can transmit the virus back to dogs. These dogs can then transmit rabies to humans as well. This is why we emphasize the need to adopt a One Health concept for rabies control since it is not restricted to the Ministry of Health. The absence of a wildlife disease surveillance system in the country is another limitation in dealing with such situations. The previous history of wildlife disease patterns is important to analyse and take appropriate actions to resolve incidents such as the one involving jackals that occurred in 2020.

It is critically important to install facilities for virus strain identification at the MRI to make sure that any possible new strain is detected early. Wildlife health authorities must rethink vaccination of animals in areas that border human settlements and the possibility of introducing safe oral bait vaccines for wild animals, selected from the products available in the market. Building a stronger link between wildlife authorities and PHVS in the Ministry of Health is important in this context. The current Sri Lankan legislation on handling and transporting wild animal specimens permits only staff of the DWC to undertake these tasks, and the general public is restricted from even entering protected forest areas. However, the DWC has inadequate numbers of staff for involvement in wild animal disease surveillance and management (Nihal *et al.*, 2020). This situation negatively influences the detection of rabies virus among wildlife populations and therefore difficulties are faced during decision making regarding rabies control in wildlife.

A good example of such a situation is the recent outbreak of rabies in jackals mentioned above, which was believed to be an activation of the sylvatic cycle and resulted in two human deaths. The jackals in the area were therefore indiscriminately killed by people. Such actions could create environmental disharmony and disturb the equilibrium of different wildlife populations in the area. The habitat in the Kaluthara district is well suited for the mixing of unvaccinated dogs with jackals. Most home gardens and rice fields are surrounded by mono-crops such as rubber, tea, cinnamon and oil palm, as well as patches of jungle, which are suitable for jackals during the daytime. Jackals are the only protected wild canine species in Sri Lanka according to the Fauna and Flora Protection Act 1938 and its amendments, No. 44 of 1964, No. 01 in 1970 and No. 49 in 1993 (Government of Sri Lanka, n.d.).

Establishing a satisfactorily vaccinated dog population around such habitats with jungles close to human habitation is one practical and feasible solution to prevent the possible transmission of rabies through a sylvatic cycle. On the other hand, if the rabies virus involved is a new strain present in wildlife, there is a risk of spreading such variants within susceptible animal populations and even into free-roaming dogs – either semi-owned or ownerless – which poses a risk to human and animal health. Therefore, it is essential to have a clear idea of the previous vaccination history of free-roaming dog populations in the area, the strain of the virus involved, and their immunity status, all of which are lacking at present. After this outbreak in jackals, all the relevant institutes and stakeholders came to one platform and planned its control and emphasized the need for a One Health approach to control rabies in Sri Lanka.

13.7 Mass Dog Vaccination and Dog Population Immunity

Dog vaccination is essential in rabies control and at least 70% of dogs must be vaccinated to achieve satisfactory herd immunity (Coleman and Dye, 1996; Coyne *et al.*, 2001; Cleaveland *et al.*, 2003; WHO, 2018). Since 1975, dog vaccination is being carried out by the PHVS of the Ministry of Health through vaccination campaigns organized by PHIs. In each of the 25 administrative districts in Sri Lanka, there are ten to 15 MOH areas, with five to six PHI areas within each such MOH area. For each PHI area, there are several fixed vaccination points (static points), and it takes at least 4–5 days to cover an entire PHI area. This activity happens under the supervision of the MOH in their regions. One such dog vaccination team is assigned to the RDHS and is deployed to the regional level on a predetermined schedule. These vaccination teams keep moving from one such fixed vaccination point to the next, to cover the entire MOH area, and finally cover the entire district – which takes an entire year. This requires completely dedicated full-time staff and other resources for full coverage to be obtained. Furthermore, in large districts, it is often difficult for one team to ensure full coverage of the entire geographical area of a district within a short time.

Vaccinators are trained by the PHVS to carry out such duties at static vaccination points, which are predetermined and publicized, with each group consisting of four to five individuals. The entire programme is supervised by the district rabies PHI and the Regional Epidemiologist (RE). Such vaccination points operate in two sessions, between 9:00 a.m. and 12:00 noon, and from 1:00 p.m. to 4:00 p.m., but only on weekdays. Wide publicity for this clinic is given via posters distributed through primary healthcare workers and on the day of the vaccination, a vehicle with a loudspeaker announcing the operation of the clinic is sent within the area. Most often, these vaccination points are held in temples, community halls and the houses of volunteers, with the maximum distance between two such points being 1.5 km. Thus, only confined dogs that can be brought to the static vaccination points would get vaccinated, while some of the semi-owned free-roaming dogs and a majority of ownerless free-roaming dogs will not get vaccinated at these points (see Fig. 13.1).

A few previous studies in Sri Lanka indicated that ownerless free-roaming dogs represent around 10% of the entire dog population (Perera *et al.*, 2000). Dog census, surveys or surveillance

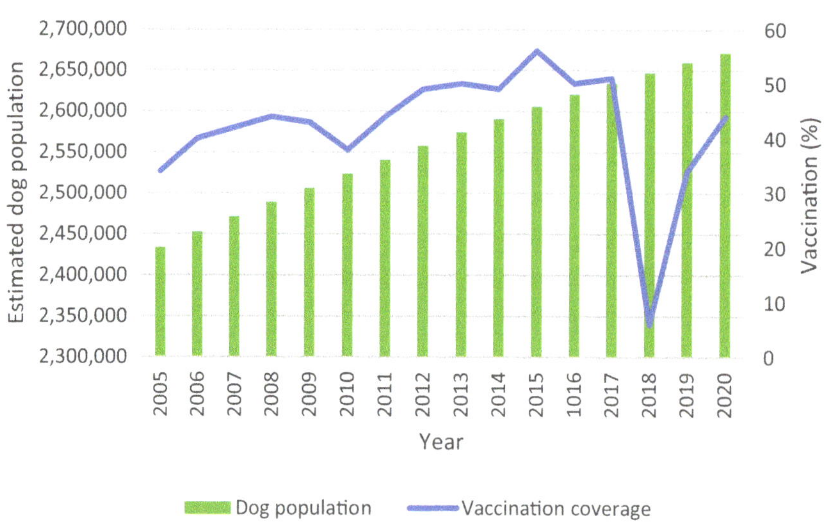

Fig. 13.1. Estimated dog population and percentage covered by annual vaccination in 2005–2020. The drop in dog vaccination in 2018 was due to the transfer of the rabies control programme from the Ministry of Health to the Department of Animal Production and Health (DAPH), and again back to the Ministry of Health. Created by Nihal Pushpakumara based on data from PHVS, Ministry of Health, Sri Lanka, 2020.

have not been conducted in Sri Lanka, and much of the data is based on estimates done by the respective PHIs using the commonly applied human:dog ratio of 8:1. The human population keeps steadily increasing, and therefore, the dog population estimations also increase accordingly, despite continuous dog population control programmes. The efficacy of rabies vaccines or the longevity of the antibodies developed in dogs have also not been consistently studied in Sri Lanka.

13.7.1 Vaccination of free-roaming dogs

There are substantial numbers of free-roaming dogs in community places such as temples, schools and marketplaces. These include both ownerless and semi-owned free-roaming dogs which are not leashed or caged. The ownerless free-roaming dogs are entirely unowned and are a critically important group of which statistics are needed to ensure adequate vaccination coverage. Some practical issues in capturing and/or restraining such dogs can be overcome by using a locally made device called the 'auto plunger', which can inject the vaccine from a distance without manual handling of the animal. A few resourceful districts employ another team of vaccinators to follow the static-point vaccination clinic to vaccinate dogs that are not brought to the clinic. However, the Ministry of Health does not employ dog catchers. Furthermore, the inability to identify previously vaccinated ownerless free-roaming dogs has always been a practical problem and the various types of ink markers that are used have not been long lasting. In addition, dog catchers, handlers and vaccinators require a list of their duties, proper education and regular refresher training.

Reports from many countries in Asia and Africa show that free-roaming dogs including ownerless dogs and unrestrained owned dogs are less likely to be vaccinated (Matter *et al.*, 2000; Kitala *et al.*, 2001; Kongkaewa *et al.*, 2004; Tang *et al.*, 2005; Menezes, 2008). The existence of the same situation can be clearly seen, in addition to improper population estimates in perusing vaccination records in each district in Sri Lanka (see Table 13.2). For example, the population of dogs vaccinated exceeded the estimated dog population in Monaragala (107% vaccination coverage) and Polonnaruwa (102% vaccination coverage).

Dog vaccinations by the Ministry of Health are done free of charge. Nevertheless, for owners of confined dogs, it is quite common, especially in cities, to get their dogs and other pets vaccinated against rabies from private veterinarians by paying for this service. Such clinics are not registered anywhere and there is no mechanism to obtain their vaccination data. Therefore, the information on total vaccinations available at the Ministry of Health is incomplete. If the government veterinary surgeons and private veterinarians are given the rabies vaccine by the Ministry of Health to be provided as a free service, they can assist in reaching 70% coverage in dogs. This could contribute immensely to reaching zero human rabies deaths by 2030.

13.7.2 Immunized dogs as a biological barrier

Dogs usually gather in small groups and demarcate their territory. The entrance of a new dog to an already identified territory, the removal of a resident dog, or the presence of a female in oestrus will increase the interactions among them. This could lead to the spread of the rabies virus unless such groups of free-roaming dogs are vaccinated (Baren, 1991; Chomel, 1993; Perera *et al.*, 2000). Therefore, one of the major challenges in rabies control in Sri Lanka is to ensure herd immunity among the semi-owned and ownerless free-roaming dog population.

For this purpose, Sri Lanka conducted a pilot project in the Panadura area in 2000 on the application of oral bait vaccine on domestic dogs that had missed the vaccine dose at static points. This was considered to be successful with certain limitations but was not continued due to logistic problems (Perera *et al.*, 2000). There was also a fear among veterinarians that the inactivated vaccinia virus used in the oral bait vaccine could potentially become active and infect the animal that consumed it, which also contributed to the cessation of this programme.

Table 13.2. Human and dog populations in the districts of Sri Lanka and vaccination coverage of dogs in 2019 at previously announced fixed places (static points), by mobile teams (free-roaming dogs), and percentages of coverage. Human population data from Central Bank of Sri Lanka (2020); dog population data were estimated based on human population data; and dog vaccination data were from PHVS, Ministry of Health, Sri Lanka (2020).

District	Human population	Estimated dog population	Dog vaccination coverage – 2019			
			Static point	Free-roaming dogs	Total	Percentage (%)
Ampara	719,000	89,875	18,877	323	19,200	21
Anuradhapura	930,000	116,250	69,685	7,843	77,528	67
Badulla	873,000	109,125	58,850	–	58,850	54
Batticaloa	570,000	71,250	14,949	–	14,949	21
Colombo	2,439,000	304,875	14,254	969	15,223	5
Galle	1,124,000	140,500	47,080	4,165	51,245	36
Gampaha	2,409,000	301,125	32,355	9,808	42,163	14
Hambanthota	65,5000	81,875	26,396	1,751	28,147	34
Jaffna	613,000	76,625	22,891	–	22,891	30
Kaluthara	1,281,000	160,125	44,704	–	44,704	28
Kandy	1,468,000	183,500	69,217	–	69,217	38
Kegalle	884,000	110,500	61,903	2641	64,544	58
Kilinochchi	126,000	15,750	1796	846	2642	17
Kurunegala	1,711,000	213,875	35,418	9894	45,312	21
Mannar	109,000	13,625	5743	–	5743	42
Mathale	519,000	64,875	50,946	2658	53,604	83
Mathara	858,000	10,7250	27,685	1132	28,817	27
Monaragala	491,000	61,375	65,785	8	65,793	107
Mulathive	96,000	12,000	7869	–	7869	66
Nuwaraeliya	763,000	95,375	23,398	106	23,504	25
Polonnaruwa	436,000	54,500	39,898	15,802	55,700	102
Puttalam	825,000	103,125	33,225	1752	34,977	34
Rathnapura	1,163,000	14,5375	36,008	–	36,008	25
Trincomalee	421,000	52,625	18,000	–	18,000	34
Vavunia	187,000	23,375	4249	51	4300	18
Grand total	**21,670,000**	**2,708,750**	**824,381**	**59,749**	**884,130**	**33**[a]

[a]Overall average percentage vaccination of dogs in 2019.

The establishment of a properly immunized dog population around patches of jungles in which wild animals and dogs interact will act as a biological barrier and prevent the spread of the virus into domestic dogs and vice versa. This approach requires wildlife authorities to work closely with the Ministry of Health based on the One Health approach. Such activities could also be used to stimulate schoolteachers and students to join and assist, using them as teaching-learning opportunities, further contributing to the groups of stakeholders trying to achieve zero human rabies deaths by 2030.

13.8 Conclusion

It is impossible to achieve zero human deaths due to rabies without controlling rabies in dogs. The essential requirement for breaking the rabies transmission cycle is achieving vaccination in at least 70% of the dog population, including all three categories of owned and confined, semi-owned free-roaming, and ownerless free-roaming dogs. A major problem in achieving this goal is the unavailability of reliable data on the numbers, distribution and ecology of the dog population in Sri Lanka. Surveys are needed to determine the dog population and their ecology in each district, including total numbers, relative proportions of the three categories of dogs specified above, and their habitats (urban, suburban, rural, etc.). Studies on these aspects should also include the immunity status and maintenance of herd immunity in different dog populations. Identification of food-waste dumping sites, habitat mapping of locations where free-roaming dogs potentially mix with wild animals, and surveillance for rabies among wildlife species that are known to have tested positive for rabies, are also important to facilitate an overall successful rabies control programme. Adequate human and physical resources must be ensured to implement the required vaccination coverage within a short period. The solutions to addressing adequate coverage of all types of dogs include providing facilities to private and government veterinary surgeons to vaccinate such dogs, using dog catchers to catch and bring them for vaccination, and the use of oral bait vaccines. A practical and long-lasting method of identifying vaccinated dogs is also essential.

The establishment of a rabies surveillance system for domestic animals as well as wildlife, backed by a geographically well-distributed rabies diagnostic laboratory network is essential. Legal aspects of controlling dog populations and rabies should be reorganized and rigorously implemented, and action should be taken to minimize free-roaming dogs (both semi-owned and ownerless). If some quota of immunized free-roaming dogs must be kept as community biological barriers, their immunity level must be assured, maintained and regularly monitored. We recommend that local government authorities must be assisted to implement, capture and detain at least a portion of unvaccinated free-roaming dogs for two weeks, giving adequate time for those who claim ownership to recover them, with a legal requirement to register the dog and keep it under their control, in accordance with the law of the country. If no one claims ownership of such dogs, the local authorities should have the power to hand them over to welfare centres, which should be established in each district. These shelters may help them to survive, confined to the centres, for their remaining natural life (this could be 4–5 years).

Legal implications on all aspects of dog ownership, registration and vaccination must be revised and enforced, while public awareness and education must encourage responsible dog ownership. The Ministry of Health must increase awareness of the public on preventing dog bites as well as the potential for rabies transmission from wildlife to dogs and other domestic animals. The importance of seeking immediate medical help in the event of exposure and the availability of PEP in all hospitals must be emphasized. The Ministry of Health must target 70% vaccination of the entire dog population, with the help of veterinarians in DAPH and private-pet animal practitioners, together with wider coverage of birth control programmes. Community physicians and Education Officers at the RDHS office must implement rigorous programmes in this context. The DWC must be strengthened with the necessary trained staff and resources for implementing wildlife disease surveillance. Local government bodies must be legally empowered with regard to registration and vaccination of all categories of dogs, as well as the capture, detaining, and transfer of unclaimed ownerless free-roaming dogs to appropriate institutions.

Zero human rabies deaths by 2030 is achievable if the issues and limitations highlighted above are addressed, and all relevant stakeholders and responsible organizations make genuine efforts and move forward with a joint national action plan based on the One Health approach.

Authors' Declaration

All authors declare that they have no conflict of interest.

All authors have approved this manuscript, agree with its submission, and share collective responsibility and accountability.

This manuscript has not been published or is not under review elsewhere.

References

Baren, G.W. (1991) Urban rabies. In: George, M.B. (ed.) *The Natural History of Rabies*, 2nd edn. CRC Press, Boston, Massachusetts, pp. 427–445.

Central Bank of Sri Lanka (2020) Central Bank of Sri Lanka report. Available at: https://www.cbsl.gov.lk (accessed 20 April 2022).

Chomel, B.B. (1993) The modern epidemiological aspect of rabies in the world. *Comparative Immunology, Microbiology and Infectious Diseases* 16(1), 11–20.

Cleaveland, S., Kaare, M., Tiringa, P., Mlengeya, T. and Barrat, J. (2003) A dog rabies vaccination campaign in rural Africa: impact on the incidence of dog rabies and human dog-bite injuries. *Vaccine* 21(17–18), 1965–1973.

Coleman, P.G. and Dye, C. (1996) Immunization coverage required to prevent outbreaks of dog rabies. *Vaccine* 14(3), 185–186.

Coyne, M.J., Burr, J.H.H., Yule, T.D., Harding, M.J., Tresnan, D.B. *et al.* (2001) Duration of immunity in dogs after vaccination or naturally acquired infection. *The Veterinary Record* 149(17), 509–515. DOI: 10.1136/vr.149.17.509.

Gongal, G. and Wright, A.E. (2011) Human rabies in the WHO Southeast Asia Region: forward steps for elimination. *Advances in Preventive Medicine* 2011, Article ID: 383870. DOI: 10.4061/2011/383870.

Goonaratna, C. (2004) A tale of two polls. *The National Medical Journal of India* 17(1), 41–42.

Government of Sri Lanka (n.d.) *Sri Lanka Fauna and Flora Protection Act 1938 and Its Amendments* (1964- No. 44, 1970- No. 01 and 1993- No. 49. Department of Government Printing, Colombo.

Gunawardena, P.S., Marston, D.A., Ellis, R.J., Wise, E.L., Karawita, A.C. *et al.* (2016) *Lyssavirus* in Indian flying foxes, Sri Lanka. *Emerging Infectious Diseases* 22(8), 1456–1459. DOI: 10.3201/eid2208.151986.

Hemachudha, T. (2005) Rabies and dog population control in Thailand: success or failure? *Journal of the Medical Association of Thailand* 88(1), 120–123.

Jackman, J. and Rowan, A. (2007) Free-roaming dogs in developing countries: the benefits of capture, neuter, and return programs. In: Salem, D.J. and Rowan, A.N. (eds) *The State of the Animals*. Humane Society Press, Washington, DC, pp. 55–78.

Kanda, K., Jayasinghe, A., Jayasinghe, C. and Yoshida, T. (2021) Public health implication towards rabies elimination in Sri Lanka: a systematic review. *Acta Tropica* 223, 106080. DOI: 10.1016/j.actatropica.2021.106080.

Kitala, P., McDermott, J., Kyulea, M., Gathuma, J., Perry, B. and Wandeler, A. (2001) Dog ecology and demography information to support the planning of rabies control in Machakos District, Kenya. *Acta Tropica* 78, 217–230.

Kongkaewa, W., Coleman, P., Pfeifferc, D.U., Antarasenaa, C. and Thiptara, A. (2004) Vaccination coverage and epidemiological parameters of the owned-dog population in Thungsong District, Thailand. *Preventive Veterinary Medicine* 65, 105–115.

Matsumoto, T., Ahmed, K., Wimalaratne, O., Yamada, K., Nanayakkara, S. *et al.* (2011) Whole-genome analysis of a human rabies virus from Sri Lanka. *Archives of Virology* 156(4), 659–669. DOI: 10.1007/s00705-010-0905-8.

Matter, H.C., Wandeler, A.I., Neuenschwander, B.E., Harischandra, L.P. and Meslin, F.X. (2000) Study of the dog population and the rabies control activities in the Mirigama area of Sri Lanka. *Acta Tropica* 75(1), 95–108. DOI: 10.1016/s0001-706x(99)00085-6.

Menezes, R. (2008) Rabies in India. *Canadian Medical Association Journal* 178(5), 564–566. DOI: 10.1503/cmaj.071488.

Nanayakkara, S., Smith, J.S. and Rupprecht, C.E. (2003) Rabies in Sri Lanka: splendid isolation. *Emerging Infectious Diseases* 9(3), 368–371. DOI: 10.3201/eid0903.020545.

Nihal, P.D.B., Dangolla, A., Hettiarachchi, R., Abeynayake, P. and Stephen, C. (2019) Surveillance opportunities and the need for intersectoral collaboration on rabies in Sri Lanka. *Journal of Veterinary Medicine* 2019, 7808517. DOI: 10.1155/2019/7808517.

Nihal, P.D.B., Dangolla, A., Hettiarachchi, R., Abeynayake, P. and Stephen, C. (2020) Challenges and opportunities for wildlife disease surveillance in Sri Lanka. *Journal of Wildlife Diseases* 56(3), 538–546. DOI: 10.7589/2019-07-181.

Perera, M.A., Harischandra, P.A., Wimalaratne, O. and Damboragama, S.N. (2000) Feasibility of canine oral rabies vaccination in Sri Lanka--a preliminary report. *The Ceylon Medical Journal* 45(2), 61–64.

Perera, I.N.G., Pinidiyapathirage, M.J. and Wickremasinghe, A.R. (2007) Rabies vaccination of domesticated dogs in the Central Province of Sri Lanka. *The Ceylon Medical Journal* 52(2), 69–70. DOI: 10.4038/cmj.v52i2.1031.

PHVS (Public Health Veterinary Services), Ministry of Health, Sri Lanka (2020) Statistics. Colombo: PHVS, Ministry of Health. Available at: https://www.rabies.gov.lk (accessed 20 April 2022).

Pimburage, R.M.S., Harischandra, P.A.L., Gunatilake, M., Jayasinhe, D.N., Balasuriya, A. *et al.* (2017) A cross-sectional survey on dog ecology and dog anti-rabies vaccination coverage in selected areas in Sri Lanka. *Sri Lanka Veterinary Journal* 64(1A), 1–7. DOI: 10.4038/slvj.v64i1A.18.

Tang, X., Luo, M., Zhang, S., Fooks, A.R., Hu, R. *et al.* (2005) Pivotal role of dogs in rabies transmission, China. *Emerging Infectious Diseases* 11(12), 1970–1972. DOI: 10.3201/eid1112.050271.

Wandeler, A.I., Matter, H.C., Kappeler, A. and Budde, A. (1993) The ecology of dogs and canine rabies: a selective review. *Revue Scientifique et Technique* 12(1), 51–71. DOI: 10.20506/rst.12.1.663.

WHO (World Health Organization) (1988) "*Report of Dog Ecology Studies Related to Rabies*" (WHO/Rab. Res/88.25). WHO, Geneva, Switzerland.

WHO (World Health Organization) (2018) WHO expert consultation on rabies - Third report. *World Health Organization Technical Report Series* 1012, 1–182. Available at: https://apps.who.int/iris/bitstream/handle/10665/272364/9789241210218-eng.pdf

Wilde, H., Khawplod, P., Khamoltham, T., Hemachudha, T., Tepsumethanon, V. *et al.* (2005) Rabies control in South and Southeast Asia. *Vaccine* 23(17–18), 2284–2289. DOI: 10.1016/j.vaccine.2005.01.030.

14 The Humane Management of Dog Populations and the Contribution to Rabies Elimination

Elly Hiby[1]*, Pankaj KC[1], Eric Brum[2] and Katie Hampson[3]
[1]The International Companion Animal Management coalition (ICAM), Cambridge, UK; [2]Emergency Centre for Transboundary Animal Diseases (ECTAD), Food and Agriculture Organization of the United Nations (FAO), Dhaka, Bangladesh; [3]Institute of Biodiversity, Animal Health & Comparative Medicine, University of Glasgow, Glasgow, UK

Abstract
Effective rabies elimination requires direct interaction with dogs, principally mass vaccination of dog populations and surveillance for suspected dog rabies cases. The most important dogs to reach are free-roaming dogs who, through their behaviour, are most epidemiologically relevant to rabies. Dog population management (DPM) uses a system of services designed to suit local dog population dynamics to improve responsible dog ownership and humanely minimize the population of unwanted and unmanaged dogs. DPM contributes to rabies control by increasing access to epidemiologically relevant dogs to reach and maintain herd immunity and ensure all dogs are monitored for effective surveillance; this has the potential to reduce the costs of rabies elimination actions. The impacts of effective humane DPM are multiple, hence the political will and funding for DPM should be accessed in addition to funding earmarked for mass vaccination. DPM should not divert resources away from vaccination.

14.1 Introduction

The prompt and effective use of post-exposure prophylaxis (PEP) in the treatment of people exposed to the rabies virus through a dog bite is a highly valued lifesaving action in rabies control. But to achieve zero human deaths demands more by ridding the world of this dog-mediated virus completely. This would eliminate the fear and anxiety of people bitten by suspect dogs, the suffering and inevitable death of people and animals from this nearly 100% fatal virus once symptoms appear, and the enormous cost to our health services around the world. It is this aim that underpins the global call to action 'Zero by 30' which sets the goal of zero human dog-mediated rabies deaths by 2030 worldwide (Minghui *et al.*, 2018).

Elimination of the rabies virus can only be achieved if we interrupt transmission within its reservoir host. For the Zero by 30 goal, this host is domestic dog (*Canis familiaris*). To achieve elimination, we need to recruit the dogs themselves and their immune systems, to kill the virus circulating within their populations by using mass dog vaccination. Thankfully in this war on rabies, we have in our arsenal a highly efficacious, long-lasting and affordable vaccine

*Corresponding author: ellyhiby@icam-coalition.org

 One Health for Dog-mediated Rabies Elimination in Asia (eds V. Slack *et al.*)
DOI: 10.1079/9781800622975.0014

against rabies in dogs. And as dogs are a domesticated species, rather than wildlife, we benefit from the dogs being an accessible population to be our soldiers.

Elimination also requires good surveillance of rabies in dogs so that any suspect cases can be promptly and appropriately managed to prevent the virus from spreading further. As a domesticated species, dogs depend on the care provided by people for their survival and hence dogs and people have a close relationship. Even in the absence of formally recognized ownership, we see free-roaming dogs congregating in and mostly reliant on human settlements. This can be a great benefit to surveillance as suspect cases are likely to be noticed by a well-informed public that knows the signs of rabies in dogs.

There are clear gains made from the rabies control journey towards elimination, with proven reductions in dog-bite prevalence (e.g. Tanzania (Cleaveland *et al.*, 2003)), costs to the health service, in terms of PEP (e.g. India (Larkins *et al.*, 2020)) and prevention of human rabies deaths (e.g. Latin America (Vigilato *et al.*, 2013)). The examples of successful rabies control projects, and the political will channelled by Zero by 30, provide the foundation for the elimination of dog-mediated rabies. But it is only by harnessing the immune power of dogs, and in particular, those that are the most epidemiologically relevant to rabies, that virus elimination can be achieved.

Any intervention involving dogs, and in particular these epidemiologically relevant dogs (see Box 14.1), will need to consider dog population management (DPM). DPM involves a cohesive set of measures to influence dog population dynamics to achieve a number of goals. These goals include a reduction in public health risks but are not limited to rabies control. In this chapter, we provide an overview of DPM, key factors for effective rabies vaccination campaigns, and then focus on the overlap between DPM and rabies control, specifically on the contribution of DPM to rabies elimination through achieving and sustaining sufficient vaccination coverage and effective surveillance.

14.2 What Is Dog Population Management (DPM)?

DPM starts with an understanding of dog population *dynamics* (Fig. 14.2). Rabies elimination may require a focus on epidemiologically relevant free-roaming dogs, but population management does not only look at the current unwanted and unowned free-roaming dogs. It also seeks to understand the sources of these dogs and aims to sustainably influence the processes that bring dogs on to the street and make them less accessible to management actions like vaccination. Interventions that work only with the current unwanted and unowned free-roaming dogs may be more accurately termed 'stray control'.

Dog population dynamics differ across communities in Asia, driven by variations in human behaviours. It may seem surprising to emphasize

Box 14.1. Epidemiologically relevant dogs.

From the perspective of rabies, not all dogs pose equal risk. Dogs that roam freely are likely to have the greatest contact with other dogs and therefore opportunity for viral transmission. Hence if vaccinated, these free-roaming dogs then have the greatest ability to act as a barrier to prevent further spread. Dogs are also not equal in their likelihood of being vaccinated. Owned adult dogs that are confined are more likely to be vaccinated through traditional vaccination delivery strategies such as central-point campaigns or through reliance on owners/carers proactively arranging vaccination with vets or animal health workers. This illustrates how the likelihood of vaccination tends to run in the opposing direction to epidemiological relevance – the most valuable dogs epidemiologically are paradoxically the least likely to be vaccinated (Fig. 14.1). Vaccination campaign approaches that target free-roaming dogs of all ages and levels of ownership bring vaccination coverage in line with epidemiological relevance. These approaches include door-to-door campaigns, engaging the support of dog feeders and carers to identify and handle community dogs or using expert dog catchers to catch unmarked and therefore unvaccinated dogs in the street.

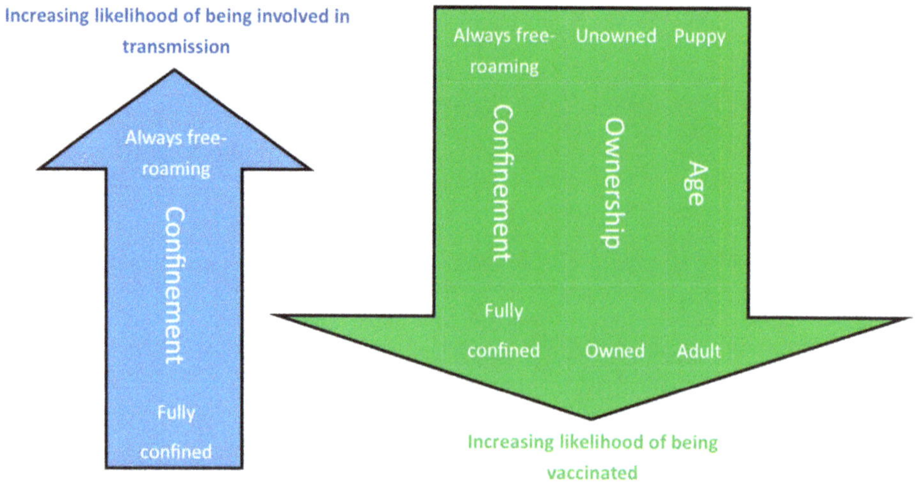

Fig. 14.1. Epidemiological relevance and opposing likelihood of vaccination against rabies for the characteristics of confinement, ownership and age of dogs. Diagram created by International Companion Animal Management coalition (ICAM).

the importance of *human* behaviour in the context of DPM, but all dog population dynamic processes are under the influence of what people *do* with dogs. The process of domestication has resulted in dogs being reliant on care provided by people for their survival and reproduction and they are unable to maintain their populations independently of the resources provided by people (Boitani *et al.*, 2017).

Some communities will be tolerant towards free-roaming dogs, offering them food and shelter and hence community dogs will make up a large proportion of their free-roaming dog population. Other communities, particularly in rural areas, will allow their owned dogs to roam freely to fulfil homestead, crop and livestock protection roles or because the physical structures of their home do not offer humane methods of confinement, resulting in most free-roaming dogs being owned roaming dogs. Some communities keep owned dogs confined within private property and are intolerant of free-roaming dogs. In other communities in Asia, there is an active trade in dog meat with consequences for animal welfare and rabies transmission. These differences in human behaviour will alter what types of dogs make up the free-roaming and epidemiologically relevant population and the processes that act as sources of these dogs. By identifying these sources, authorities can then select which DPM services will be most effective.

A DPM system is a cohesive programme of services, supported by a foundation of legislation, political will and social motivation (ICAM, 2019a, pp. 31–78) (Fig. 14.3). This system works to establish a positive relationship between dogs and their owners or carers ('carers' in the case of community animals without a single referral household). This positive relationship is characterized by responsible human behaviour to maintain good welfare and mitigate risks that dogs may present to other animals, the environment and people.

The services which make up the DPM system include four fundamental services which are critical to all DPM systems; and five context-dependent services (ICAM, 2019a, pp. 48–76) (Fig. 14.3). Context-dependent services are not always required but there will be some communities in which they become important to implement and also possible to enforce. For example, introducing identification and registration into a community where animal health services have neither the resources nor capacity for inserting and reading microchips would be a

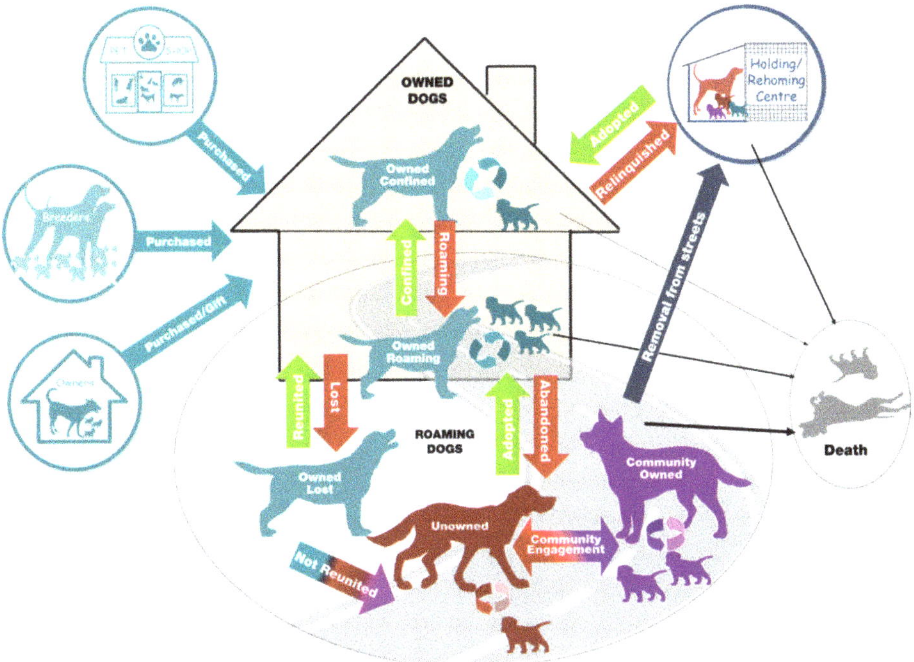

Fig. 14.2. Simplified representation of dog population dynamics. Dog states are represented by labelled dog icons and dynamic processes by labelled arrows. From ICAM, 2019a.

waste of resources that could be better spent on greater vaccination efforts.

Fundamental services:

1. Promoting responsible behaviour – This involves a combination of legislation, behaviour change communication and social pressure to increase responsible, safe and compassionate behaviour towards dogs.
2. Strengthening DPM professional capacity – Recognizing and investing in relevant professions, including veterinarians, animal handling/care and enforcement agencies is vital so that the services required for responsible ownership are accessible and of good quality.
3. Reproduction control – Access to reproduction control services is fundamental to all population management as the humane way to limit population size (rather than increased mortality) and limit the production of puppies to those that are wanted. Ensuring access requires overcoming barriers to reproduction control, which may include cost, veterinary capacity and skills, and transporting animals to clinics. Catch, neuter, vaccinate and return (CNVR) is an approach to control the reproduction of unowned animals that are to be managed *in situ* on the streets. (See discussion at the end of this section for how this can be used as an alternative/in combination with rehoming services.)
4. Veterinary care – Access to veterinary care ensures owners and carers can keep their animals in a good state of welfare and limit the risk of zoonotic diseases. It brings them into contact with animal health workers and veterinarians who can encourage responsible ownership behaviours and provide an opportunity for disease surveillance. Ensuring access may require subsidies for owners with limited resources and outreach services for those owners unable to reach static facilities. Veterinary care includes timely rabies vaccination to protect animals

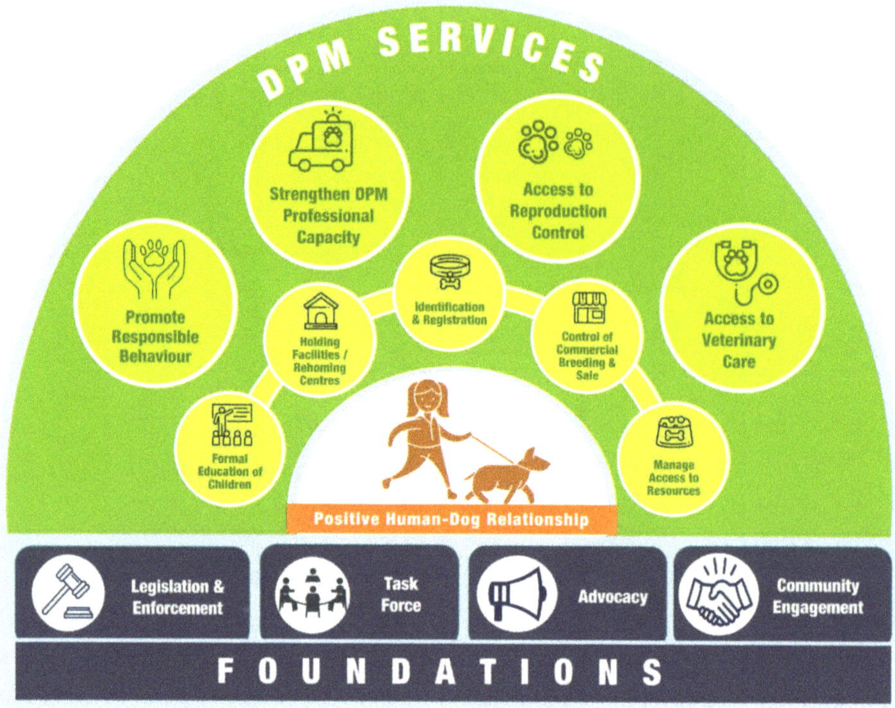

Fig. 14.3. Diagrammatic representation of a dog population management (DPM) system. From the 'Humane dog population management guidance' in ICAM (2019a)

and people from this virus and to mitigate inappropriate reactions to rabies outbreaks.

Context-dependent services:

1. Formal education of children – This includes education about safe and compassionate interaction with dogs and preventative measures for both children and animals to protect health, including prompt dog-bite treatment.
2. Holding facilities and rehoming – These can be provided through shelters or foster networks, achieving efficient and reliable reuniting of lost dogs and a humane and responsible option of relinquishment (rather than abandonment) followed by efficient rehoming to new suitable homes. (See discussion at the end of this section for how this can be used as an alternative/in combination with CNVR services.)
3. Identification and registration – Central or interoperable regional/local databases support efficient reuniting of identified and registered lost owned animals. This also provides proof of ownership for enforcement, traceability and tracking/reminders relating to health measures (where health records are linked to registration). Widespread adoption may be achieved through voluntary systems but mandatory legislation may be required.
4. Control of commercial breeding and sale – Puppies should be healthy, well socialized and habituated, leading to increased retention by new owners. Breeders, sellers and consumers are aware of expected standards and those below standard are identified and improved/penalized.
5. Managing access to resources – This is important to reduce conflict with free-roaming dogs. Improved waste management has been touted as a solution to stray control.

However, if free-roaming dogs are reliant on this food source, a sudden reduction would lead to starvation and is hence not humane and can lead to increased conflict as free-roaming dogs search for alternative sources. Rather than *reducing* food sources, these can be *manipulated* by moving food sources away from high-conflict areas to more appropriate locations, for example, using feeding stations. Poor waste management causes other more significant problems, including water contamination and air pollution, so an improvement in waste management over time is to be expected regardless of population management motivations. It is the role of population management to prepare the dog population for the inevitable improvement in waste management driven by other motivations, ensuring free-roaming dogs are appropriately resourced by owners and carers and not reliant on waste for their nutrition.

Core to the population management system is addressing the source of future unwanted and unowned free-roaming animals. Owners that have the knowledge, motivation, and ability to access services to practise responsible ownership behaviours can prevent their dogs from contributing to population management problems. This includes the services of affordable reproduction control and veterinary care for owners to invest in managing their dog's reproduction and health. The animal health services may also have the capacity to support owners by implementing an identification and registration system to reunite lost owned dogs and to enforce regulations on dog breeding and sale to ensure owners can acquire dogs knowing they are safe and healthy.

But this DPM system must also address the current unwanted and unowned free-roaming dogs. This requires careful consideration of how to use two relevant services. These are not mutually exclusive and can be used in combination:

- Holding facilities/rehoming centres or 'shelters' for relinquished pets and unowned dogs removed from the streets for reuniting or rehoming. Factors to consider include:
  - Rehoming 'centres' do not need to be physical structures. This service can be delivered through foster networks with potential economic and animal welfare benefits.
  - For rehoming centres to be a functional service, there must be a culture of adoption, providing a reasonable chance that an animal will be rehomed, otherwise centres fill to capacity or become places where most dogs will be euthanized. CNVR may provide a management option while building adoption capacity.
- Reproduction control and veterinary care provided to unowned animals using CNVR leading to the management of unowned animals *in situ* on the streets. Factors to consider include:
  - Tolerance and acceptance of free-roaming dogs by the majority of the public is necessary for dogs to maintain good welfare following CNVR.
  - For an individual animal to be considered a suitable candidate for CNVR, there should be evidence that the animal can maintain an acceptable level of welfare living on the streets. Very young animals or those with health issues may be best suited to rehoming. Where CNVR is used, ideally, rehoming is also available as an option for individual animals that are not suitable for release.
  - Community dogs that enjoy some care from local people may be able to breed successfully on the street. CNVR will also limit their breeding as a source of future unowned dogs.
  - There is a risk that when legislation prohibits the abandonment of owned animals, authorities may interpret the release of sterilized dogs as a form of abandonment. In such situations, agreement with the authorities on the difference between 'abandonment' and 'release' will need to be sought before starting CNVR.

The assessment of which of these two options is most appropriate for managing unwanted and unowned dogs will differ between communities and over time. Both shelters and CNVR have been misunderstood as equating to population management when used in isolation. A service focused only on the current free-roaming population is not comprehensive population management as it does not address the sources of unwanted and unowned dogs; conceptually akin to continuously refilling a leaking bucket without first fixing the leak. Although it should be noted that CNVR does have the benefit of addressing one important source: the birth of unowned puppies on the street by currently unowned or community dogs. Rehoming and CNVR function as part of a DPM system and must be used in combination with other services that also address other sources of future unwanted and unowned dogs.

14.3 Culling

Culling of dogs as a method of stray control, and as part of rabies control, has been used many times throughout history but has been repeatedly found to be ineffective and unpopular. Reasons for its lack of effect and unpopularity are listed in Table 14.1.

From a rabies perspective, one may argue that when faced with a free-roaming unowned or community-owned dog, killing it removes the dog as a potential transmission vector. However,

Table 14.1. Reasons for not culling dogs.

Reason	Explanation[a]
Population dynamics	• Minimal impact on the source of future unwanted and unowned free-roaming dogs and culling reduces competition over resources for the remaining dogs, hence populations quickly recover.
Disease (rabies) transmission	• Mistaken assumption that fewer dogs will mean less rabies, but rabies transmission is largely density independent, so there is only a minimal reduction in transmission with reduced density (Hampson *et al.*, 2009). • Where vaccinated dogs are either not marked, or these markers fail over time, culling may disproportionally remove vaccinated dogs as they tend to be more accessible. Culled dogs quickly replaced by owners (through purchase or birth) are also likely to be unvaccinated. Together this reduces herd immunity. • Culling can inadvertently increase contact between dogs and lead to rabies spread to other populations due to social perturbation or people moving their dogs to avoid culling teams.
Social acceptability	• Support for culling is limited and may be actively protested against in some communities, in particular when alternatives such as vaccination or CNVR are possible. Reports of owned and vaccinated dogs being culled are frequent and lead to further mistrust between government services and local communities. • Acceptability will be particularly low where inhumane methods of culling are used (WOAH lists recommended and unacceptable methods of killing dogs in Chapter 7.7 of the *Terrestrial Animal Health Code*; (WOAH, 2022b)).
Cost	• Culling appears straightforward and cheap, but costs per dog can be higher than rabies vaccination, not least because vaccination campaigns can recruit widespread public action in support of campaigns, which is not the case for culling.
Ethics	• Non-lethal and effective alternatives to culling for DPM and rabies control exist, and dogs are sentient beings with the capacity to suffer, hence ethical arguments do not support culling.

[a]CNVR, catch, neuter, vaccinate and return; DPM, dog population management; WOAH, World Organization for Animal Health.

this dog will be quickly replaced through birth and immigration by a new dog, almost certainly unvaccinated, resulting in no benefit to rabies control. Instead, there is the option to vaccinate and release such dogs, maintaining population stability and creating a biological barrier to virus spread with an immunity 'trap' of rabies antibodies in these vaccinated and released dogs. To be an effective biological barrier, 70% of the population needs to be vaccinated to achieve herd immunity, as described in Section 14.6.1 'Sustaining herd immunity', later in this chapter.

In recent history, the attempt to control a rabies outbreak in Bali was wholly unsuccessful when mass culling was used, but showed immediate and significant improvements in reducing dog bites, dog rabies cases and human rabies cases once mass dog vaccination was used instead (Putra *et al.*, 2013; Suseno *et al.*, 2019). The expert consultation by WHO reports:

> Mass dog vaccination has repeatedly been shown to be effective for controlling dog-mediated rabies, whereas removal of dogs does not decrease dog density or control rabies in the long run. Mass culling of dogs should therefore not be a part of a rabies control strategy: it is ineffective and may be counterproductive to vaccination programmes, particularly when they target free-roaming dogs.
>
> (WHO, 2018a, pp. 79–80)

Culling is the killing of an animal for purposes other than its own welfare, while euthanasia is killing an animal to prevent its further suffering. Although culling has no place in DPM or rabies control, all organizations that have direct interaction and responsibility over dogs must have a euthanasia policy in place and the capacity for humane methods of euthanasia (the World Organization for Animal Health (WOAH) lists recommended and unacceptable methods of killing dogs in Chapter 7.7 of the *Terrestrial Animal Health Code*; WOAH, 2022b). This includes organizations responsible for implementing mass rabies vaccination campaigns, holding and rehoming centres and CNVR services. This policy should be founded on animal welfare principles, appropriate to national legislation and local regulations, and realistic to implement within their veterinary, physical and staff capacity. The goal is that euthanasia is only used for those dogs that are suffering from an incurable illness, an injury, or have an unmanageable behaviour problem that prevents them from being rehomed or released, or are not coping well enough with rehoming facilities to maintain reasonable welfare. For some countries with limited rehoming potential and limited resources, this threshold for euthanasia may not be achieved immediately, but is the goal to work towards. The use of euthanasia for dogs showing signs of rabies is encouraged to prevent their suffering and protect human health and should be followed by prompt testing for laboratory confirmation of a rabies diagnosis. Where feasible, quarantine may be used to observe the dog for the progression of signs of rabies. If signs of rabies become evident, dogs should be euthanized, while any dog that remains healthy can be confirmed as not rabid. Further details about quarantining in the event of an animal bite are discussed in Section 14.6.2 'DPM contribution to rabies surveillance' and in other chapters on Integrated Bite Case Management (e.g. see Chapter 6, this volume).

In addition to culling, another common practice in parts of Asia that is both relevant to animal welfare and rabies control is the dog meat trade (Asia Canine Protection Alliance, 2013). We purposely do not cover the dog meat trade in any detail here, since it is beyond the scope of this chapter. However, we note that the dog meat trade is renowned for inhumane practices and has no place in humane DPM (FOUR PAWS, 2020). Hence our discussion is only to highlight potential rabies risks inherent in the trade. In countries where the dog meat trade is large and purposeful breeding of dogs for meat is limited, the removal of dogs for meat may impact owned and free-roaming populations. The turnover of dog populations in settings with an active dog meat trade is expected to be particularly high, creating large susceptible populations in which rabies can easily spread. The movement of dogs for the meat trade is frequent, often over large distances, and even across national borders, making rabies incursions into new areas a risk. These factors, together with the complete lack of regulations to reduce transmission through quarantine or vaccination means that the trade presents an obstacle to rabies elimination. Moreover, in communities with dog meat consumption, the risk of transmission between dogs and people, at all points of handling, killing and during the slaughter process is high (Wertheim *et al.*, 2009). For these reasons, there is a need for engagement with the dog meat trade for the joint aims of improved animal welfare and for eliminating dog-mediated rabies.

14.4 Key Factors for Successful Vaccination

Successful dog vaccination campaigns for progressive control and elimination of rabies share the key factors outlined in Table 14.2.

14.5 Intersection Between DPM and Rabies Control

Rabies control and DPM share a focus and a need to access epidemiologically relevant dogs. Although they have aims specific to their particular scope, they also share the overlapping aims of protecting animal and human health. Hence there is an opportunity for synergy and mutual benefit between these interventions. In this section, we return to the concept of rabies elimination as a war against the rabies virus and dogs as our soldiers to hunt down and kill the virus with their immune systems. We look specifically at how DPM helps us to leverage the capacity of these dog soldiers towards rabies elimination through achieving and sustaining vaccination coverage and effective surveillance (Fig. 14.4). The question we aim to answer here is: 'If I am responsible for vaccinating the greatest possible proportion of the dog population, or I am responsible for spotting and quickly responding to suspect rabies cases, how can DPM help me do that?'

14.6 Vaccination Coverage Where It Matters Most

We started this chapter with the concept of epidemiologically relevant dogs and their value in the war on rabies. But we know that many rabies vaccination campaigns achieve their greatest coverage in populations of dogs that are confined and are at limited risk of rabies transmission as well as in populations already covered by previous campaigns. This is not to say that vaccination of these dogs should not be done, but coverage in only this subpopulation will not achieve rabies elimination. It is the epidemiologically relevant dogs that need to be recruited, and yet they are often missed by vaccination campaigns year after year, providing an ongoing reservoir of susceptible hosts. DPM can increase the accessibility of these dogs through improved responsible dog ownership, community engagement, CNVR and improved staff capacity and skill. Finally, accessing these dogs for vaccination breaks the cycle of transmission within their populations and supports virus elimination.

The promotion of responsible dog ownership is a fundamental service of DPM and is achieved through legislation, behaviour change communication, social pressure and legislation working together to change human behaviour. A feedback loop of owner behaviour improvement can be supported through DPM services by increasing the value of their dog; the more management investment made in the dog, including sterilization and parasite control, the more it may be valued by an owner and the more effort they may expend in maintaining their dog in good health and welfare through future management investment, including regular vaccination. In practical vaccination campaign terms, an owner that values their dog is more likely to make an effort to catch and physically transport, or at least handle, their dog for vaccination teams, reducing campaign effort and therefore resources. They may even take the step of proactively having their dog vaccinated through animal health services outside of vaccination campaigns, eliminating the cost of vaccinating this dog entirely from campaign budgets. Conversely, an owner that has amassed a perceived excess of dogs through unwanted breeding, may not value his/her dogs sufficiently to spend their time and resources on vaccination, or may only expend effort on a preferred few.

Accessing more than 70% of the dog population during an annual mass vaccination campaign is not feasible without community engagement and support. Some

Table 14.2. Factors for successful vaccination campaigns.

Factor	Description[a]
Preparation for vaccination	1. Prepare and document vaccination plan in consultation with stakeholders, including required financial, human and material resources. 2. Provide pre-exposure prophylaxis and training in humane dog capture and handling, appropriate vaccine storage and handling, vaccine administration, and managing exposure to suspect rabid animals. 3. Procure and prepare all required materials and equipment. 4. Inform communities ahead of planned vaccination activities to increase community participation and reduce the time required to complete vaccination. 5. See further guidance in the WOAH *Terrestrial Animal Health Code* Chapter 4.18 (WOAH, 2022a).
Vaccination	1. A 3 year vaccine should be used for mass dog vaccination and stray dog vaccination, including during CNVR activities. The cold chain should be maintained at all levels of vaccine storage until administration, but importantly, efforts to maintain the cold chain should also ensure no vaccine is frozen, which is more likely to reduce vaccine efficacy. 2. A minimum coverage level of at least 70% of the dog population should be achieved in all administrative units of the area targeted for control. For accelerating time to elimination, higher coverage levels may be targeted. 3. For maximal impact on rabies virus transmission, epidemiologically relevant dogs should be prioritized. Methods for safe capture and handling (e.g. dog nets) should be used when dogs cannot be readily caught or safely restrained by hand. 4. Long-term markers (e.g. permanent collars) should be applied to all vaccinated adult dogs for monitoring vaccination coverage and enabling communities to differentiate vaccinated from unvaccinated dogs. Short-term markers (e.g. non-toxic paint marks) should be used for vaccinated puppies and applied to adult dogs only when there are insufficient resources for the purchase of long-lasting collars. 5. Vaccination should be applied on at least an annual basis. For accelerating viral elimination, 6-monthly vaccination campaigns may be considered and/or strategies to target new puppies in the interval between campaigns. 6. To slow the decline in vaccination coverage levels over time, population turnover should be minimized by utilizing the DPM tools described in this chapter.
Monitoring vaccination coverage	1. Vaccination data should be recorded following the vaccination of each individual animal and stored within an electronic database (resolved to the smallest available administrative unit). 2. Vaccination coverage should be monitored at the smallest administrative level possible (e.g. sub-village) by either post-vaccination surveys or by comparing to vaccination tally from the previous campaign in which the vaccination target was achieved.
Monitoring vaccination efficacy	1. Rabies cases in humans and animals should be monitored. 2. Integrated Bite Case Management may be used for increasing the sensitivity of detection of animal rabies cases (WHO, 2018a, p. 126).

Continued

Table 14.2. Continued

Factor	Description[a]
Response to suspect rabid dogs	1. Rapid detection and removal of rabid animals are crucial to reduce the spread of rabies and community impact. 2. All persons dealing with suspect rabies cases or entering a field situation where there may be rabid animals should receive a full course of pre-exposure prophylaxis before initiating activity (WHO, 2018a, p. 58). 3. Suspect animals should be evaluated and animals demonstrating signs of rabies should be humanely euthanized and submitted for testing as per the WOAH *Manual of Diagnostic Tests and Vaccines for Terrestrial Animals* Chapter 3.1.18 (WOAH, 2018). 4. Refer to the WHO (World Health Organization), 2018c position paper on rabies vaccines and immunoglobulins for guidance on humans exposed to suspect rabid animals (WHO, 2018c). 5. A positive rabies case in an area is an indication for immediate vaccination of the area, including revaccination of previously vaccinated dogs (note that reactive vaccination will have limited impact if not undertaken rapidly, i.e. within the week of the detected animal case, and is not a replacement for comprehensive routine vaccination in endemic settings). Culling should not be carried out.

[a]CNVR, catch, neuter, vaccinate and return; DPM, dog population management; WHO, World Health Organization; WOAH, World Organization for Animal Health.

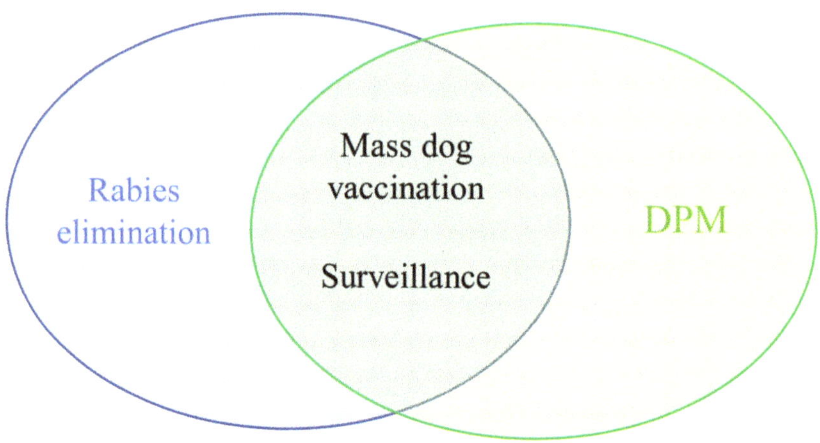

Fig. 14.4. Mass dog vaccination and surveillance activities are at the intersection of rabies elimination and dog population management (DPM) programmes. Diagram created by ICAM.

epidemiologically relevant dogs will be particularly difficult for vaccination teams to reach without the help of community members known and trusted by these dogs to find, catch and handle them. As community engagement and action are a foundation of an effective DPM system, locations with already functioning DPM are likely to have stronger communication and support from local communities. In particular, where DPM systems have involved CNVR, which have the opportunity to recruit community members into accessing and monitoring community dogs, these same individuals can support mass vaccination campaigns of community dogs (see Box 14.2 Case Study 1 about the ManuMitra programme).

By definition, CNVR accesses only epidemiologically relevant dogs that are always free

Box 14.2. Case studies of synergy between DPM and rabies control.

Case study 1

ManuMitra ('friend of human') is a comprehensive programme to manage dog populations and control rabies in Kathmandu, Nepal (ICAM, 2019b). One key aspect is the engagement of animal management assistants (AMAs) – local volunteers with a track record of caring for animals in the community. AMAs join a peer-supported training programme covering rabies prevention and control, veterinary first aid and post-operative care for sterilized dogs. They work closely with ManuMitra veterinarians and animal welfare officers to ensure all their local dogs are vaccinated by helping to find, identify and catch dogs that they know. They also identify dogs for sterilization, first gaining consent from an owner or carer for every dog. AMAs provide ongoing monitoring of free-roaming dogs, informing ManuMitra staff when new dogs needing vaccination or sterilization arrive, treating those with minor ailments and skin disease *in situ* and remaining vigilant for suspect rabies cases. AMAs are the experts in their local dogs, and as local residents are best placed to ensure saturation of rabies vaccination and DPM effort in their ward while being continuous advocates for responsible dog ownership and animal welfare.

Case Study 2

Help In Suffering has been running a CNVR programme in Jaipur, India, since 1994 complying with the national Animal Birth Control (Dog) Rules, 2001. This involves spaying and vaccinating female dogs, castrating and vaccinating young males and vaccination only of adult males, as castration is expected to make little difference to their behaviour and vaccination only uses fewer resources; each dog is vaccinated once in its lifetime. The vaccine used conveys 3 years of immunity. With a high coverage of the population (> 75%) and these dogs' relatively short lifespan of around 3 years (Reece *et al.*, 2008), this is estimated to maintain immunity in >40% of the dog population. An economic assessment of the period 1994–2017 estimated CNVR had averted over 360,000 dog bites at a cost of US$5.62 million. In addition, nearly 500 human rabies deaths had been averted by animal birth control (ABC) over the 23 year period. When added to the US$5.62 million saved from bites averted, this becomes a total societal economic benefit, estimated to be US$38.48 million. A sizeable financial benefit compared to the US$658,744 cost of Help In Suffering's ABC work over 23 years; for every US$1 spent on ABC, US$8.50 were saved in dog-bite treatment and US$58.40 in total societal economic losses from both rabies and bites (Larkins *et al.*, 2020).

Case Study 3

A fishing community on the outskirts of Karachi, Pakistan, called Ibrahim Hyderi, became the site of Rabies Free Karachi's pilot One Health project. This project involved mass dog vaccination and CNVR of free-roaming dogs, as well as distinctive yellow collars to provide a visual indication that dogs had been treated. The Rabies Free Karachi team also conducted workshops and engaged local people to gain their support in implementing the activities and explain the meaning of the collars. An exploratory follow-up survey of a small sample of local people revealed satisfaction with the project, a more positive perception of the collared dogs and visually more relaxed and friendly behaviour towards these specific treated dogs. Replication of this pilot in other areas of Pakistan is under way with similar results. However, it should be recognized that some people have strongly entrenched opinions of dogs based on previous negative experiences, and for them, the presence of a collar and knowledge of prior treatment may be insufficient to change their perceptions. DPM requires long-term engagement and high population coverage to achieve the changes in the free-roaming dog population required to meet the needs of all citizens. (More details of this case study are provided by Salahuddin *et al.* in Chapter 15, this volume and in WHO, 2018b).

roaming. CNVR is usually running throughout the year rather than as an annual pulse, steadily building the proportion of the free-roaming dogs that are vaccinated and preventing the birth of unvaccinated puppies on the street. Although regular revaccination of 'CNVRed' dogs is the ideal approach, the short lifespan of these free-roaming unowned or community-owned dogs makes a single vaccination at the time of sterilization sufficient for effective rabies control (Reece and Chawla, 2006) (see Box 14.2 Case Study 2 about Help In Suffering in Jaipur). However, a potential conflict between CNVR and rabies control can occur where vaccination is tied to sterilization, as this can limit the geographical scope and coverage of mass vaccination. A preferred approach is for vaccination and sterilization to be decoupled, allowing for vaccination only when appropriate, for example with male dogs and during mass vaccination campaigns, with CNVR following up the rest of the year as resources allow.

Some epidemiologically relevant dogs will be behaviourally difficult to catch and handle for vaccination. DPM services present a need and an opportunity to invest in the recruitment and skills development of dog handlers on an ongoing basis; these expert staff can then be utilized for mass vaccination campaigns.

14.6.1 Sustaining herd immunity

The critical percentage of the dog population that must be immune for disease control is related to the basic reproductive number (R_0) of the disease; for rabies R_0 is typically between 1 and 2, resulting in a critical percentage of 20–40% (Hampson *et al.*, 2009; Brum, 2019). However, the recommended threshold coverage for an annual vaccination campaign is 70% (Coleman and Dye, 1996; WHO, 2018a, p. 79). This higher target ensures that population turnover between annual campaigns does not allow herd immunity to fall below the critical percentage. While incursions of rabies virus still occur, herd immunity must be sustained above the critical percentage to prevent transmission. DPM can support herd immunity by reducing population turnover and providing services for the vaccination of puppies and newly acquired dogs in the period between annual vaccination campaigns (Table 14.2).

Targeted reproduction control services can reduce the birth of puppies that were likely to go unvaccinated and would have driven herd immunity down. This includes sterilization of community and unowned dogs, and owned dogs identified by their owners as unwanted for breeding. A focus on the spaying of female dogs is likely to have the greatest impact on the production of such 'at risk' puppies, as females are the limiting factor in dog population growth. Sterilization coverage should be monitored to check these targets are being met and avoid reliance on dogs that are easier to access and sterilize but less important for 'at risk' puppy production, such as castration of owned and confined male dogs.

The other population turnover process important for herd immunity is the survival of dogs that have been vaccinated. Sterilization may support the survival of vaccinated dogs by removing the energetic costs of reproduction, increased contact rates, and associated disease transmission risk of breeding behaviours. The previously described feedback loop of improved owner behaviour with increasing management investment may also increase survival as the amount and quality of care provided to dogs is likely to be positively correlated with their perceived value.

A critical service of the DPM system is access to veterinary care. Unlike annual mass vaccination campaigns, these services provide year-round access to basic healthcare including rabies vaccination. This provides owners with the opportunity to have puppies and newly acquired dogs vaccinated in the period between vaccination campaigns, an action that should be encouraged through responsible dog ownership education. Veterinary services should also offer humane euthanasia for individual dogs that are suffering from an incurable illness, injury, or a behaviour problem that their owners find unmanageable. This provides an important alternative to abandonment and protects dogs from further suffering. These veterinary services may be private, government, or a combination through government subsidies of private veterinary care.

14.6.2 DPM contribution to rabies surveillance

When rabies outbreaks do occur, their control, and therefore the prevention of human and animal deaths, relies on surveillance and prompt quarantine or humane euthanasia and testing of suspect cases. Where DPM systems exist, the proportion of the dog population that is not under the management of an owner or community carer should be reduced, therefore dogs with signs of rabies are less likely to go unnoticed and accessing dogs for revaccination in response to an outbreak will be easier. Rabies spread in unmanaged populations of dogs is more likely to go unnoticed until deaths occur in either people or livestock.

Integrated Bite Case Management, or IBCM, is an approach that can improve surveillance and human rabies prevention, including delivery of PEP resulting in faster and more effective management of outbreaks (Suseno *et al.*, 2019; Swedberg *et al.*, 2022). This requires functioning communication channels and a working relationship between human and animal health services, with animal health alerting human health when suspected animal cases occur, and human health alerting animal health when they treat a bite from what is considered to be a symptomatic rabid dog. These alerts include the location and date/time to allow action to be taken by the relevant services. Where DPM systems are in place, animal health services will have greater knowledge of the dog population across their jurisdiction and contacts with owners and carers, allowing them to respond faster and with greater accuracy.

14.6.3 Funding dog-mediated rabies elimination – DPM contribution and conflict

The elimination of dog-mediated rabies requires widespread vaccination of epidemiologically relevant dogs in all rabies-endemic countries. Resourcing this effort requires funding from local and national governments, supported by donor agencies, pharmaceutical companies and non-governmental organizations (NGOs). In this section, we discuss how DPM systems can contribute to these resources by reducing the costs of rabies control. But there is also the potential for conflict over limited resources and hence resources need to be carefully prioritized, and opportunities are taken to bring in additional funding sources through the wider goals of DPM.

Dogs resort to biting people for a number of reasons. This is usually motivated by fear and has nothing to do with a rabies virus infection. But in an abundance of caution, human health services in countries with endemic rabies may treat these bites as a suspect rabies exposure, because of the fatal implication of untreated rabies exposure. DPM can help to reduce this costly bite treatment and wastage of PEP by working to remove some of the motivations and contexts that can lead to dog bites. This includes a reduction in maternal defensive aggression (Reece *et al.*, 2013) and a reduction in breeding behaviours, such as competition between males over females in oestrus, which can spill over into aggression towards people. Further, DPM systems may include community and school education initiatives that include bite prevention strategies leading to an avoidance of contexts and human behaviours that may provoke a dog to bite. The use of clear identifying markers for vaccinated free-roaming dogs, such as long-lasting collars, can also support the avoidance of fear-motivated bites. As community members begin to perceive these vaccinated dogs as less of a threat to their health they may treat them with less outward aggression, subsequently avoiding situations where dogs may feel they need to defend themselves by biting (see Box 14.2, Case Study 3 on vaccination collars in Karachi, Pakistan).

We have previously described how DPM can increase vaccination coverage where it matters most through responsible dog ownership and community engagement. Increasing the engagement and the actual action of owners and carers in catching and bringing dogs for vaccination reduces the reliance on expert catchers and handlers, and the cost of hiring these professionals. Payment for vaccination of owned dogs during vaccination campaigns has been shown to reduce coverage (Dürr *et al.*, 2008) and any barriers to achieving high coverage should be avoided during vaccination campaigns. However, outside of vaccination campaigns, owners should be encouraged to access and pay for vaccination through available animal health services, in particular for puppies and newly acquired dogs that come into their care between vaccination campaigns. For owners that can afford vaccination, proactively and independently accessing animal health services for their dogs, leaves government-funded vaccination campaign resources available for those owners that cannot afford these services. Promoting the benefits of other population management interventions may make owners' investment in vaccination more likely, as part of a package of sterilization and parasite control for their dog.

Conflict can occur where resources that could be used for vaccination alone appear to have been limited to a smaller geographical area by requiring associated DPM measures such as sterilization. Vaccination for rabies should be a priority, as this protects both human and animal health. However, DPM has the potential to achieve additional aims, including: (i) reduced dog bites (in addition to rabies exposures); (ii) improved dog welfare; (iii) reduced free-roaming dogs and associated public safety and public perceptions; and (iv) negative impacts on wildlife and livestock. Hence DPM has the potential to bring in *additional* funding, political support and community engagement that would not otherwise have been available for rabies elimination alone. But this can only be achieved through effective stakeholder engagement and partner collaboration, making clear the collective goals to be achieved and pre-empting their potential for conflict.

14.7 Conclusion

Whether the goal of Zero by 30 is achieved will be driven by rabies elimination success in Asia, the continent with the highest burden of human rabies deaths and PEP costs (Hampson *et al.*, 2015). The elimination of dog-mediated rabies is possible, but only by accessing epidemiologically relevant dogs for vaccination and through effective surveillance actions that are supported by DPM. Reflecting on rabies control success in Latin America, we can see this was achieved through widespread and high-coverage dog vaccination, apparently without additional DPM investment (Vigilato *et al.*, 2013). However, dogs are valued in many parts of Latin America by their owners and community carers (Yue, 2019), and we hypothesize there was already greater investment and action in DPM when rabies control was launched in earnest in the 1980s. This made free-roaming epidemiologically relevant dogs more accessible for vaccination by their owners and carers. In Asia, the lack of DPM systems and associated responsible dog ownership makes these dogs harder to access. This does not mean rabies elimination will fail, but that investment in DPM is likely to be needed, and indeed wanted, by politicians and communities to support rabies elimination and bring additional benefits. It should also be recognized that the scale, investment and political will behind rabies vaccination in Latin America was far greater than anything seen in Asia to date and this greater commitment underpins their success.

Looking long-term, government-funded annual mass vaccination campaigns should not be required as the vast majority of the dog population should be proactively vaccinated by their responsible owners through accessible veterinary services throughout the year. This is the current situation in most high-income countries. DPM systems help achieve this long-term vision of sustainable herd immunity by raising owner expectations of their responsibilities along with investment in the veterinary profession required to service these owned dogs. Achieving this across Asia will be a long process for some countries, but this journey needs to begin with investment in DPM systems.

Authors' Declaration

All authors declare that they have no conflict of interest.

All authors have approved this manuscript, agree with its submission, and share collective responsibility and accountability.

This manuscript has not been published or is not under review elsewhere.

References

Asia Canine Protection Alliance (2013) Risk assessment: the risk the dog meat trade poses to rabies transmission and the ASEAN plus 3 countries' pledge to eliminate rabies by 2020. Available at: http://www.acpagroup.org/images/resources/Risk%20Assessment_TheIllegalTradeinDogsandRabiesTransmission_ACPA.pdf (accessed 6 February 2023).

Boitani, L., Francisci, F., Ciucci, P. and Andreoli, G. (2017) The ecology and behaviour of feral dogs: a case study from central Italy. In: Serpell, J.A. (ed.) *The Domestic Dog: Its Evolution, Behaviour, and Interactions with People*. Cambridge University Press, Cambridge, pp. 342–368.

Brum, E. (2019) The epidemiological principles of rabies control and the RACE against rabies. In: *Presentation at the Third International Conference on Humane Dog Population Management,* 18–20 October 2019. Available at: https://www.icam-coalition.org/conferences/international-conference-dog-population-management-2019/

Cleaveland, S., Kaare, M., Tiringa, P., Mlengeya, T. and Barrat, J. (2003) A dog rabies vaccination campaign in rural Africa: impact on the incidence of dog rabies and human dog-bite injuries. *Vaccine* 21(17–18), 1965–1973. DOI: 10.1016/s0264-410x(02)00778-8.

Coleman, P.G. and Dye, C. (1996) Immunization coverage required to prevent outbreaks of dog rabies. *Vaccine* 14, 185–186.

Dürr, S., Meltzer, M.I., Mindekem, R. and Zinsstag, J. (2008) Owner valuation of rabies vaccination of dogs, Chad. *Emerging Infectious Diseases* 14(10), 1650–1652. DOI: 10.3201/eid1410.071490.

FOUR PAWS (2020) The dog and cat meat trade in Southeast Asia: a threat to animals and people. Available at: https://media.4-paws.org/8/0/0/3/80039a8956751c7b9bf934c35993858592182db3/FOURPAWS_Big_DCMT_Report_GB.pdf (accessed 6 February 2023).

Hampson, K., Dushoff, J., Cleaveland, S., Haydon, D.T., Kaare, M. *et al.* (2009) Transmission dynamics and prospects for the elimination of canine rabies. *PLoS Biology* 7(3), e53. DOI: 10.1371/journal.pbio.1000053.

Hampson, K., Coudeville, L., Lembo, T., Sambo, M., Kieffer, A. *et al.* (2015) Estimating the global burden of endemic canine rabies. *PLoS Neglected Tropical Diseases* 9(4), e0003709. DOI: 10.1371/journal.pntd.0003709.

ICAM (International Companion Animal Management coalition) (2019a) Humane dog population management guidance. Available at: https://www.icam-coalition.org/download/humane-dog-population-management-guidance/ (accessed 6 February 2023).

ICAM (International Companion Animal Management coalition) (2019b) Manu mitra: an example of community engagement in Kathmandu, Nepal. Available at: https://www.icam-coalition.org/wp-content/uploads/2019/08/4-Community-engagement-Kathmandu-Manu-Mitra.pdf (accessed 6 February 2023).

Larkins, A.J., Reece, J.F., Shaw, A.P.M. and Thrusfield, M.V. (2020) An economic case study of the control of dog-mediated rabies by an animal welfare organisation in Jaipur, India. *Preventative Veterinary Medicine* 183, 105120. DOI: 10.1016/j.prevetmed.2020.105120.

Minghui, R., Stone, M., Semedo, M.H. and Nel, L. (2018) New global strategic plan to eliminate dog-mediated rabies by 2030. *The Lancet. Global Health* 6(8), e828–e829. DOI: 10.1016/S2214-109X(18)30302-4.

Putra, A.A.G., Hampson, K., Girardi, J., Hiby, E., Knobel, D., *et al.* (2013) Response to a rabies epidemic, Bali, Indonesia, 2008–2011. *Emerging Infectious Disease* 19, 648–651.

Reece, J.F. and Chawla, S.K. (2006) Control of rabies in Jaipur, India, by the sterilisation and vaccination of neighbourhood dogs. *Veterinary Record* 159, 379–383.

Reece, J.F., Chawla, S.K., Hiby, E.F. and Hiby, L.R. (2008) Fecundity and longevity of roaming dogs in Jaipur, India. *BMC Veterinary Research* 4, 6. DOI: 10.1186/1746-6148-4-6.

Suseno, P.P., Rysava, K., Brum, E., De Balogh, K., Ketut Diarmita, I. *et al.* (2019) Lessons for rabies control and elimination programmes: a decade of One Health experience from Bali, Indonesia. *Revue Scientifique et Technique (International Office of Epizootics)* 38(1), 213–224. DOI: 10.20506/rst.38.1.2954.

Swedberg, C., Mazeri, S., Mellanby, R.J., Hampson, K. and Chng, N.R. (2022) Implementing a One Health approach to rabies surveillance: lessons from integrated bite case management. *Frontiers in Tropical Diseases* 3, 829132. DOI: 10.3389/fitd.2022.829132.

Reece, J.F., Chawla, S.K. and Hiby, A.R. (2013) Decline in human dog-bite cases during a street dog sterilisation programme in Jaipur, India. *The Veterinary Record* 172(18), 473. DOI: 10.1136/vr.101079.

Vigilato, M.A.N., Clavijo, A., Knobl, T., Silva, H.M.T., Cosivi, O. *et al.* (2013) Progress towards eliminating canine rabies: policies and perspectives from Latin America and the Caribbean. *Philosophical Transactions of the Royal Society B* 368(1623), 20120143. DOI: 10.1098/rstb.2012.0143.
Wertheim, H.F.L., Nguyen, T.Q., Nguyen, K.A.T., De Jong, M.D., Taylor, W.R.J. *et al.* (2009) Furious rabies after an atypical exposure. *PLoS Medicine* 6, 0265–0268. DOI: 10.1371/journal.pmed.1000044.
WHO (World Health Organization) (2018a) WHO Expert Consultation on Rabies. *Technical Report Series* 1012. DOI: 10.1007/s13398-014-0173-7.2.
WHO (World Health Organization) (2018b) Eliminating dog-mediated human rabies: Karachi determined to overcome a public health hazard. *Departmental News*. Available at: https://www.who.int/news/item/26-02-2018-eliminating-dog-mediated-human-rabies-karachi-determined-to-overcome-a-public-health-hazard (accessed 6 February 2023).
WHO (World Health Organization) (2018c) Rabies vaccines and immunoglobulins: WHO position April 2018. *Weekly Epidemiological Record* 16(93), 201–220.
WOAH (World Organisation for Animal Health) (2018) Rabies (infection with rabies virus and other lyssaviruses) (Chapter 3.1.18). In: *Manual of Diagnostic Tests and Vaccines for Terrestrial Animals*. Available at: https://www.woah.org/en/what-we-do/standards/codes-and-manuals/terrestrial-manual-online-access/ (accessed 14 February 2023).
WOAH (World Organisation for Animal Health) (2022a) Vaccination (Chapter 4.18). In: *Terrestrial Animal Health Code*. Available at: https://www.woah.org/en/what-we-do/standards/codes-and-manuals/terrestrial-code-online-access/?id=169&L=1&htmfile=chapitre_vaccination.htm (accessed 14 February 2023).
WOAH (World Organisation for Animal Health) (2022b) Dog population management (Chapter 7.7). In: *Terrestrial Animal Health Code*. Available at: https://www.woah.org/en/what-we-do/standards/codes-and-manuals/terrestrial-code-online-access/?id=169&L=1&htmfile=chapitre_aw_stray_dog.htm (accessed 14 February 2023).
Yue, Y. (2019) Exploring attitudes and behaviour towards roaming dogs around the world. In: *Presentation at Third International Conference on Humane Dog Population Management,* 18–20 October 2019. Available at: https://www.icam-coalition.org/conferences/international-conference-dog-population-management-2019/ (accessed 6 February 2023).

15 Mass Dog Vaccination and Animal Birth Control: A One Health Pilot Project in Karachi, Pakistan

Naseem Salahuddin[1]*, Naila Baig-Ansari[2], M. Aftab Gohar[3], Rabia M. Khan[1], Wajiha Javed[4], S. Anzal Abbas Jaffari[4], Qamer Mahmood[5], Manzoor Ahmad[6], Muhammad Rafi Ullah[6], Fiaz Ahmad[6] and Waqas Ahmad[6]

[1]Infectious Diseases, Indus Hospital and Health Network (IHHN), Karachi, Pakistan; [2]Indus Hospital Research Center, Indus Hospital and Health Network (IHHN), Karachi, Pakistan; [3]Rabies Prevention Center, Indus Hospital and Health Network (IHHN), Karachi, Pakistan; [4]Public Health, Getz Pharma Private Limited, Karachi, Pakistan; [5]Veterinary Epidemiology Unit, Department of Internal Medicine, Reproduction, and Population Medicine, Faculty of Veterinary Medicine, Ghent University, Ghent, Belgium; [6]Department of Clinical Sciences, University of Veterinary and Animal Sciences, Lahore, Narowal Campus, Narowal, Pakistan

Abstract

The Sindh province in Southern Pakistan reports staggering numbers of dog bites. Not only to treat but to prevent, dog-bite casualties, in January 2018, the Indus Hospital launched a proof of concept of One Health. The pilot project 'Rabies Free Pakistan' (RFP) consisted of the performance of mass dog vaccination (MDV) and animal birth control (ABC) in a fishing village near the hospital. The initial team included a physician, an epidemiologist and some research assistants, but it soon expanded to veterinarians, members of the civil society and the city municipality, and received support from the World Health Organization (WHO). This project, in which MDV and ABC were performed programmatically for the first time in Pakistan, demonstrated that MDV and ABC are technically feasible if the political will is secured. RFP is now sharing the lessons learnt with decision-makers in other Pakistani provinces to ensure the scaling up of this approach.

15.1 Introduction

Rabies is one of the 20 diseases included in the World Health Organization (WHO)'s list of neglected tropical diseases (WHO, 2018), and it is endemic in low- and middle-income countries such as Pakistan, where free-roaming dogs pose a health risk to humans and other animals (Rupprecht and Salahuddin, 2019). According to estimates, there are around 12 million free-roaming dogs in Pakistan, with approximately thousands of dog bites annually, while data pertaining to the requirements of human rabies vaccines are not available (Ahmad *et al.*, 2021). Pakistan is a federation of four provinces, a Capital Territory and Federally Administered Tribal

*Corresponding author: naseem.salahuddin@tih.org.pk

 One Health for Dog-mediated Rabies Elimination in Asia (eds V. Slack *et al.*)
DOI: 10.1079/9781800622975.0015

Areas with a total population of 225 million. The Sindh province has an even distribution of urban and rural population (Nishtar, 2010). Karachi is the capital of Sindh and the commercial hub of Pakistan. It is a sprawling city of 17 million inhabitants, with >24,000 cases of dog bites annually reported in a 2-year study period conducted in private and public medical hospitals, while the range of these dog-bite injuries varies among different hospitals of the city (Zaidi *et al.*, 2013). Some bite victims seek care at one of the three large city hospitals where post-exposure prophylaxis (PEP) is free, but many cases (e.g. the ones managed by secondary and tertiary care facilities, private practitioners and spiritual healers) are not recorded.

Despite the massive network of public and private healthcare facilities in Pakistan, rabies prevention receives little attention. Unfortunately, many healthcare providers have not kept pace with the newer developments in PEP recommended by WHO. As happens in other low- and middle-income countries, emergency room physicians are often unaware of the updated rabies vaccination protocol, are reluctant to infiltrate rabies immunoglobulin (RIG) into wounds, or consider it too expensive (Salahuddin *et al.*, 2014). Many do not perform thorough wound washing and, being unfamiliar with whether and how dog bites and rabies are managed in their area, refer dog-bite victims to rabies prevention centres in another city (Salahuddin *et al.*, 2011, 2014). Establishing functioning PEP centres in every district hospital with trained healthcare providers and available and affordable PEP is fundamental to human rabies prevention in Pakistan and beyond.

The social and cultural norms of Pakistani communities influence people's attitudes towards dogs, hence dog-keeping practices and the perception of rabies risk in dogs and humans. Orthodox Muslims, especially in cities, consider dogs unclean and avoid any interaction and contact with them, particularly when it comes to free-roaming dogs. In rural areas, the human–dog relationship can be different with people sometimes keeping dogs to protect their cattle (Ahmad *et al.*, 2021). In both cases, dog-bite events that involve free-roaming dogs often result in the public demanding the local authorities to take immediate and drastic action. Mass dog poisoning is the response that the Karachi government has traditionally adopted, as a supposedly quick fix to a disease that has never been considered a priority disease for the Pakistani government.

Yet, scientific evidence shows that indiscriminate dog culling is not the answer. It is ineffective as a long-term measure because the dog population will eventually bounce back, as well as counterproductive, because if any dog has been vaccinated, they may get killed too. Additionally, it may cause the spreading of rabies, if dog owners relocate their (rabid) dogs elsewhere to save them from culling. Finally, some sections of the population may disapprove of such an extreme act.

15.2 The One Health Pilot Model

In this context, in 2017, some human health professionals at the Indus Hospital and Health Network (IHHN) decided that their efforts for rabies prevention through PEP were insufficient, as they addressed the issue of rabies at a late stage, after the bite from a potentially rabid dog. We wanted to develop an effective strategy for controlling rabies at the source through a holistic One Health approach, as recommended by the Asia Pacific Regional Tripartite partnership between WHO, the Food and Agriculture Organization of the United Nations (FAO) and the World Organization for Animal Health (WOAH) (Gongal and Wright, 2011). Nevertheless, the Tripartite's approach towards rabies does not address the problem of dog population control in clear terms. This is understandable because research has demonstrated that reducing dog density does not have an impact on rabies transmission. Yet, based on our living in Karachi and witnessing the human–dog conflict that unfolds daily on the streets, we realize that, at least in Pakistan, most people consider the overpopulation of free-roaming dogs as a major hindrance in their daily lives, and demand that municipalities deal with it swiftly. To meet this local need, our One Health strategy simultaneously addressed rabies, through mass dog vaccination (MDV), and dog overpopulation, through animal birth control (ABC).

In 2018, the Indus Hospital launched the pilot project 'Rabies Free Pakistan' (RFP) to demonstrate to the local and global community that

Pakistani health professionals and civil society are committed to eliminating dog-mediated human rabies. The objective was to provide proof of concept to the government of Pakistan, and foreign government bodies as well, that vaccinating and sterilizing free-roaming dogs is a feasible, effective and ethical alternative to dog culling to address situations of high human–dog conflict.

The team was led by an infectious disease specialist at the Indus Hospital and included physicians and researchers, as well as Karachi's leading veterinarians, animal rights activists, members of the civil society, community engagement experts and epidemiologists. A fishing village, Ibrahim Hyderi, was chosen as the project site because of its proximity to the Indus Hospital and the outbreaks of severe dog bites often reported from within the area. The seashore near Ibrahim Hyderi also serves as an unofficial landfill site where the waste collected from various parts of Karachi is dumped. The disposal of organic waste from the houses in the village adds up, creating an environment that provides free-roaming dogs with plenty of food for their population to grow (Wright *et al.*, 2021).

Recognizing the importance of community engagement, the RFP team adopted the participatory action research approach and held a series of meetings with stakeholders in the Ibrahim Hyderi community to discuss the proposed MDV and ABC plan. Stakeholders identified the presence of waste in the area as a key issue in relation to dog rabies and population control and requested effective and long-term action to manage this problem. Other meetings were also organized with the union council members, as well as the general public.

The issue of waste management, another important aspect in the control of dog-mediated rabies, is discussed in Box 15.1.

15.3 The Actions Taken

15.3.1 Training phase

A budget was created by the RFP team and the local government pledged the initial funds necessary to begin the project. Money was also raised through crowdfunding and corporate and personal donations. Interestingly, it is to be noted that the Indus Hospital receives a consistent share of its funding through zakat, a form of almsgiving that is considered a religious obligation in Islam. A large shipping container donated by a private donor was custom designed as a surgical room, equipped with an air conditioner, water supply and drainage, as well as all the basic equipment required by veterinarians for surgical sterilization. Later in the year, a purpose-built surgical van was donated by another private donor, and this allowed us to increase the number of surgical procedures performed.

The Ibrahim Hyderi community provided the space necessary to establish the project's field office. The surgical container, cages, holding pens and field equipment were kept on the premises. This secure rent-free space served as the base for all our field MDV and ABC activities. The entire RFP field personnel was employed from within the Ibrahim Hyderi community, which included dog vaccinators, caretakers at the site, and community engagement persons who visited all households and sensitized residents regarding the RFP activities, collecting feedback throughout the duration of the pilot project.

WHO provided support with dog vaccines and sponsored an 11-day training workshop held in Karachi by an experienced animal behaviourist from South Africa. The purpose of the workshop was the training of the initial, local dog-catching team. It included classroom lectures and hands-on activities on dog behaviour and body language, humane dog catching, handling, tagging and transportation, choice and management of equipment, dog vaccination protocols, and dog vaccination campaign planning and implementation (including the community awareness-raising component). The poles and nets necessary for dog catching were made locally after seeing the prototypes available online. Two weeks before the field training, all trainees were given rabies pre-exposure prophylaxis. Dog vaccines were carried to the field in coolers to maintain the cold chain.

In the field, hands-on MDV training also included the use of the two Rabies Data Collector devices that RFP had purchased from the Global Alliance for Rabies Control. These devices are used to enter the exact location (using the Global Positioning System), gender and age of each vaccinated dog, take a picture of the dog, and transmit this information to the central database

Box 15.1. The environmental component of rabies control: waste management.

The global approach to dog-mediated human rabies control mainly consists of MDV to reduce transmission at its source and PEP to stop infection in exposed individuals. Awareness regarding dog-bite prevention and wound management complement this strategy. Waste mismanagement is acknowledged as a factor that facilitates and sustains the presence, growth and survival and breeding rate of the dog population (Taylor *et al.*, 2017). Furthermore, it influences dog behaviour and the interactions among dogs and with people (e.g. competition and defence), increasing the risk of rabies transmission among dogs and from dogs to people (especially vulnerable groups such as street dwellers and waste collectors – Herbert *et al.*, 2012) and the need for PEP (Chandran and Azeez, 2016). Nevertheless, waste management is very rarely part of rabies control programmes (Taylor *et al.*, 2021). While so far, RFP has not addressed this environmental component either, we are very much aware of the need to include waste management in our One Health approach to dog-mediated human rabies control in Karachi and all over Pakistan. In the following paragraphs, we wish to highlight this issue in order to facilitate sustainable and One Health-informed action towards the global Zero by 30 goal.

Solid waste comes from residential, commercial and industrial sources. Organic items, such as kitchen scraps, restaurant and canteen leftovers, and garbage from open-air markets and meat shops are a consistent part of it. The composition of solid waste, including its edible portion, is different in every community depending upon the lifestyles of the people living in it and the overall geographical environment. In most Asian countries, the waste management system is not effective, or even existent, throughout all the phases of collection, transport, recycling, treatment and disposal (Guerrero *et al.*, 2013). This is due to a mix of financial, organizational, geographical and political reasons. Unmanaged and unsorted waste is usually dumped on roadsides, open land patches and vacant plots, poorly collected and transported to often overflowing landfills. When available, trash containers are often broken, full, smelly or dirty and this discourages their use (Kansal, 2002; Saree *et al.*, 2021).

In Pakistan, a 60% increase in the amount of municipal solid waste has been observed from 2015 to 2022 and the vast majority of this waste is completely unmanaged (Abbasi *et al.*, 2015). Ibrahim Hyderi provides a telling example. Located in a geographically marginal and low-income area of Karachi, it serves as the unregulated dumping ground for the waste collected elsewhere in the city. Incidentally, this highlights very well the unfair distribution of the rabies burden on poor, underserved and socially marginalized communities (Cleaveland *et al.*, 2017). Moreover, since most members of the local community work in the fishing sector, fish scraps resulting from fish processing and local fish consumption provide a valuable source of food for the dogs who live in the area and the adjacent ones.

Despite the complex and broad challenges involved in waste management, we argue that this component of the rabies problem deserves more attention and action by all rabies stakeholders. A complete One Health approach to rabies should include discussion with environmental authorities and part of the design of a rabies plan should address the problem of waste management, by looking at its root causes and striving for sustainable solutions. Community engagement actions and rabies awareness campaigns can be a good starting point to increase public awareness and create a solid base for advocacy both with the population at large and with government bodies (Kato *et al.*, 2003).

that the RFP team uses to monitor and evaluate performance. Later on, the RFP team created a similar program on Android smartphones, using the Open Data Kit software to collect the necessary information more easily and economically. This data collection system is what allows the central RFP team to plan the daily activities of the field teams and direct where more MDV and ABC needs to be performed.

Six veterinarians from some local veterinary institutions were trained on ABC by selected private veterinarians. The trainees were taught the fundamentals of surgical sterilization methods (ovariohysterectomy and castration), anaesthesia techniques, sterilization of instruments, pre- and post-operative care of dogs, and monitoring and management of post-operative complications. Four months after the initial training, a consultant veterinarian was hired for refresher training, as well as for exploring an economical but safe alternative to ABC (Leoci *et al.*, 2014; Ibrahim *et al.*, 2016). An experienced veterinarian taught the team about the chemical castration of male dogs using calcium chloride to minimize time and reduce costs while maximizing the output of sterilization procedures per day.

15.3.2 Implementation phase

Every street within the target area of a given team on a given day is combed for dogs of all ages. Each MDV team consists of three people with dog catching and vaccinating duties and one person responsible for handling the vaccines, marking dogs with spray paint, recording MDV data on the device and submitting it to the central field supervisor at the end of each workday. Dogs are restrained by hand or caught using a lightweight butterfly net, vaccinated subcutaneously with Rabisin dog rabies vaccine, and marked with non-toxic paint along the top and back of the head for identification and to prevent revaccination during the same vaccination drive. When dog collars became available through donations from Boehringer Ingelheim, collars of a certain colour were used to identify vaccinated dogs and collars of another colour to identify vaccinated *and* sterilized dogs. Vaccination teams continue to work in the same area on consecutive days until the maximum number of dogs are vaccinated, and then they move to the next area. The typical MDV daily target varies from 20 to 30 dogs in areas with fewer dogs and 50–60 in high-count areas.

Non-sterilized dogs are also brought in for ABC. Both males and females are selected for surgery but, as recommended by WHO and WOAH, priority is given to females. The dogs are vaccinated and held overnight in the holding pen, and on the day of surgery, two veterinarians and one assistant use standard surgical protocols for sterilization. A V-shaped ear notch is performed while the dog is still under anaesthesia as a permanent means of sterilization identification. Operated dogs are kept under observation until they fully recover and then are released back to the area where they were picked up from.

In May 2018, the team explored chemical castration as a potential alternative to male orchidectomy (Leoci *et al.*, 2014; Ibrahim *et al.*, 2016). As this was our first experience with chemical castration, the team started with five dogs (four adults and one juvenile dog), who underwent intratesticular calcium chloride injection. The dogs were then observed for 45 days, checking for any signs of pain, testicular size, serum testosterone levels and complications such as abscesses. The testosterone level in the four adult dogs reduced significantly at day 23 from baseline. No reduction was observed in the juvenile dog. The testicular size in all dogs decreased visibly by day 45 and this was considered a proof of success.

With practice, the ABC team is able to perform an average of 12 surgeries and ten chemical castrations per day. A record of morbidity or mortality cases, with related causes, is maintained, so that future errors or mishaps could be rectified. To date, we found chemical castration to be safe, rapid and considerably less expensive than surgical castration, although a 2% morbidity from testicular abscess was observed. We do not have data yet on the long-term efficacy of this procedure.

15.4 Lessons Learnt for Ourselves and Others

Each of the four stakeholder groups involved in RFP proved essential: (i) the Ibrahim Hyderi community; (ii) the Indus Hospital project team; (iii) the local government; and (iv) WHO.

The Ibrahim Hyderi community was at the forefront of this project, both by providing the space and the human resources necessary for the field operations, and by being willing to allow the use of their locality to pilot-test the effect of MDV on dogs and their human community. Some members of the community were crucial to increase rabies awareness, especially during global events such as World Rabies Days (28 September).

After the use of collars was introduced, we observed an improvement in the human attitude towards dogs. In particular, children feel that collared dogs are safe to hold or play with, while adults consider them 'respectable' animals. In a community survey run by RFP, 60% of the respondents reported that they react to uncollared dogs with hostility by teasing, running away or throwing stones at them. Once collared, 93% of them reported adopting neutral behaviour. Nevertheless, a downside related to collaring was also observed, which is that children often remove the collar for amusement. A boy was seen removing the collar from a dog and putting it on an unvaccinated dog. MDV teams report that only a limited number of collars remain in place, on the right dog, after 1 or

2 weeks after vaccination. They assume that either children pull them off, or that the collars simply get unlatched and fall off. Either way, it is important to improve the latch mechanism, to make intentional or accidental removal of collars difficult.

The Indus Hospital project team was innovative in many areas. First, the whole idea of extending rabies prevention beyond human PEP by vaccinating dogs was promoted by a human healthcare professional. This represents the essence of the One Health model and paradigm shift that it demands. Second, we engaged with veterinarians and animal rights activists to both create a robust social media presence and set up a coalition meant to convince the public, civil society and members of local and national government bodies that mass dog culling is not the right answer to dog–human conflict and dog-mediated human rabies. This coalition promotes a well-rounded approach that includes: (i) adequate waste management; (ii) dog-bite prevention through awareness campaigns; (iii) prompt PEP for exposed individuals; (iv) MDV; and (v) ABC. We strive to make sure that each complaint from the public about free-roaming dogs is properly addressed, to assure Karachi residents that action is being taken and avoid retaliatory measures being adopted, which may jeopardize MDV if vaccinated dogs are involved in mass culling.

Third, the Indus Hospital project team created their own data collection system using an open-source software and easily available Android phones instead of relying on imported, expensive devices. Using cheap and locally available tools and resources is crucial to the financial sustainability of long-term programmes like MDV and ABC. Fourth, we looked into alternative methods of dog sterilization and gathered initial data to show that chemical castration can take place with minimal side effects if properly carried out.

In hindsight, we recognize the importance of performing a baseline dog survey to determine the impact of MDV and ABC campaigns. This survey does not necessarily need to be conducted before starting the campaign but can occur simultaneously by doing a visual count of the dogs sprayed with paint after the intervention, and the others.

The local government of Sindh missed the target expected of them since they only provided 25% of the funding initially pledged. This severely affected the pace of RFP operations. Furthermore, they failed to provide staff to be trained from within their field workers, or any veterinarians from the government animal health sector.

Support from WHO proved essential to secure the dog vaccines needed for FRP and learn from experts and other countries' experiences how to perform MDV in a science-based and efficient way.

At the end of the pilot project and its assessment, we increased the number of MDV and ABC teams and expanded our One Health-based approach to other areas beyond Ibrahim Hyderi. That said, we are conscious that implementation needs to be sustained, and the tangible outcomes of our action will not be evident for at least 5 years of continued effort. So far, we have demonstrated that the systematic implementation of MDV and ABC is possible in Pakistan because we have the technical capacity to plan and implement MDV campaigns.

Throughout the setting up and implementation of this pilot project, RFP met several administrative and organizational challenges. The hardest was the rising cost of petrol, vaccines and anaesthetics, coupled with the increase in pay scales requested by Pakistani government regulations. Moreover, over 2 years of coronavirus 2019 (COVID-19) severely restricted the movement of MDV teams, while a heat wave in 2018 and heavy rains and flooding in 2019 furthered hampered field activities. As a result, in some areas of Karachi, the city government and even individuals started culling dogs, including those who had been vaccinated or sterilized by us. Upon protests from animal rights activists, including those who are part of RFP, the culling was called off.

At present, our immediate goal is to ensure that the trained and motivated MDV staff created by RFP continue their work and systematically expand it to all the areas of Karachi. However, complete scale-up and success mainly depend on dog-mediated human rabies elimination occupying a solid position in the agenda of the Pakistani government at the central, province and city levels. Through RFP advocacy, the government of the Sindh province recently announced a budget for the Rabies Control Programme of Sindh, aimed at vaccinating and sterilizing free-roaming dogs across 30 districts of the Sindh province in 3 years. They co-opted

RFP as their technical advisers. If agreed upon, such a public–private partnership has the potential to make a historical change in rabies control in the Sindh province and, if expanded to the rest of the county, the whole of Pakistan. We need to continue to promote and highlight our work so that political commitment materializes, and more public and private partners emerge to make scaling up possible. We hope to earn recognition from the international community that Pakistanis can contribute towards the global Zero by 30 target.

As rabies control is a team effort, the RFP team recently shared the collected experience at Pakistan's leading veterinary school, the University of Veterinary Animal Sciences (UVAS), in Lahore, in the Punjab province. A 5-day workshop was organized to provide theoretical and practical knowledge on MDV to undergraduate and postgraduate veterinary students, the staff of a local animal rescue and welfare organization and members of the Lahore city government. At present, the Punjab government is planning to follow the example of the Sindh province and introduce city-wise MDV and ABC. We hope all other provinces will follow.

15.5 Conclusion

Pakistan's goal of eliminating dog-mediated rabies by 2030 is a tall order, but the prevention of this agonizing and lethal disease must be prioritized and achieved by tackling the root cause of the problem. An interdisciplinary and intersectoral group consisting of physicians, veterinarians, researchers, the community, the city government and civil society initiated the One Health-inspired RFP pilot project. We credit ourselves with having raised community awareness, brought together One Health champions who can help increase advocacy for rabies control and proved to the government that MDV and ABC, which were previously unknown in Pakistan, can be done and is cheaper and more effective than culling in reducing the burden of rabies in dogs and, as a result, in humans. We invite other cities to establish PEP centres that provide up-to-date and free PEP and – at the same time – vaccinate free-roaming dogs and control their population humanely, especially through adequate waste management. Finally, dog and human rabies should be made notifiable and rabies surveillance initiated if Pakistan is to attain Zero by 30 (Welburn *et al.*, 2015).

Authors' Declaration

All authors declare they have no conflict of interest.

All authors have approved this manuscript, agree with its submission, and share collective responsibility and accountability.

This manuscript has not been published or is not under review elsewhere.

References

Abbasi, H., Xiwu, L. and Zhao, G. (2015) An overview of Karachi solid waste disposal sites and environs. *Journal of Scientific Research and Reports* 6(4), 294–303. DOI: 10.9734/JSRR/2015/11125.

Ahmad, W., Naeem, M.A., Akram, Q., Ahmad, S. and Younus, M. (2021) Exploring rabies endemicity in Pakistan: major constraints & possible solutions. *Acta Tropica* 221, 106011. DOI: 10.1016/j.actatropica.2021.106011.

Chandran, R. and Azeez, P.A. (2016) Stray dog menace: implications and management. *Economic and Political Weekly* 51, 58–65.

Cleaveland, S., Sharp, J., Abela-Ridder, B., Allan, K.J., Buza, J. *et al.* (2017) One Health contributions towards more effective and equitable approaches to health in low- and middle-income countries. *Philosophical Transactions of the Royal Society of London. Series B Biological Sciences* 372, 1725. DOI: 10.1098/rstb.2016.0168.

Gongal, G. and Wright, A.E. (2011) Human rabies in the WHO Southeast Asia Region: forward steps for elimination. *Advances in Preventive Medicine* 2011, 383870. DOI: 10.4061/2011/383870.

Guerrero, L.A., Maas, G. and Hogland, W. (2013) Solid waste management challenges for cities in developing countries. *Waste Management* 33(1), 220–232. DOI: 10.1016/j.wasman.2012.09.008.

Herbert, M., Basha, R. and Thangaraj, S. (2012) Community perception regarding rabies prevention and stray dog control in urban slums in India. *Journal of Infection and Public Health* 5(6), 374–380. DOI: 10.1016/j.jiph.2012.05.002.

Ibrahim, A., Ali, M.M., Abou-Khalil, N.S. and Ali, M.F. (2016) Evaluation of chemical castration with calcium chloride versus surgical castration in donkeys: testosterone as an endpoint marker. *BMC Veterinary Research* 12(1), 46. DOI: 10.1186/s12917-016-0670-3.

Kansal, A. (2002) Solid waste management strategies for India. *Indian Journal of Environment Protection* 22(4), 444–448.

Kato, M., Yamamoto, H., Inukai, Y. and Kira, S. (2003) Survey of the stray dog population and the health education program on the prevention of dog bites and dog-acquired infections: a comparative study in Nepal and Okayama Prefecture, Japan. *Acta Medica Okayama* 57(5), 261–266. DOI: 10.18926/AMO/32829.

Leoci, R., Aiudi, G., Silvestre, F., Lissner, E.A., Marino, F. *et al.* (2014) A dose-finding, long-term study on the use of calcium chloride in saline solution as a method of nonsurgical sterilization in dogs: evaluation of the most effective concentration with the lowest risk. *Acta Veterinaria Scandinavica* 56(1), 63. DOI: 10.1186/s13028-014-0063-1.

Nishtar, S. (2010) *Choked Pipes. Reforming Pakistan's Mixed Health System.* Oxford University Press, Oxford.

Rupprecht, C.E. and Salahuddin, N. (2019) Current status of human rabies prevention: remaining barriers to global biologics accessibility and disease elimination. *Expert Review of Vaccines* 18(6), 629–640. DOI: 10.1080/14760584.2019.1627205.

Salahuddin, N., Jamali, S., Ibraheem, K. and Sardar, S. (2011) Awareness about rabies post exposure prophylaxis in Pakistan among patients and health care workers: results from an Asian rabies expert bureau study. *Journal of the College of Physicians and Surgeons--Pakistan* 21(8), 491–494. DOI: 08.2011/JCPSP.491494.

Salahuddin, N., Mubashar, K. and Baig-Ansari, N. (2014) Use of rabies immune globulin in seven urban emergency rooms in Pakistan. *Asian Biomedicine* 8(1), 61–65. DOI: 10.5372/1905-7415.0801.262.

Saree, A., Klinpikul, A., Kongchu, A., Wongaree, A., Navykarn, B. *et al.* (2021) Stray dog feeding and self-reported stray dog-related problems in Thailand. *In Review*. DOI: 10.21203/rs.3.rs-900524/v1.

Taylor, E., Del Rio Vilas, V., Scott, T., Coetzer, A., Prada, J.M. *et al.* (2021) Rabies in the Middle East, Eastern Europe, Central Asia and North Africa: building evidence and delivering a regional approach to rabies elimination. *Journal of Infection and Public Health* 14(6), 787–794. DOI: 10.1016/j.jiph.2021.02.009.

Taylor, L.H., Wallace, R.M., Balaram, D., Lindenmayer, J.M., Eckery, D.C. *et al.* (2017) The role of dog population management in rabies elimination – a review of current approaches and future opportunities. *Frontiers in Veterinary Science* 4, 109. DOI: 10.3389/fvets.2017.00109.

Welburn, S.C., Beange, I., Ducrotoy, M.J. and Okello, A.L. (2015) The neglected zoonoses: the case for integrated control and advocacy. *Clinical Microbiology and Infection* 21(5), 433–443. DOI: 10.1016/j.cmi.2015.04.011.

WHO (World Health Organization) (2018) WHO expert consultation on rabies - third report. *World Health Organization Technical Report Series* 1012, 1–182. Available at: https://apps.who.int/iris/bitstream/handle/10665/272364/9789241210218-eng.pdf (accessed 5 December 2022).

Wright, N., Subedi, D., Pantha, S., Acharya, K.P. and Nel, L.H. (2021) The role of waste management in control of rabies: a neglected issue. *Viruses* 13(2), 225. DOI: 10.3390/v13020225.

Zaidi, S.M.A., Labrique, A.B., Khowaja, S., Lotia-Farrukh, I., Irani, J. *et al.* (2013) Geographic variation in access to dog-bite care in Pakistan and risk of dog-bite exposure in Karachi: prospective surveillance using a low-cost mobile phone system. *PLOS Neglected Tropical Diseases* 7(12), e2574. DOI: 10.1371/journal.pntd.0002574.

16 Rabies in China: The Role of Rabies Ecologies and Pet Activism

Suzanne Barber[1] **and Michael J. Hathaway**[2*]
[1]*Perdue University, West Lafayette, Indiana, USA;* [2]*Simon Fraser University, Burnaby, British Columbia, Canada*

Abstract
Despite phenomenal efforts to control other diseases, the People's Republic of China still has the world's second-highest amount of human rabies cases after India. This chapter explores several major challenges that remain in existing strategies to reduce or eliminate rabies. We argue for looking not only at dog–human relations but taking a One Health approach, we advocate for understanding the underappreciated roles of 'rabies ecologies' that include wild animals. Also, this chapter explores the impact of regional differences, problems with vaccine reliability, animal governance and the rise of animal activist movements that are trying to reimagine dog–human relations. Activists, for example, are challenging the prevalence of large-scale dog culls as an effective measure of rabies prevention. As China's dog numbers boom, however, there are growing difficulties in a country with many diverse social situations, where one finds low rates of dog registration, desexing and vaccination.

16.1 Introduction

While rabies is a particularly stigmatized and fatal disease, it is a potentially manageable disease that not only affects humans but at least several dozen mammal species. This book emphasizes that Asia plays a major role in global human rabies deaths, with almost two-thirds of cases, although it could be noted that this is also approximately in relation to global population ratios. Of these, India constitutes overwhelmingly the largest number of cases, and China has the second largest, varying from 3300 in 2007 to 202 in 2020 (Feng *et al.*, 2021). This chapter describes rabies in China while also drawing attention to several aspects of rabies transmission that might limit the effectiveness of rabies control campaigns. We look at three aspects: (i) the role of wild animal species in 'rabies ecologies'; (ii) regional variation; and (iii) issues around animal governance and health reporting. Also, we examine dog culling and rabies vaccination campaigns in this country.

We would also like to point out that since 2019 China has received much unwanted attention as the global epicentre of coronavirus 2019 (COVID-19) and ground zero in a series of zoonotic diseases (forms of bird flu and severe acute respiratory syndrome (SARS)). This has heightened the state's sense of being blamed and watched by other countries (Dionne and Turkmen, 2020). This sense of being observed and judged might shape the kinds of statistics that China produces for international consumption. As the world has seen, in China there is a tendency to foster a rigid and sweeping eliminationist logic as seen in the 'Zero COVID' approach. Yet, an outside observer might not

*Corresponding author: mjhath@gmail.com

 One Health for Dog-mediated Rabies Elimination in Asia (eds V. Slack *et al.*)
DOI: 10.1079/9781800622975.0016

know that this extreme approach oscillates with a pattern of uneven or haphazard management. We can see such oscillation when it comes to managing rabies. In China, cities often create onerous and expensive procedures for dog registration, so rates of compliance (and thus linked rates of desexing (spaying and neutering) and vaccination) remain quite low, sometimes under 1% (*China Daily*, 2009). Also, there remains high rates of suspicion about the effectiveness of Chinese rabies vaccines, especially with increasing discoveries of illegal production methods, or counterfeit vaccines that may render them ineffective (Wu *et al.*, 2009; Patranobis, 2018). This creates a challenging context for China's recently created bureau looking at rabies issues on a national basis. Historically, cities have carried out massive dog culls under the name of disease management. Yet by the early 2000s, China also had a growing animal rights movement that was advocating for non-violent means of rabies and disease control.

16.2 Rabies Ecology: Dogs Are One of Many Mammalian Hosts

We argue for the importance of a One Health perspective that pays attention to a wide range of species that exist as rabies reservoirs (Miao *et al.*, 2021), especially in countries such as China where such wild species play an underappreciated role. In China, it is often said that 95% of all rabies transmission to humans comes from dogs (Yin *et al.*, 2021) and globally, the World Health Organization (WHO) statistics claim that dog bites account for 99% of human rabies. Thus, it is totally understandable that dogs should be a major focus. We wish, however, to draw attention to a number of other mammals that may be important reservoirs of rabies, and the concept of viral reservoirs or interspecies 'spillovers' has risen to prominence in the aftermath of the COVID-19 pandemic (Enserink, 2020; Dimonaco *et al.*, 2020).

In certain areas other animals might mediate the persistence and spread of rabies. In China, the main term for rabies is 狂犬病, literally fierce or mad dog disease. As in a commonly used English euphemism (mad dog disease), dogs are often assumed to be the only carriers of the disease to humans and hence dogs are blamed for all cases. In terms of efficacy, campaigns that focus exclusively on dogs may be less successful in countries with high rates of rabies among wild animals that might transmit rabies directly to humans, or might act as vectors to dogs. In China, based on the term, people can reasonably assume that only dogs carry rabies. Thus, if people are scratched or bitten by wild animals, they are less likely to pursue treatment in clinics or laboratories. It is often incredibly difficult to catch wild animals. In rural areas in China, where dogs can be free roaming, they have more opportunities to encounter wild animals, and thus possibilities for transmission can be higher than urban dogs typically kept within apartments. We should also acknowledge that in many places in rural China, in particular, low rates of dog vaccination mean that dogs can potentially infect wild animals, thus playing both roles and we shouldn't assume a unidirectional flow. In sum, these ecologies are highly variable based on regional specifics, and therefore moving from generic national approaches to more regionally salient campaigns is more likely to be effective.

We can see this issue of wild reservoirs even in countries with very high rates of dog vaccination such as the USA and Canada, where rabies still exists among wild animals such as raccoons and foxes. These were some of the world's largest wild animal vaccination campaigns, mainly carried out by aerial or hand delivery of baits with oral vaccines. Although expensive, such campaigns have often been quite successful in reducing rabies transmission between wild and domestic animals. Even if dog vaccination rates near 100%, there are still other species that can exist as a reservoir. When the issue of rabies is not just viewed as a human–dog relation but, instead is seen as a kind of ecology, then a whole range of potential wild hosts become visible and might even in certain areas become the most important species for the survival of rabies. Within the last five decades, in some places in the world such as Mexico, rigorous dog-based vaccination campaigns have been so effective that there are no longer dog-mediated human rabies deaths (Velasco-Villa *et al.*, 2017). These places might have been more effective not only because the state amassed a large amount of resources to carry out rabies vaccination in dogs (and carried out and monitored vaccinations in consistent ways over a significant period of time), but also because vaccination created a biological barrier between dogs and wild animal hosts.

Thus, we encourage greater research and awareness of the role of wild animals in the ecology of rabies, both in China and elsewhere. However, we imagine that in places such as China (and other countries where there is little public awareness of the role of wild animals as rabies carriers), wild animals probably play a larger role in rabies transmission to humans, dogs and other domesticated and wild animals than has been typically understood. Scientists have found that rabies can easily flow from wild animals to unvaccinated dogs (Velasco-Villa *et al.*, 2008). As shown by Feng *et al.* (2021) in some regions of China, such as the North-west, rabid wild foxes 'play a pivotal role in animal rabies epizootics', so thinking carefully about regional variations in reservoir species would prove more useful than national approaches that only target dogs.

Taking a human-centred perspective, if we look at low-income rural farmers whose animals are exposed to rabies and, like humans, consequently experience a high mortality rate, this can become a serious social problem. Interestingly, in some places in China this issue is acknowledged, and some are finding that vaccines developed for dogs have also been effective in cattle and Bactrian camels, showing that different mammal species share similarities in terms of immune response. This also means that vaccinating a range of animals might be more affordable than otherwise assumed based on the notion that each species needs its own custom vaccine (Liu *et al.*, 2016).

16.3 Regional Variation

As mentioned in the section above on 'rabies ecologies', there are critical differences between regions in the vast and diverse terrain now known as 'China'. There can also be significant variations between first tier and second tier cities, and major differences between urban and rural contexts. In terms of cities, there are often big differences in the availability of reliable vaccines and veterinary services.

Currently, the majority of human rabies cases occur in south-east China (Zhou *et al.*, 2016), which overall represents an area with high human population densities and large cities, but still 65% of all cases involve economically disadvantaged rural populations (Yin *et al.*, 2021). Thus, while rabies control efforts are still important in urban companion pet populations, different methods might be necessary in rural areas with a higher rabies risk.

16.4 Issues Around Pet Popularity, Activism and Animal Governance

In terms of rabies and dogs in urban China, one of the most important issues is the recent explosion of interest in pet dogs. The Chinese state has a history of state-based antagonism towards dogs that has only started to recently subside (Barber and Hathaway, 2022). For decades, in urban areas it was illegal to have a dog as a pet. Even after it was legalized, it was extremely expensive for people to purchase pet registration. In some places, such as Guangzhou, the registration fee was 10,000 yuan in 1997, a substantial fee, greater than the annual income of many teachers. By 2009, these high fees meant that although only 800 dogs were registered in the city, there were an estimated 100,000 dogs (*China Daily*, 2009). In a number of places, registered dogs had to be relatively small (under 35 cm), and some cities enforced a 'one dog' policy. As some scholars have argued, dog vaccination and desexing is often tied to registration, but such high registration costs and annual fees mean that relatively few dogs are registered. According to the International Center for Veterinary Services, a China-based organization, about 5% of urban stray, feral and sheltered dogs and cats are spayed or neutered and this is almost 0% for their rural compatriots (International Center for Veterinary Services, 2022).

The recent spread of dogs as pets and overall pet culture in China is relatively new. As this book describes, many places in the world have different relations to dogs compared to expectations in the West that tend to see dogs as family members, who exist mainly as coddled pets, but also as working or guard dogs. Such dogs are also regarded as owned by individual people, who are responsible for vaccinating their dogs, and taking care of them including providing dogs with their complete needs for food, water and exercise. In this context, dog breeding is almost entirely controlled by professionals, and it is seen as normal and desirable that puppies are spayed or neutered at a young age. Such treatments are not universal within the West, and in Europe there are strong arguments against 'desexing' dogs at such a young age because it might cause physical problems in later life.

The flip side of such Western notions is that free-roaming dogs are often vilified, quickly captured by animal control officers and placed into dog pounds, where overall many are killed by humane methods. Yet, in many places in the world, there are street dogs or village dogs, and in these places people often show some degree of dispersed care of these dogs, even though it is also acknowledged that some can be a threat to people, especially children, in terms of bites or as carriers of disease. In China, some animal rights activists and veterinarians explained that the low desexing rate of dogs was due to many people thinking that spaying and neutering is cruel, that dogs have some right to breed and to have their own puppies (Barber, 2018). There is a widespread public sense that it is wrong to desex an animal, even their own pet. Until recently, veterinary training in China has focused primarily on livestock. Thus, a veterinarian's main concern and training was often to increase breeding of livestock, not decrease animal numbers. This resulted in less training and experience with spaying and neutering. In addition, with the recent pet boom, there is now a large unmet demand for veterinarians that specialize in dogs and cats. Furthermore, activists pointed to a widespread scepticism in China that veterinarians are more motivated by money over a sense of care, so this also results in a greater suspicion towards what might be seen as unnecessary surgery (Barber, 2018).

Along with the re-emergence of the urban pet-keeping culture, the Chinese companion animal rights movement has also begun to grow. The Chinese companion animal rights movement is primarily made up of urban, middle-class people in their 20s–40s. It consists of both local and international non-governmental organizations (NGOs), smaller unofficial local groups, and many individuals who have formed important social networks through social media. The movement has been heavily involved in the conversations and actions surrounding the elimination of rabies as it has proved to be a powerful rhetorical tool through which to argue for the establishment of domestic animal protection laws. Furthermore, activists regard some of the rabies control methods such as dog culls as inhumane and have worked to find and promote alternatives.

One way activists and animal protection groups have worked to reduce the number of dog culls is to reduce the number of dog bites by producing and distributing educational literature. While these materials do not directly address rabies, the hope is that by reducing dog bites, there will be fewer injuries, and fewer government-led dog culls (or fewer excuses for them in some cases). Within urban areas, the educational material can often take the form of brochures and posters reminding people to keep their dogs on a leash. However, the largest number of dog bites involve the elderly and children under the age of 15 (Yin *et al.*, 2021), and thus much of the educational literature is now being addressed towards elementary school-aged children. ACTAsia is one organization that has been at the forefront of this development, and their 'care for life education' is now part of the official curriculum in many Guangdong schools (ACTAsia, 2022). This programme is currently 1 week long and is taught in the first year of primary school. During this week children are taught about the importance of valuing and respecting other forms of life. As part of this curriculum, children are taught how to understand canine body language in order to avoid being bitten. This curriculum is spread throughout both the urban and the rural areas of Guangdong Province. While other organizations' educational programmes are rarely as large or as formal as ACTAsia's, there is a growing recognition both within the animal rights movement as well as within local governments that such programmes demonstrate a lot of potential in reducing dog bites. Unlike other rabies control methods, education campaigns such as this move away from focusing on dogs as disease vectors that need to be controlled, and instead recentre the dog as a potential victim of human ignorance. It is human behaviour, both in terms of pet-care responsibility and human–dog interactions, that needs to be moderated. Such bite reduction, combined with increasing vaccination, and a focus on culling only infected dogs, might offer the potential promise of reducing dog-induced rabies cases.

16.5 Dog Culls

Within China, dog culls are perhaps the most common rabies control method carried out by the state, and certainly the most visible and contentious. They have generated a lot of public anger and helped mobilize some of the activists. Due to the lack of official records and the emotional reaction towards dog culls, it is difficult to get an accurate number of the amount of dogs culled in China. What is clear though, is that they can be quite large (where tens of thousands of dogs are killed during a short-lived campaign) and are often controversial (Watts, 2006; *Reuters*, 2009). Supporters of dog culls argue that large numbers of stray and loose dogs are responsible for the high rate of rabies in China. Vaccination numbers, as we will discuss, while improving, are still low, and some argue that these low numbers, combined with little or no sterilization, means dog culls are the best method to continue to reduce rabies numbers in China and reach the Zero by 30 goal. Although the exact method of culling varies by location, it is frequently carried out by police or people hired to perform the cull and most often involves beating the dogs, drowning them, or sometimes poisoning them. Other experts argue that mass culling of stray dogs is ineffectual, and instead culls should be focused only on dogs that carry rabies (Qiang *et al.*, 2012).

While dog culls are often rationalized as a way to help control or prevent future rabies outbreaks, a number of animal rights activists argue that the majority of culls have little to do with rabies control. Many of these culls, they argue, occur in urban areas where the risk of rabies transmission is comparatively low. Specifically, the culls most often appear in gentrifying areas of a city, particularly when a 'beautifying campaign' has been launched, suggesting that the culls have more to do with image and economic gain than rabies control. Furthermore, activists have pointed to the timing of these culls as another clue that they are more about international appearance than disease eradication. Culls, they are quick to point out, are common before important international or regional events where an extensive amount of news coverage is expected.

16.6 Vaccination

Activists, public health experts and government officials have long recognized that canine vaccination is probably the most effective method of rabies control. However, the vaccination rate, especially in rural areas, remains relatively low, falling far below the estimated 70% vaccination rate required to reach herd immunity and the zero-rabies goal (Yin *et al.*, 2021). There are a number of challenges to increasing the vaccination rate in China. Prior to 2021, there were no national regulations for the management of dogs. As a result, some places, particularly major cities, had strict management laws focusing on things such as vaccinations and leash laws, but other areas had little or no management requirements. A revised Animal Epidemic Prevention Law of the People's Republic of China was passed in 2021, which helped to unify requirements such as rabies vaccinations. While this law is an important step towards the goal of rabies elimination, the actual implementation of the law continues to face a number of challenges, particularly in terms of enforcement, lack of access to the rabies vaccine, and lack of veterinarian training.

Chinese animal rights activists have long pointed to the unpredictable enforcement of animal quarantine laws meant to help control zoonotic diseases. In 2016, some activists initiated a social media campaign where different form letters (i.e. standardized letters written from a template) were sent to various party officials, pointing out the different problems that occur when anti-epidemic and food safety laws are not followed, and urging them to make the sale of dog meat illegal. One such letter, sent to a high-ranking provincial official, specifically referenced the danger of rabies:

> The formal inclusion of the rabies disease prevention quarantine system of dogs and cats is

> in name only. Guangdong and Guangxi are the main importers of dogs and cats, it is also the area with the highest prevalence of rabies in the whole country, which is closely linked to both the warm and humid climate in the south and the illegal operation of dog and cat smuggling.
>
> (Anonymous poster's comment on WeChat, 2016)

Guangdong Province did have rabies prevention laws prior to 2021, but it was frequently discussed that local police either did not consider the issue serious enough to enforce or were being paid off to look the other way. Vaccination laws for companion dogs were also very rarely enforced in cities, and almost never in rural areas.

The challenge of raising the vaccination rate goes beyond lack of enforcement, as people living in rural areas of the country where rabies is the most prevalent often do not have a reliable way to vaccinate their dogs. This can be due to there being no veterinarians in the area, the high cost of the vaccines, or a combination of both. One veterinarian working with an animal rights group in Sichuan Province explained:

> We go out into the local areas to alert them to vaccinate their pets, we don't at the moment provide a clinic for local people just because we don't have the resources yet ... We just need some more money and more staff ... People do see the benefits of the vaccines. Their pets aren't going to get sick, and they [the owners] aren't going to get sick [with a zoonotic disease]. Especially with rabies being a huge problem in China.
>
> (Anonymous veterinarian, Sichuan Province, 2016, personal communication)

In this particular instance, while the organization was able to educate people on the importance of vaccinations, the lack of money and trained staff prevented them from carrying out what they hoped would be the next step: rural vaccination clinics. Some groups, such as the NGO ACTAsia, have been successful in working with local government officials and veterinarians to offer free vaccinations in high-risk areas that normally do not have access to these services (ACTAsia, 2022). A particularly important component of ACTAsia's programing is their focus upon educating local veterinarians and working with local government officials, thus allowing for broader understandings of the importance of vaccination campaigns with an understanding of the regional conditions. While programmes like these are still limited, they offer a compelling blueprint of what is possible.

16.7 Conclusion

We argue that three key aspects limit the effectiveness of rabies control methods in China: (i) the role that wild species potentially play in creating 'rabies ecologies'; (ii) the regional variation of these ecologies that can render generic national programming difficult to implement; and (iii) the problems associated with inconsistent governance and accurate reporting. Chinese rabies control methods are focused upon dog management, with the majority of the work centred around reducing stray dog populations, usually through dog culls, and increasing vaccination rates. Some of these methods, particularly in the case of large-scale dog culls, have been met with resistance by the animal rights movement and some experts who argue this method is ineffective and often appears to be connected to other motives. While experts and activists alike agree that vaccinations and desexing are a better long-term solution for rabies control, countrywide laws and campaigns requiring vaccination and registration of dogs have proved to be difficult to implement in the areas most prone to rabies breakouts. However, when animal protection organizations have been able to work closely with local governments and veterinarians to create programming designed for specific regional conditions, they have proved to be successful. These campaigns are still small in number and size, but they offer a compelling example of what is possible in rural areas which have traditionally been unable to access vaccination and education campaigns.

Authors' Declaration

All authors declare they have no conflict of interest.

All authors have approved this manuscript, agree with its submission, and share collective responsibility and accountability.

This manuscript has not been published or is not under review elsewhere.

References

ACTAsia (2022) How ACTAsia is helping to combat rabies in animals, adults, and children. Available at: https://www.actasia.org/news/how-actasia-is-helping-to-combat-rabies-in-animals-adults-and-children/ (accessed 3 January 2023).

Barber, S. (2018) Persistence is power: modes of action in the Chinese animal protection movement. PhD thesis, Indiana University, West Lafayette, Indiana.

Barber, S. and Hathaway, M. (2022) China's pet activists: using moral arguments and epidemic concerns to make space for animal rights. *International Review of Environmental History* 8(1), 65–82.

China Daily (2009) One-dog rule in effect. Available at: http://www.chinadaily.com.cn/bw/2009-07/20/content_8447345.htm (accessed 6 February 2023).

Dimonaco, N.J., Salavati, M. and Shih, B.B. (2020) Computational analysis of SARS-CoV-2 and SARS-like coronavirus diversity in human, bat and pangolin populations. *Viruses* 13(1), 49. DOI: 10.3390/v13010049.

Dionne, K.Y. and Turkmen, F.F. (2020) The politics of pandemic othering: putting COVID-19 in global and historical context. *International Organization* 74(S1), E213–E230. DOI: 10.1017/S0020818320000405.

Enserink, M. (2020) Coronavirus rips through Dutch mink farms, triggering culls. *Science* 368(6496), 1169. DOI: 10.1126/science.368.6496.1169.

Feng, Y., Ma, J., Sun, S., Chi, L., Kou, Z. *et al.* (2021) Epidemiology of animal rabies – China, 2010–2020. *China CDC Weekly* 3(39), 815–818. DOI: 10.46234/ccdcw2021.202.

International Center for Veterinary Services (2022) Spay/Neuter Sponsorship for Stray/Rescued/Shelter Dogs and Cats. Available at: http://www.icvsasia.com/index.php?option=com_content&view=category&layout=blog&id=62&Itemid=99&lang=en (accessed 6 February 2023).

Liu, Y., Zhang, H.-P., Zhang, S.-F., Wang, J.-X., Zhou, H.-N. *et al.* (2016) Rabies outbreaks and vaccination in domestic camels and cattle in Northwest China. *PLoS Neglected Tropical Diseases* 10(9), e0004890. DOI: 10.1371/journal.pntd.0004890.

Miao, F., Li, N., Yang, J., Chen, T., Liu, Y. *et al.* (2021) Neglected challenges in the control of animal rabies in China. *One Health* 12, 100212. DOI: 10.1016/j.onehlt.2021.100212.

Patranobis, S. (2018) China tight-lipped about recall of fake rabies vaccines exported to India. *Hindustan Times*. Available at: https://www.hindustantimes.com/world-news/china-tight-lipped-about-recall-of-fake-rabies-vaccines-exported-to-india/story-ZD7opnULIBNJ3HvkYUUK0M.html (accessed 3 January 2023).

Qiang, H., Jin, Z. and Ruan, S. (2012) Dynamics of rabies epidemics and the impact of control efforts in Guangdong Province, China. *Journal of Theoretical Biology* 300, 39–47. DOI: 10.1016/j.jtbi.2012.01.006.

Reuters (2009) China city kills 36,000 dogs after rabies deaths. *Reuters*. Available at: https://www.reuters.com/article/idUSPEK35560 (accessed 6 February 2023).

Velasco-Villa, A., Reeder, S.A., Orciari, L.A., Yager, P.A., Franka, R. *et al.* (2008) Enzootic rabies elimination from dogs and reemergence in wild terrestrial carnivores, United States. *Emerging Infectious Diseases* 14(12), 1849–1854. DOI: 10.3201/eid1412.080876.

Velasco-Villa, A., Escobar, L.E., Sanchez, A., Shi, M., Streicker, D.G. *et al.* (2017) Successful strategies implemented towards the elimination of canine rabies in the western hemisphere. *Antiviral Research* 143, 1–12. DOI: 10.1016/j.antiviral.2017.03.023.

Watts, J. (2006) Chinese county culls 50,000 dogs in crackdown on rabies. *The Guardian,* 2 August.

Wu, X., Hu, R., Zhang, Y., Dong, G. and Rupprecht, C.E. (2009) Reemerging rabies and lack of systemic surveillance in People's Republic of China. *Emerging Infectious Diseases* 15(8), 1159–1164. DOI: 10.3201/eid1508.081426.

Yin, W., Fu, Z.F. and Gao, G.F. (2021) Progress and prospects of dog-mediated rabies elimination in China. *China CDC Weekly* 3(39), 831–834. DOI: 10.46234/ccdcw2021.205.

Zhou, H., Vong, S., Liu, K., Li, Y., Mu, D., *et al.* (2016) Human rabies in China, 1960-2014: a descriptive epidemiological study. *PLoS Neglected Tropical Diseases* 10(8), 1–12. DOI: 10.1371/journal.pntd.0004874.

17 Comparison of the Different Brain Collection Techniques and Evaluation of Mixed Brain Tissues as a Specimen for Rabies Diagnosis

Daria Llenaresas Manalo* and Ma. Ricci Resurreccion Gomez
Research Institute for Tropical Medicine, Muntinlupa City, Philippines

Abstract

The opening of the skull is the standard method used for the collection of brain specimens for rabies diagnosis. It requires several tools, strict biosafety protocols, intensive staff training, and an exclusive dissection room. Three alternative methods of brain collection (i.e. the straw method, the hook method and the throat approach) were compared to the standard technique using criteria based on immunofluorescence readings and dissection procedures. The direct fluorescent antibody test (dFAT) results of the mixed brain tissues, collected using the alternative methods, had no significant difference as compared with the brain stem collected using the standard method (P-value < 0.001). However, the dissection variables of the alternative methods differ significantly from the standard technique. Brain collection techniques using the straw and hook method are less hazardous, cheaper, faster and easier to perform. Application of these techniques in animal and human diagnosis can strengthen the One Health approach to human rabies elimination.

17.1 Introduction

The Philippines is one of the South-east Asian countries where rabies is endemic, causing a major health and economic issue. There is a 462% increase in the number of animal-bite cases reported in the country from 2009 (206,253 bite cases) to 2018 (1,159,711 bite cases) while a 13.5% increase was reported in clinically diagnosed positive human cases based on the recorded clinical case at 243 in 2009 and 276 in 2018 (DOH *et al.*, 2019, p. 16).

The zoonotic nature of rabies complicates its prevention and control, challenging both the animal and the human health sectors. The global consortium on rabies emphasized the One Health approach, encompassing collaboration between the veterinary and medical public health departments with programmes that focus on the strategies for controlling animal-to-human rabies transmission and outbreak prevention (Wallace *et al.*, 2017, p. 2; Acharya *et al.*, 2020, p. 5). In the Philippines, the Anti-Rabies Act of 2007 was created to strengthen rabies control and elimination efforts in the country through the One Health approach with the goal of zero human deaths from dog-mediated rabies by 2030 (DOH *et al.*, 2019, pp. 17–18).

Accurate epidemiologic data are significant in streamlining rabies control and prevention.

*Corresponding author: darsky1566@gmail.com

 One Health for Dog-mediated Rabies Elimination in Asia (eds V. Slack *et al.*)
DOI: 10.1079/9781800622975.0017

Hence the scarcity of prevalence and incidence data poses a significant challenge. Assessment of control and prevention of the disease requires support from locally equipped laboratory services with reliable, affordable techniques that are easy to perform. Increasing surveillance data for both human and animal rabies cases starts with sample collection (Acharya *et al.*, 2020, p. 5; Haselbeck *et al.*, 2021).

The standard method of brain collection is the opening of the skull. This method extracts the whole brain and specifically collects the medulla or brain stem, Ammon's horn or hippocampus, and cerebellum, all of which are recommended tissues for rabies diagnosis (Velleca and Forrester, 1981, pp. 24–30; Rupprecht *et al.*, 2018, pp. 68–69; WOAH, 2018, p. 580). Because of the hazards associated with specimen collection, safety is of paramount importance when working with rabies. All persons involved in rabies testing should receive pre-exposure immunization with regular serological tests and booster immunizations as necessary (Kaplan, 1996, pp. 3–7; Rupprecht *et al.*, 2018, pp. 31, 67; WHO, 2018, pp. 57–58). The collection of brain samples requires training and a separate room designated exclusively for the purpose. In general, appropriate biosafety practices are required to avoid direct contact with potentially infected tissues or fluids (Velleca and Forrester, 1981, pp. 5–11; Barrat, 1996, p. 425; Rupprecht *et al.*, 2018, pp. 28–31). However, some of the existing rabies laboratories in the Philippines do not have a special room for extracting brain tissues. In most cases, the rabies dissection room is the same room used for other procedures aside from brain collection. In addition, the opening of the skull method is very dangerous and hazardous if the technicians are not fully trained (Shankar, 2009, p. 75). Some of these hazards include accidental injuries from possible rabies-contaminated blades or sharp edges of the skull (Iamamoto *et al.*, 2011, p. 30; Rupprecht *et al.*, 2018, pp. 28–31). Moreover, the standard procedure of opening the skull to collect brain samples is a difficult task to perform by other staff who are not properly trained and not used to performing the procedures.

Alternative methods of brain tissue collection such as the straw method through the occipital route and retro-orbital route have already been endorsed by the World Health Organization (WHO) and by the World Organization for Animal Health (WOAH), especially for their application in the field setting. These procedures simply require minimum barrier protection such as safety glasses, gloves and masks since the brain tissues are collected through a small incision only (Hirose *et al.*, 1991, p. 291; Barrat, 1996, pp. 425–428; WHO, 2018, pp. 67–68; WOAH, 2018, p. 580). Moreover, these alternative methods can lessen biological hazards and can be used in basic field conditions or in a simple benchtop laboratory with appropriate biosafety practices to avoid direct contact with potentially infected tissues or fluids. Despite the ease and safety, these alternative methods have not been implemented in the Philippines, most likely due to lack of specific materials, skills and a quality assurance system. There is currently no quantitative comparison between these alternative techniques and the standard method.

Thus, this study aimed to evaluate three alternative brain collection techniques, including the drinking straw method, the hook method and the throat approach, by comparing them to the standard method. The quality of the mixed brain tissue for rabies diagnosis by direct fluorescent antibody test (dFAT) was also evaluated.

17.2 What We Did

17.2.1 Specimen and data collection

Two hundred and one (201) dog and cat head specimens submitted to the rabies laboratory of the Research Institute for Tropical Medicine (RITM), Department of Health, Philippines from 2016 to 2018 for routine rabies diagnosis were included in this study. The submitted specimens were suspected rabid animals that had shown signs of rabies, had been involved in biting cases, and had simply died for no apparent reasons. The people submitting the specimen were the owners of the dogs or, in the case of unowned dogs, bite victims or relatives of the victims, or local government officials. These people also provided information through a questionnaire (rabies diagnosis form)-guided interview. The form included questions on the health history, current health status, and demographic data (age, sex, specific address) of the biting animal. The reason behind the biting incident and the treatment received by the bite victim was also recorded. If the bite victim was not present during the submission of the rabies sample, he or she was contacted and interviewed via phone. The data collected was then encoded using FileMaker Pro Advanced.

The 201 samples included in this study were equally divided into the three alternative collection techniques; therefore, yielding a total of 67 specimens for each alternative method (straw, hook and throat). After collecting the mixed brain tissues using the assigned alternative method, the skull was opened to collect the anatomical brain tissues using the standard method. This study only used the brain stem, which is reported to contain the highest load of viral antigen (Bingham and van der Merwe, 2002, p. 93; Beck *et al.*, 2017, pp. 6–7) as the standard reference material.

17.2.2 Dissection procedures

All the brain samples included in this study were collected using two dissection procedures in a designated dissection laboratory. The first method used was a pre-assigned alternative, less-invasive method of brain collection, followed by the second procedure, which is the standard method. After dissection, the technician accomplished an Outcome Measure Form, which contained a scoring system designed to objectively assess the performance of each procedure for every sample processed. A 'buddy system' was practised while performing the dissection procedure wherein one was the prosector and the other person assisted and recorded the data. The dissection methods were scored using the following criteria: (i) cost, defined as the price of tools used for each procedure; (ii) time, the length in minutes of every procedure; (iii) biosafety consideration, the required personal protective equipment (PPE) and room for biosafety; and (iv) the difficulty of the procedure, defined as the number of tools needed to get the specimen. The score given to each collection parameter decreases as the procedure becomes more costly, more time-consuming, or requires more PPE and tools for sample collection as presented in Table 17.1.

17.2.2.1 Collection using a drinking straw

The brain samples were collected following the procedure described by Barrat (1996, pp. 425–432) with few modifications. Briefly, the animal was placed in a lying position, face down. The head was bent to expose the occipital region while palpating for the atlanto-occipital joint or the joint between the C0 and C1 vertebra. Just above the joint, a small incision (about 5 cm) was made to the skin following the muscles as far as the vertebrae behind the atlanto-occipital protuberance. The atlanto-occipital joint was opened by cutting the dorsal atlanto-occipital membrane and the meninges exposing the occipital foramen. The samples were obtained by inserting the straw into the occipital foramen with a slight twisting movement towards one of the eyes (see Fig. 17.1a). Depending on the size of the animal, around 5–10 cm (2–4 inches) of the straw was inserted into the head, or until resistance was felt. The straw was pinched upon removal and the collected sample of the brain tissue, comprising a mixture of the medulla oblongata, the base of the cerebellum, the hippocampus and the cerebral cortex, was transferred into a container using an applicator stick.

17.2.2.2 Collection using a hook

A small incision was made in the same way as the previous method. A hook (a long slender bar with a scoop at the end) was inserted into the occipital foramen towards the direction of the eye (see Fig. 17.1b), and the mixed brain tissue was scooped and placed in a sterile container.

17.2.2.3 Brain collection via the throat

The animal head was laid on the table in a dorsal position and a midline incision was made from the throatlatch region. The musculature attachments of the tongue were severed towards the nose until beyond the throat to free the larynx, trachea and oesophagus. The muscles were retracted to expose the throat, the surface of the spinal column, and associated muscles. Upon identification of the atlanto-occipital joint, the connective tissue overlying the spinal

Table 17.1. Scoring criteria for the dissection procedure of rabies samples. Created by the authors.

Parameter	Score	Description[a]
Cost	3	The method required minimum PPE and one to two dissection tools
	2	The method required moderate PPE and three dissection tools
	1	The method required full PPE and several (more than four) dissection tools
Time	4	The method was performed in less than 1 min per sample
	3	The method was performed in 1–3 min per sample
	2	The method was performed in 4–10 min per sample
	1	The method was performed in more than 10 min per sample
Biosafety consideration	3	Laboratory shoes, scrub suits, gloves, hair cap, mask and goggles
	2	Laboratory shoes, scrub suits, gloves, hair cap, mask, goggles and laboratory gown
	1	Scrub suits, gloves, hair cap, mask, face shield, laboratory gown, waterproof apron, hair cap and boots
Difficulty of the procedure	3	The method required one to two tools to collect the specimen
	2	The method required three tools and a slight force to collect the specimen
	1	The method required four to five tools and a force to collect the specimen

[a]PPE, personal protective equipment.

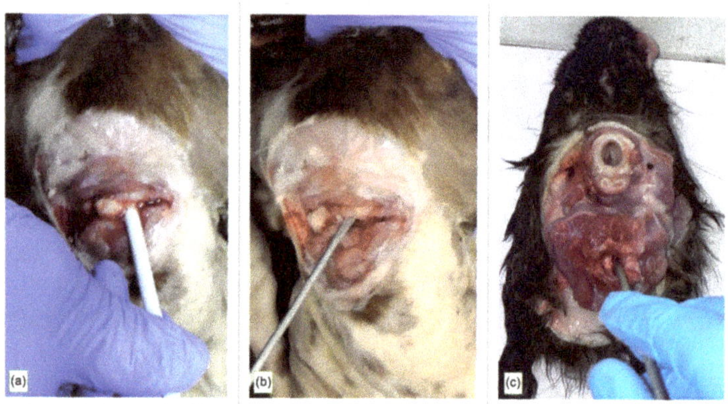

Fig. 17.1. The three alternative methods of brain sample collection showing the location where the tool was inserted using (a) the straw method; (b) the hook method; and (c) collection via throat or ventral approach. Photographs (a–c) are copyright of Daria Manalo.

cord was carefully cut with a scalpel to expose the brainstem (see Fig. 17.1c). From here, the mixed brain sample was scooped using a hook for rabies testing.

17.2.2.4 Opening of the skull (standard method)

After extracting the samples using the alternative method, the brain is collected using the standard method (Lepine and Atanasiu, 1996, pp. 68–74). The ventral side of the head was laid on the working area and a midline cut was made through the skin, the connective tissue, and the muscle on top of the head. A cut was made from behind the eyes to the animal's back of the neck. The skin was peeled just above the level of the ears and was carefully folded away to reduce the amount of hair that might contaminate the specimen. The remaining muscle tissue was dissected along the lines where cuts in the bone were made. Consequently, cuts through the skull were made: (i) behind the eye sockets; (ii) laterally just above both ears; and (iii) at the back of the skull just above the area where the spinal cord enters the cranial cavity. After making a clean cut through the skull, the skullcap was removed, and the brain was freed from the covering membrane. The hippocampus, cerebellum and brain stem were identified and collected for routine diagnosis. However, only the brain stem was used as a specimen for this study.

17.2.3 Touch impression and fixation

From the tissue samples obtained, a small section at approximately 0.5 cm^2 of the mixed brain tissue from alternative methods and brain stem from the standard method was used for touch impression. Two glass slides for each sample type were prepared. For each slide, duplicate touch impressions of the brain tissue samples were made. The touch impressions were air-dried for 5 min at 27°C and were fixed with cold acetone for 16–18 hours.

17.2.4 Rabies testing by direct fluorescent antibody test (dFAT)

The brain touch impressions were processed for dFAT as described by Dean *et al.* (1996, pp. 88–95), which took about 2 hours using fluorescein isothiocyanate-labelled antibody (FITC)-conjugated Anti-Rabies Monoclonal Globulin (Fujirebio Diagnostic Inc.). To determine the quality of the obtained mixed brain samples for dFAT, the result was compared to that of the brain stem, which is the standard brain sample. The scores of the dFAT reading were recorded by the technician in the outcome measure form. A score of '0' was given if the dFAT result of the mixed brain tissues was in contrast with the dFAT results of the standard method and '55' was given if the results were similar. Moreover, the presence or absence of non-specific staining, fluorescence staining intensity, distribution of rabies virus antigen, presence of dust-like particles, and presence of large oval masses were graded for each sample following the criteria in Table 17.2. The scoring used in this study was devised by the authors based on the Centers for Disease Control and Prevention (CDC) *Minimum Standard for Rabies Diagnosis* (CDC, n.d.; Rupprecht *et al.*, 2018, pp. 116–118).

17.2.5 Data management

All data collected were encoded using FileMaker Pro Advanced. Demographic data were analysed using descriptive statistics. The four methods of brain collection techniques were described using mean and standard deviation. A paired *t*-test was used to compare the immunofluorescence readings and dissection scores of each of the three techniques (drinking straw, hook method and throat approach collection) with the standard procedure. A *P*-value of less than 0.05 was considered to be statistically significant.

Table 17.2. Scoring criteria for the direct fluorescent antibody test (dFAT) reading of rabies samples. Created by the authors.

Parameter	Score	Description
Non-specific staining	4	Absence of non-specific staining
	0	Presence of non-specific staining
Staining intensity of rabies antigen	10	Glaring apple-green fluorescence
	8	Bright apple green fluorescence
	6	Dull apple-green fluorescence
	4	Very dim but detectable apple-green fluorescence
	0	No fluorescence was observed (negative)
Distribution of rabies antigen	10	Large, medium and dust-like fluorescing particles profusely distributed in all fields of the smear
	8	Medium and dust-like fluorescing particles dispersed in each field of the smear
	6	Small and dust-like fluorescing particles dispersed in each field of the smear (10–50% of the microscopic fields)
	4	Small and dust-like fluorescing particles present only in 10% of the smear
	0	No particles observed (negative)
Dust-like particles	4	Distributed in all fields of the smear
	3	Present in 60–90% of microscopic fields
	2	Present in 10–50% of the microscopic fields
	1	Present in only 10% of the smear
	0	No particles observed
Large, oval masses	4	Distributed in all fields of the smear, antigen present in all fields
	3	Present in 60–90% of microscopic fields
	2	Present in 10–50% of the microscopic fields
	1	Present in only 10% of the smear
	0	No masses observed

17.3 What We Discovered

17.3.1 Animal demographic data

The laboratory received 180 dog and 21 cat head samples for the study. Decomposed or autolysed brain samples that were unfit for rabies dFAT were not included. From these collected specimens, 85 were dFAT-positive for rabies, and 116 were negative. All the positive cases were dogs (100%). The cases were mostly males (52.2%), under the age of 1 year (44.3%), and owned but free roaming (44.8%). The animals were observed to be sick 2 weeks before death (68.7%) and 13.4% of these were euthanized while 85.1% were found dead. Moreover, 58.2% of the submitted animals were not vaccinated against rabies. In some cases, incomplete data was provided by the bearer, especially for stray animals where age, vaccination status and health status were unknown or unobserved.

Samples mostly came from the neighbouring provinces of Muntinlupa City, where RITM is located, particularly from Metro Manila, Region 4A, and Region 4B. Human exposure was recorded in all submitted specimens, through a bite, scratch or licking of intact skin. Victims of the rabies-positive animals either had received or were scheduled to receive post-exposure prophylaxis (PEP) at the time of the interview.

17.3.2 Alternative methods versus standard method

A comparison of each of the three alternative methods to the standard technique showed no false negative or false positive dFAT result. This means that the mixed brain sample was in 100% agreement when tested with dFAT. A more in-depth examination of the immunofluorescence readings showed that the mean score of the samples collected by the standard method was higher than the three alternative methods except for the absence of non-specific staining and staining intensity. On the other hand, the dissection procedure parameters were statistically different between the three alternative methods and the standard method. In particular, the straw method had a higher mean score in cost, biosafety consideration, and difficulty parameters while the hook method had a slightly higher mean score in the time variable (see Table 17.3).

17.3.3 Comparison between the three alternative methods

Analysis of the collection techniques based on the outcome measures is shown in Table 17.3. The immunofluorescence readings represented the reliability of the brain tissues for dFAT. Rabies virus in suspected rabid animals produces intra-cytoplasmic inclusions of various shapes. A single microscope field may contain dust-like particles of <1 μm in diameter, round to oval masses, and strings 2–10 μm in diameter. Results showed no significant difference in the value in dFAT readings: the presence of non-specific staining, intensity, dust particles, and oval mass in all the collection techniques (standard, straw, throat or ventral approach, and hook) (see Fig. 17.2). A significant difference in the distribution of rabies antigen was observed in the drinking straw (P = 0.006) and hook method (P = 0.010) when compared to the standard method (see Table 17.3). The dissection variables including cost, time, biosafety consideration, and difficulty of the procedure, were statistically different among the four collection techniques with P-values of less than 0.001 (see Fig. 17.3). The straw technique had the highest mean score among the alternative collection techniques in terms of cost, biosafety procedure, and difficulty of the procedure, whereas the hook technique had the highest mean score for time. The data suggest that the hook method was the fastest to perform as compared to the other three methods of collection. Evaluating the overall outcome of the three alternative brain collection techniques shows that the straw method had the highest mean score while the throat or ventral approach got the lowest mean score (see Table 17.3).

17.4 Lessons Learned

Of the three alternative methods presented in this study, the straw procedure was found to be the cheapest and safest alternative method used. This procedure applies suction to obtain a small sample from the brain until transferred to the specimen container, minimizing the exposure of the brain tissue in the open environment. The hook method, which avoids the need for applied pressure to get the sample, was found to be the fastest and easiest to perform. The samples in both hook and straw methods were obtained through the atlanto-occipital joint wherein the neck muscles, dorsal membrane and meninges were cut open. In contrast, collection through the throat required a bigger opening and involved cutting off the tougher connective tissue. This method showed the lowest mean score of the three alternative methods. This disparity in the anatomical entry point may also have been attributed to the difference in rabies antigen distribution readings in samples collected by the straw and hook. Tissue samples collected from hook and straw were from different parts of the brain (including the brainstem, base of the cerebellum, hippocampus, thalamus and frontal cortex), while brain tissue collected through the throat was more homogenous.

The alternative methods of brain collection are cheaper because these procedures only need minimal tools and PPE to perform, unlike the

Table 17.3. Comparison of scores for each alternative diagnosis technique with the dFAT standard method. [a]Created by the authors.

Parameters	Standard ($n = 67$)		Drinking straw ($n = 67$)			Standard ($n = 67$)		Ventral approach ($n = 67$)			Standard ($n = 67$)		Hook method ($n = 67$)		
	Mean	SD	Mean	SD	*P*-value	Mean	SD	Mean	SD	*P*-value	Mean	SD	Mean	SD	*P*-value
dFAT results	55	0	55	0	NA	55	0	55	0	NA	55	0	55	0	NA
Non-specific staining	3.34	1.49	3.40	1.44	0.321	3.34	1.49	3.46	1.37	1.000	3.52	1.31	3.46	1.37	0.321
Staining intensity of rabies antigen	9.82	0.67	9.91	0.42	0.182	9.85	0.67	9.82	0.58	0.709	9.97	0.24	9.82	0.67	0.058
Distribution of rabies antigen	9.82	0.67	9.46	1.02	0.006*	9.64	0.67	9.37	1.17	0.095	9.88	0.77	9.52	1.21	0.010*
Dust-like particles	3.84	0.45	3.73	0.64	0.128	3.79	0.45	3.82	0.49	0.727	3.84	0.54	3.79	0.62	0.410
Large, oval masses	3.25	1.16	3.27	1.24	0.859	3.36	1.16	3.45	1.20	0.484	3.54	0.99	3.45	1.05	0.358
Cost	1.60	0.55	2.72	0.52	< 0.001*	1.46	0.66	1.93	0.61	< 0.001*	1.69	0.58	2.54	0.53	< 0.001*
Time	2.67	1.24	3.78	0.60	< 0.001*	3.04	0.93	3.60	0.72	< 0.001*	3.13	0.83	3.79	0.45	< 0.001*
Biosafety consideration	1.00	0.00	1.94	0.85	< 0.001*	1.04	0.37	1.28	0.45	< 0.001*	1.04	0.27	1.66	0.75	< 0.001*
Difficulty of the procedure	1.58	0.63	2.64	0.62	< 0.001*	1.42	0.65	1.78	0.69	0.002*	1.66	0.64	2.46	0.66	< 0.001*
Total	91.93	3.09	95.85	3.07	< 0.001*	92.07	3.14	93.33	3.25	< 0.001*	93.27	2.49	95.49	3.43	< 0.001*

*$P < 0.05$ statistically significant.
[a]NA, not applicable; SD, standard deviation.

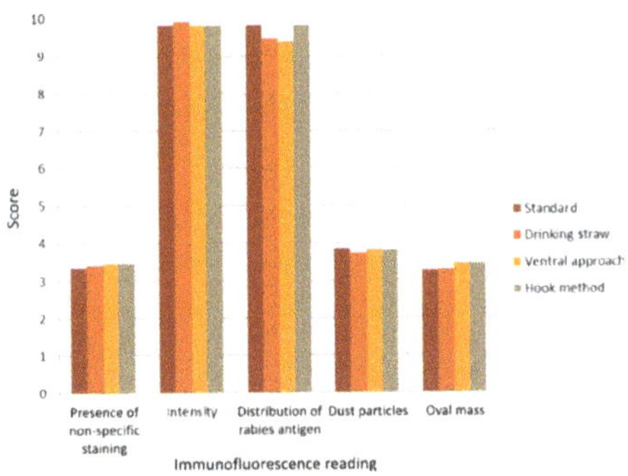

Fig. 17.2. Comparison of three alternative methods versus standard collection technique in terms of immunofluorescence reading. Higher scores denote better suitability of samples for the direct fluorescent antibody test (dFAT). Created by the authors.

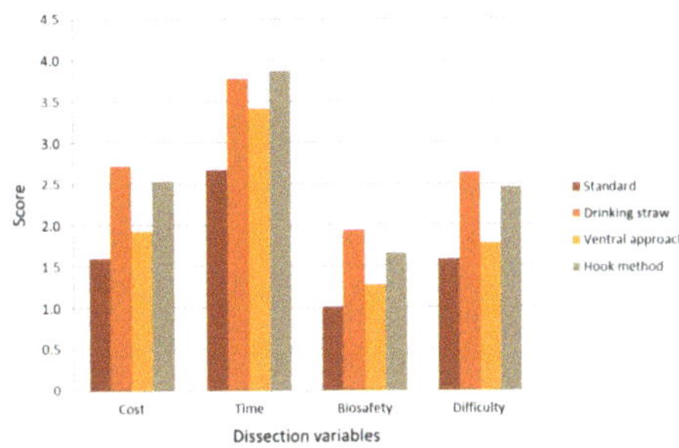

Fig. 17.3. Comparison of three alternative methods versus standard collection technique in terms of dissection. A high score means the dissection procedure is cheaper, faster, safer and easier to perform. Created by the authors.

standard method which requires several tools such as a butcher's knife, a rubber mallet and full PPE to open the skull (Lepine and Atanasiu, 1996, pp. 68–69; Rupprecht *et al.*, 2018, pp. 28–29). Tools that are used in rabies dissection also need to be disinfected and sometimes disinfection may not work, which can cause cross-contamination (Aiello *et al.*, 2016, p. 76). Considering all the materials that are not required in these techniques, this study has estimated a total of US$20–30 can be saved per specimen for rabies diagnosis. This is a major advantage when considering financial constraints in low-income countries like the Philippines (Acharya *et al.*, 2020, p. 5). Moreover, the alternative methods are faster and easier to perform than the standard method. In the hook and straw methods, the shortest time is less than 5 s to collect the brain (data not shown) unlike in the standard method where it takes a minimum of 5 min. However, the throat approach for collecting the brain tissues is slightly difficult because several organs and tissues need to be severed before reaching the atlanto-occipital joint.

WOAH and WHO have recommended the straw method of sample collection for more than 25 years. However, as a paradox, the application of this technique has not been widely practised, probably due to a lack of knowledge, training and a quality assurance system. The goal of rabies elimination has pushed the need for rapid and effective new techniques that can be used for initial screening in rural areas and to help ease the burden of testing in central laboratories. Studies in India of the samples collected using the straw method and the retro-orbital route have shown remarkable results when tested with lateral flow assays (Gandhimathy and Asokkumar, 2017; Yale *et al.*, 2019). Brain collection through the foramen magnum using a hook or brain scoop was studied in Sri Lanka, although the applicability of the collected sample in laboratory diagnosis was not investigated (Gunawardena, 2007, pp. 155–156). Similarly, a sample spoon was also used to collect brain tissue in a study conducted in Chad that yielded high diagnostic accuracy when tested with lateral flow devices (Naïssengar *et al.*, 2021, p. 221). In the Philippines, a recent study on rabies diagnosis by lateral flow assay using samples collected by the straw method showed promising results, with 94.3% sensitivity and 100% specificity (Mananggit *et al.*, 2021, pp. 1–6).

The results of this study showed a 100% positive predictive value 100% of the time, but there could be severe consequences if an alternative method did not correctly identify a positive result in a positive specimen because different sections of the brain were not collected. As with any diagnostic procedure, collecting the history of the biting animal through an interview with the sample bearer is important to support the laboratory diagnosis of rabies. The data gathered here can provide critical information, especially when laboratory results are uncertain and need to be repeated or when an alternative diagnostic test needs to be performed (WHO, 2018, pp. 78–79).

Several disadvantages when using alternative methods can also be noted. In the hook and straw methods, there are times when the incision of the skin through the muscles is not on top of the atlanto-occipital joint, making it difficult to locate the joint, thus, the procedure could take longer, and the brain tissue collection is harder. Identification of the specific parts of brain material is also impossible when using the alternative methods therefore these techniques are not applicable if the study will need to focus on specific parts of the brain. Moreover, there are losses of brain tissue collected via the alternative methods which account for the differences in the rabies antigen distribution in the alternative techniques versus the standard technique (Iamamoto *et al.*, 2011, p. 30). In addition, there are times that brain tissue is covered with blood and requires blotting before performing touch impressions. A laboratory may also choose not to use the alternative methods because of the unavailability of tools (hook and straw) in their area or when other brain parts are required for other diagnostic laboratory procedures.

Control in animal populations is ultimately the most important part of prevention. Mass parenteral vaccination of dogs and cats is still the most important measure for controlling animal rabies, especially in dogs. Aside from mass vaccination, epidemiologic surveillance and dog population control are still basic elements in the control programme (WHO, 2018, pp. 78–124). The majority of rabies cases in the Philippines are unreported, particularly in rural areas. To achieve the country's goal of 'Zero human rabies deaths by 2030', a definitive laboratory-based diagnosis must be implemented. Specimen collection techniques shown in this study are also applicable to human rabies diagnosis especially since there are societal, cultural and religious barriers that may impede the performance of autopsy for brain collection (Hemachudha *et al.*, 2002, p. 101; Smith *et al.*, 2003, p. 150). The straw and hook collection methods have been successfully used in one of the rabies patients at RITM, and the sample was processed for dFAT. Other than the hook and straw method, brain tissue can be obtained by brain biopsy or by needle (e.g. Vim-Silverman or Tru-cut needle) aspiration with entry via orbit or foramen magnum (Tong *et al.*, 1999; Sudarsanam *et al.*, 2008, pp. 163–164; Duong *et al.*, 2016; WHO, 2018, p. 24). A study conducted in Senegal reported a 95.6% positivity in samples aspirated through sub-occipital aspiration and tested by indirect fluorescent staining (Sow *et al.*, 1996).

Rabies diagnosis is a major component of the surveillance of the disease especially if the country is aiming for elimination. It is also crucial for the detection of imported cases or the emergence of rabies in non-endemic areas. In endemic areas, rabies diagnosis is essential to guide surveillance systems as well as the management of animal-bite exposures by helping physicians. It also provides reliable, laboratory-confirmed data to estimate the burden of the disease and prioritize resources towards its control (Duong *et al.*, 2016, pp. 107–114; WHO, 2018, pp.21–22, 80–81). As a result, in the years leading up to 2030, countries aiming for rabies eradication must establish and strengthen laboratory diagnosis, particularly in rural areas where most rabies cases are underdiagnosed.

17.5 Conclusion

To end dog-mediated human rabies deaths by 2030, a definitive laboratory diagnosis is essential. Laboratories must be equipped with techniques that can be used even with limited resources. The alternative methods of sample collection were significantly cheaper, easy to perform, less hazardous, and effective in collecting brain samples without sacrificing the integrity of the tissues for testing. This study also shows that mixed brain tissues are appropriate and reliable for dFAT rabies diagnosis, and the straw and hook methods are the best alternative techniques for brain collection.

These recommended techniques should be replicated and validated to confirm our findings before employing these techniques as standard procedures in diagnostic laboratories. In addition, the laboratory diagnosis can be further supported by taking the history of the biting animal and bite victims. A laboratory result can guide physicians in prioritizing PEP administrations, especially in patients bitten by laboratory-confirmed positive animals. Integrating these techniques in human and animal diagnosis through the One Health approach, government-supported training, and policies on rabies diagnosis can eventually lead to the goal of human rabies elimination by 2030.

Acknowledgements

The authors are grateful to the Veterinary Research Department of RITM for the technical support given to this project and Ms Ma. Kristine Joy S. Calvario for the statistical analysis.

This study was funded by the Department of Science and Technology – Philippine Council for Health Research and Development DOST-PCHRD (project no. FP 160013).

Authors' Declaration

All authors declare they have no conflict of interest.

All authors have approved this manuscript, agree with its submission, and share collective responsibility and accountability.

This manuscript has not been published or is not under review elsewhere.

References

Acharya, K.P., Acharya, N., Phuyal, S., Upadhyaya, M. and Lasee, S. (2020) One-health approach: a best possible way to control rabies. *One Health*10, 100161. DOI: 10.1016/j.onehlt.2020.100161.

Aiello, R., Zecchin, B., Tiozzo Caenazzo, S., Cattoli, G. and De Benedictis, P. (2016) Disinfection protocols for necropsy equipment in rabies laboratories: safety of personnel and diagnostic outcome. *Journal of Virological Methods* 234, 75–79. DOI: 10.1016/j.jviromet.2016.03.017.

Barrat, J. (1996) Simple technique for the collection and shipment of brain specimens for rabies diagnosis. In: Meslin, F.X., Kaplan, M.M. and Koprowski, H. (eds) *Laboratory Techniques in Rabies*, 4th edn. World Health Organization, Geneva, Switzerland, pp. 425–432.

Beck, S., Gunawardena, P., Horton, D.L., Hicks, D.J., Marston, D.A. *et al.* (2017) Pathobiological investigation of naturally infected canine rabies cases from Sri Lanka. *BMC Veterinary Research* 13(1), 99. DOI: 10.1186/s12917-017-1024-5.

Bingham, J. and van der Merwe, M. (2002) Distribution of rabies antigen in infected brain material: determining the reliability of different regions of the brain for the rabies fluorescent antibody test. *Journal of Virological Methods* 101(1–2), 85–94. DOI: 10.1016/s0166-0934(01)00423-2.

CDC (Centers for Disease Control and Prevention) (n.d.) Protocol for postmortem diagnosis of rabies in animals by direct fluorescent antibody testing: A minimum standard for rabies diagnosis in the United States. Available at: https://www.cdc.gov/rabies/pdf/rabiesdfaspv2.pdf (accessed 20 September 2022).

Dean, D.J., Albelseth, M.K. and Atanasiu, P. (1996) The fluorescent antibody test. In: Meslin, F.X., Kaplan, M.M. and Koprowski, H. (eds) *Laboratory Techniques in Rabies*, 4th edn. World Health Organization, Geneva, Switzerland, pp. 88–95.

Department of Health (DOH), Department of Agriculture (DA), Department of Education (DepEd) and Department of Interior and Local Government (DILG) (2019) *National Rabies Prevention and Control Program: Manual of Procedures*. Department of Health, Government of the Philippines. Available at: https://doh.gov.ph/sites/default/files/publications/Rabies%20Manual_MOP_2019%20nov28.pdf (accessed 5 December 2022).

Duong, V., Tarantola, A., Ong, S., Mey, C., Choeung, R. *et al.* (2016) Laboratory diagnostics in dog-mediated rabies: an overview of performance and a proposed strategy for various settings. *International Journal of Infectious Diseases* 46, 107–114. DOI: 10.1016/j.ijid.2016.03.016.

Gandhimathy, S. and Asokkumar, M. (2017) Advances in brain specimen collection for rabies diagnosis. *International Journal of Current Microbiology and Applied Sciences* 6(11), 148–151. DOI: 10.20546/ijcmas.2017.611.019.

Gunawardena, G.S.P. de S. (2007) Brain scoop: a simple post-mortem instrument to collect brain samples for diagnosis of rabies. *Proceedings of the Peradeniya University Research Sessions, Sri Lanka* 12(1), 155–156.

Haselbeck, A.H., Rietmann, S., Tadesse, B.T., Kling, K., Kaschubat-Dieudonné, M.E. *et al.* (2021) Challenges to the fight against rabies – the landscape of policy and prevention strategies in Africa. *International Journal of Environmental Research and Public Health* 18(4), 1736. DOI: 10.3390/ijerph18041736.

Hemachudha, T., Laothamatas, J. and Rupprecht, C.E. (2002) Human rabies: a disease of complex neuropathogenetic mechanisms and diagnostic challenges. *Lancet Neurology* 1(2), 101–109. DOI: 10.1016/s1474-4422(02)00041-8.

Hirose, J.A., Bourhy, H. and Sureau, P. (1991) Retro-orbital route for brain specimen collection for rabies diagnosis. *The Veterinary Record* 129(13), 291–292. DOI: 10.1136/vr.129.13.291.

Iamamoto, K., Quadros, J. and Queiroz, L.H. (2011) Use of aspiration method for collecting brain samples for rabies diagnosis in small wild animals. *Zoonoses and Public Health* 58(1), 28–31. DOI: 10.1111/j.1863-2378.2009.01282.x.

Kaplan, M. (1996) Safety precautions in handling rabies virus. In: Meslin, F.X., Kaplan, M.M. and Koprowski, H. (eds) *Laboratory Techniques in Rabies*, 4th edn. World Health Organization, Geneva, Switzerland, pp. 3–8.

Lepine, P. and Atanasiu, P. (1996) Histopathological diagnosis. In: Meslin, F.X., Kaplan, M.M. and Koprowski, H. (eds) *Laboratory Techniques in Rabies*, 4th edn. World Health Organization, Geneva, Switzerland, pp. 66–79.

Mananggit, M.R., Manalo, D.L., Saito, N., Kimitsuki, K., Garcia, A.M.G. *et al.* (2021) Lateral flow devices for samples collected by straw sampling method for postmortem canine rabies diagnosis. *PLoS Neglected Tropical Diseases* 15(12), 6–11. DOI: 10.1371/journal.pntd.0009891.

Naïssengar, K., Oussiguere, A., Madaye, E., Mbaipago, N., Mindekem, R. *et al.* (2021) Challenges to improved animal rabies surveillance: experiences from pilot implementation of decentralized diagnostic units in Chad. *Acta Tropica* 221, 105984. DOI: 10.1016/j.actatropica.2021.105984.

Rupprecht, C.E., Fooks, A.R. and Abela-Ridder, B. (2018) *Laboratory Techniques in Rabies*, 5th edn., Vol. 1. World Health Organization, Geneva, Switzerland. Available at: https://apps.who.int/iris/handle/10665/310836 (accessed 10 September 2021).

Shankar, B.P. (2009) Advances in diagnosis of rabies. *Veterinary World* 2(2), 74–78.

Smith, J., McElhinney, L., Parsons, G., Brink, N., Doherty, T. *et al.* (2003) Case report: rapid ante-mortem diagnosis of a human case of rabies imported into the UK from the Philippines. *Journal of Medical Virology* 69(1), 150–155. DOI: 10.1002/jmv.10253.

Sow, P.S., Diop, B.M., Ndour, C.T., Soumare, M., Ndoye, B. *et al.* (1996) La ponction aspiration sous-occipitale : technique de prelevement cerebral post-mortem pour le diagnostic virologique de l'encéphalite rabique humaine a Dakar. *Medecine et Maladies Infectieuses* 26(5), 534–536. DOI: 10.1016/S0399-077X(96)80237-X.

Sudarsanam, T.D., Chacko, G. and David, R.D. (2008) Postmortem trucut transnasal brain biopsy in the diagnosis of encephalitis. *Tropical Doctor* 38(3), 163–165. DOI: 10.1258/td.2007.070091.

Tong, T.R., Leung, K.M., Lee, K.C. and Lam, A.W.S. (1999) Trucut needle biopsy through superior orbital fissure for the diagnosis of rabies. *Lancet* 354(9196), 2137–2138. DOI: 10.1016/s0140-6736(99)95093-6.

Velleca, W. and Forrester, F. (1981) *Laboratory Methods for Detecting Rabies*. US Government Printing Press Office, Washington, DC.

Wallace, R.M., Undurraga, E.A., Blanton, J.D., Cleaton, J. and Franka, R. (2017) Elimination of dog-mediated human rabies deaths by 2030: needs assessment and alternatives for progress based on dog vaccination. *Frontiers in Veterinary Science* 4, 9. DOI: 10.3389/fvets.2017.00009.

WHO (World Health Organization) (2018) WHO expert consultation on rabies - Third report. *World Health Organization Technical Report Series* 1012, 1–182. Available at: https://apps.who.int/iris/bitstream/handle/10665/272364/9789241210218-eng.pdf (accessed 5 December 2022).

WOAH (World Organisation for Animal Health) (2018) Rabies (infection with rabies virus and other Lyssaviruses). In: *Manual of Diagnostic Tests and Vaccines for Terrestrial Animals*. WOAH, Paris, pp. 578–612. Available at: https://www.oie.int/fileadmin/Home/eng/Health_standards/tahm/3.01.17_RABIES.pdf (accessed 5 December 2022).

Yale, G., Gibson, A.D., Mani, R.S., Harsha, P.K., Costa, N.C. *et al.* (2019) Evaluation of an immunochromatographic assay as a canine rabies surveillance tool in Goa, India. *Viruses* 11(7), 649. DOI: 10.3390/v11070649.

18 High-Throughput Techniques to Understand Evolution and Transmission Trends of Rabies Virus in Asian Countries

Niti B. Jadeja[1] and Abi Tamim Vanak[1,2,3*]
[1]Ashoka Trust for Research in Ecology and the Environment, Bengaluru, India; [2]University of KwaZulu-Natal, Pietermaritzburg, South Africa; [3]DBT-Wellcome Trust India Alliance, Hyderabad, India

Abstract

The 12 kb single-stranded negative-sense rabies virus (RABV) genome exhibits unusual genomic plasticity owing to which this virus often dodges immune surveillance and is highly adaptative in new host animals. Attempts to link genome evolution, transmission and prevalence of RABV have been greatly facilitated by genomic approaches. Previously reported large-scale phylodynamic studies attempting to identify the origin of diverse RABV lineages and spill-over events have been deficient in representation from Asian countries where most human deaths due to rabies are recorded. Advancements in high-throughput DNA sequencing techniques have enabled low-cost and rapid analysis of trends in RABV genome evolution, sustenance and transmission. Here, we discuss the improvements and requirements for genome-based surveillance methods and their limitations. Ultimately, a prevention programme can be more effective if high-risk areas are identified through phylogenetic and spatiotemporal studies of human and animal rabies.

18.1 Introduction

Rabies is caused by a single-stranded, negatively polar genome-bearing species of the *Lyssavirus* genus in the family *Rhabdoviridae*. While a total of 17 *Lyssavirus* species, classified under four phylogroups, have been identified from various hosts, the Rabies *Lyssavirus* remains the species of concern as it is the only species known to infect and persist in multiple host reservoirs through independent transmission cycles (Reddy *et al.*, 2018; Walker *et al.*, 2018). In the past decade, several distinct rabies virus (RABV) strains have been reported based on phylogenetic analysis of genes and RABV genomes. These findings provide growing evidence of RABV genome evolution that governs the virulence and transmission of this virus in diverse hosts (Troupin *et al.*, 2016, pp. 1–4). The World Health Organization (WHO) enlists carnivorous animals such as domestic dogs (*Canis lupus familiaris*), red foxes (*Vulpes vulpes*), ferret badgers (*Melogale moschata*) and golden jackals (*Canis aureus*) as the most common host of this virus in the Asian region (WHO, 2021b).

Despite it being a preventable disease, rabies is an endemic zoonosis in several developing countries. RABV remains widely distributed across host populations and maintains its abundance via sylvatic and urban life cycles (Troupin

*Corresponding author: avanak@atree.org

DOI: 10.1079/9781800622975.0018

et al., 2016, pp. 2–8). The sustenance of RABV in diverse animal hosts and host-influenced evolution of the virus in these animal reservoirs has led to the emergence of critical mutations and novel phenotypes that further alter the virulence capacities of the virus (Luo *et al.*, 2020, p. 481).

Given the frequent emergence and re-emergence of the infectious diseases H1N1 influenza (2009), Chikungunya (2014), Zika (2015), and severe acute respiratory syndrome coronavirus 1 (SARS-CoV1) and SARS-CoV2 (2002 and 2019, respectively), it might not be wrong to believe that this is indeed an era of pandemics. Understanding pathogen evolution and phylodynamics can help identify the emergence and early pathogen dissemination in animal and human populations (Gardy and Loman, 2018, p. 9). Technological advances in DNA sequencing now provide affordable sequencing of genomes of pathogenic microbes in disease epidemiology. Genomics is a sophisticated means of pathogen surveillance that can be used to inform human, animal and environmental health in a One Health framework. This chapter describes the genome characteristics of RABV, gene functions, tests for rabies diagnosis, high-throughput sequencing (HTS) techniques, and gene and genome sequencing-based surveillance of human and animal rabies in Asian countries.

18.2 The RABV Genome

The RABV genome is 11,928 nucleotides in length and consists of five genes that encode distinct structural proteins namely, nucleoprotein (N), phosphoprotein (P), matrix protein (M), glycoprotein (G) and large protein (L). The bullet-shaped virus is about 170 nm in length and 80 nm in diameter (Fig. 18.1.a) (Du *et al.*, 2008, p. 260). A typical RABV cell structure consists of two components, a core component formed by the protein products of *N*, *P* and *L* genes and a lipid bilayer envelope that is made up of the structural proteins produced by *G* and *M* genes (Nagaraja *et al.*, 2008, p. 451). All five genes have an equal role in causing infection and subsequent phylogenetics of RABV (Du *et al.*, 2008, pp. 261–262). Functions of the gene products in the life cycle and infection of host cells have been described in detail by Fisher *et al.* (2018). In brief, the nucleoprotein

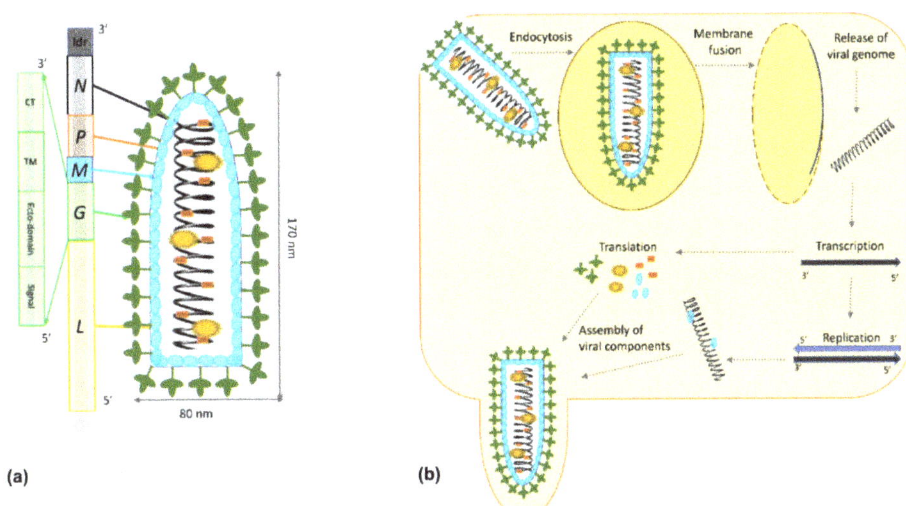

Fig. 18.1. (a) Gene arrangement and proteins encoded in the rabies virus (RABV) genome. (b) Mechanisms involved in the cellular life cycle of RABV and multiplication in host cells. CT, C terminal; TM, transmembrane; Idr, messenger RNA; N, nucleoprotein; P, phosphoprotein; M, matrix protein; G, glycoprotein; L, large protein. Adapted from Fisher *et al.*, 2018, pp. 4-5.

is responsible for encapsulation of the viral genome and assisting the viral genome transcription and replication. As the virus is incapable of penetrating intact skin, an infection most commonly occurs through direct contact with saliva through a break in the skin caused by a scratch or bite from an infected animal. Rarely, infectious materials such as saliva or nervous tissues may come in contact with mucous membranes, or an open wound/lesion. Three types of animal contact ranging from the licking of animals to transdermal bites by suspected rabid animals have been categorized (WHO, 2021a). Upon infection, the phosphoprotein facilitates the transport of the core genome components to the cytoplasm within the axon of a neuron. The matrix protein is involved in the transcription and inhibiting host response. It is also responsible for the bullet shape of this virus as it maintains the helical arrangement of the core genome. The glycoprotein is critical to cell attachment and entry into the host neuron cell. The L protein codes for a ribonucleic acid (RNA) polymerase which performs viral RNA transcription and replication.

The gene arrangement and the mechanisms involved in the cellular life of RABV are depicted in Fig. 18.1. The virus enters the host cell wall through endocytosis by initially adhering to the host cell wall receptors through the glycoprotein (Fisher *et al.*, 2018, pp. 243–244). Membrane fusion occurs across the viral and the endosomal membranes to release the viral core components in the cell cytoplasm, where the viral polymerase transcribes the genes followed by a translation step (Fig. 18.1.b). An anti-genomic RNA strand is derived from a further replication process which acts as the template for producing more copies of the viral RNA. The newly synthesized genes assemble, and the virus proliferates by repetition of these steps. The entry of the virus into the nervous system is facilitated by the nicotinic acetylcholine receptor molecule found on the muscle cell wall (Lafon, 2005, p. 82).

Methods based on animal and cell assays, antigen and antibody testing have been used in diagnoses of RABV infection and evaluation of immunity against RABV in humans and animals. The diagnosis has been extended to molecular virology using gene sequencing techniques that have better characterized the infection dynamics, dispersal and sustenance of RABV in terrestrial (domestic, livestock and wildlife) and aerial (bats) animals.

18.3 Rabies Diagnosis

There are no pre-clinical diagnosis methods for rabies infection in the case of suspected exposure to the virus. Ante-mortem diagnosis, in the case of human rabies, includes skin biopsies containing hair follicles, a biopsy of the central nervous system, and salivary tests. The diagnosis of rabies infection usually requires testing of the brain tissue in case of post-mortem diagnosis in humans and animals. For primary diagnosis, antigen detection by fluorescent antibody test (FAT) is the recommended standard by the WHO and the World Organization for Animal Health (WOAH, founded as OIE) for rabies diagnosis in animals and humans (Rupprecht *et al.*, 2018; WOAH, 2021). The WOAH recommended diagnostic techniques for rabies include: (i) antigen tests such as FAT, direct rapid immunohistochemical test, and enzyme-linked immunosorbent assay (ELISA); (ii) virus detection using cell culture technique and mouse inoculation test; and (iii) molecular techniques like reverse transcription-polymerase chain reaction (RT-PCR) and real-time RT-PCR (WOAH, 2021, p. 586).

In the case of rabies infection in humans, multiple tests are required for intra-vitam diagnosis as no single test is sufficient (Garg, 2014, p. 37). Intra-vitam or post-mortem diagnosis of rabies in humans is carried out by testing for either the whole virus, viral antigens or viral genes in infected tissues using diagnostic techniques. A molecular technique such as RT-PCR has been used for intra-vitam diagnosis and is known to be highly sensitive as it detects the copies of gene or gene fragments of the RABV genome. RT-PCR is becoming popular for detection of rabies in suspected or probable cases in several centralized and reference laboratories globally (Fooks *et al.*, 2009, p. 530; Dacheux *et al.*, 2010, p. 765).

Saliva, nuchal skin, hair follicles, urine or cerebrospinal fluid specimens can be tested for viral nucleic acid in this test. Other methods relying on the nucleic acid analysis of biological samples include nucleic acid sequence-based amplification and loop-mediated isothermal amplification which are found to be highly specific and efficient in the ante-mortem diagnosis of human rabies (Wacharapluesadee and Hemachudha, 2001; Muleya *et al.*, 2012, pp. 160–166). These molecular assays have proved to be effective in cases of decomposed biological samples and rabies infection acquired through organ transplants (Araújo *et al.*, 2008, pp. 1–2; Chen *et al.*, 2018, p. 86). However, false positives are common in these assays owing to an error in standards, contamination in the reaction mix, or non-specific amplification in the reactions. Hence RT-PCR tests are recommended for confirmatory tests (Mani and Madhusudana, 2013). Although RT-PCR-based diagnostic methods have been widely adapted for the diagnosis of other viral pathogens, there is no standard RT-PCR assay for the detection of all lyssaviruses (Gigante *et al.*, 2018, p. 17).

Besides antigen detection, viral antibodies are estimated for monitoring the immune response against rabies. These serology-based assays are not useful in rabies diagnosis as the immune response is varied and the RABV antibodies in serum appear relatively late after the onset of clinical signs (Mani *et al.*, 2014, pp. 1–4). Virus neutralization assays like rapid fluorescent focus inhibition test and fluorescent antibody virus neutralization test are generally used to estimate antibody titres (Rupprecht *et al.*, 2018). The direct FAT for G and M immunoperoxidase inhibition assay and the immunoelectrophoresis test is also used to measure the immune response against rabies post-vaccination (Rupprecht *et al.*, 2018). ELISA is another serological test that eliminates the need to manage live RABV and is used to measure specific antibodies against the viral glycoprotein or nucleoprotein (Rupprecht *et al.*, 2018). This test demonstrated 78.38–100% sensitivity and 75.76–96.77% specificity compared to the direct FAT in the case of human ante-mortem diagnosis of rabies indicating its suitability in assessing disease exposure status (Realegeno *et al.*, 2018). There is also evidence of rabies neutralizing antibodies in unvaccinated animals like bats and cattle as confirmed by rapid fluorescent focus inhibition tests, suggesting the need for rabies surveillance in animals other than dogs (Mani *et al.*, 2014, p. 1804; Bharti *et al.*, 2018). The presence of antibodies in unvaccinated animals could be an outcome of cross-reactivity to unknown lyssaviruses and further confirmation requires genetic mapping of viral isolates.

Variability of methods in rabies diagnosis and interpretation of serological test results (cut-offs) often lead to uncertainty in the surveillance of RABV (Moore *et al.*, 2013, pp. 449–451). The sensitivity of the various tests described here varies significantly owing to factors like autolysis of the sample, distinct stages of infections, and diverse immune responses. These complexities are magnified in cases of rabies-specific antibodies being detected in sera of humans, animals and wildlife that were healthy and unvaccinated (Gold *et al.*, 2020, p. 1). For ecological and evolutionary analysis of RABV, traditional molecular tools like PCR and gene sequencing, and more recently genomic techniques have been recommended by WOAH, (2021) and Rupprecht *et al.* (2018).

18.4 RABV Gene Analysis

Gene-based phylogenetic analysis has facilitated our understanding of selection and evolution in microbes. Genetic characterization of RABV genes including the leader fragment and the five genes has been carried out in rabies samples from animals and humans in various molecular studies (Nagarajan *et al.*, 2006, pp. 3218–3221; Du *et al.*, 2008, p. 260; Reddy *et al.*, 2018). Gene diversity analyses from various countries have provided a phylogenetic classification of *Lyssavirus* at the strain level which further resulted in the identification of region-specific lineages in diverse geographies (Jiao *et al.*, 2011, pp. 1-2; Reddy *et al.*, 2018). The viral gene sequences generated from such independent studies have been deposited in public repositories (e.g. the National Library of Medicine; available at: https://www.ncbi.nlm.nih.gov/ (accessed 1 June 2021)), thus making the genetic information available for large-scale analysis of gene evolution in RABV and other species. Such large-scale phylogenetic analysis and gene or genome divergence studies have resulted in the identification of African, Asian, Arctic-like, Cosmopolitan and Indian subcontinent (Sri Lankan) lineages of RABV from distinct geographies around the world (Troupin *et al.*, 2016, p. 4).

Sequencing of partial *N* gene for a cow sample from southern India revealed 96–99% homology with the Sri Lankan RABV lineage indicating transmission of the virus across these neighbouring countries (Aravindhbabu *et al.*, 2011, p. 138). Another gene-based study of rabies-positive dogs and other animals for RABV phylogenies in India revealed that the Arctic-like lineage was dominant (Reddy *et al.*, 2018). In addition to the Arctic-like lineage, the partial gene sequencing of the nucleoprotein gene and subsequent phylogenetic analysis confirmed the presence of Cosmopolitan lineage in the Indian region between 2013 and 2016 from rabies-positive animals such as buffaloes, camels, cats, dogs and others. *G* gene, *L* gene and non-coding *G–L* intergenic region sequences derived from the brain samples of animals like dogs, buffaloes, cows and goats, from various states in India during 2001–2005 confirmed the predominance of dog rabies in India and led to the identification of multiple epidemiological markers unique to Indian RABV which could be used in tracing the origin of viral strains, especially in cases of travel-related rabies (Nagarajan *et al.*, 2006, pp. 3220–3222). Similar amino acid mutations could be identified in the first genome sequence of an infected street-dog sample in China (Ming *et al.*, 2009, pp. 8–10).

To confirm the lineage classification and divergence across various countries, the partial gene-based phylogenies were replaced by complete gene sequencing and whole-genome sequencing (WGS) of the RABV (Nagaraja *et al.*, 2008, p. 449; Ming *et al.*, 2009, pp. 6–10; Cherian *et al.*, 2015, p. 333). Complete *G* gene sequencing of rabies-positive samples from the northern and southern regions of India inclusive of six host species confirmed the presence of Arctic-like lineage across the country (Cherian *et al.*, 2015, p. 333). In this study, distinct clustering of two groups termed the North Indian (Uttar Pradesh, Madhya Pradesh, Delhi, Rajasthan) and the South Indian groups (Kerala, Karnataka, Gujarat) was observed within the lineage identified.

In recent years, advanced high-throughput sequencing (HTS) techniques have resulted in increased knowledge of mechanisms of spread, the role of wildlife, distinctions in disease burden on the urban and rural community, and efficacy of control strategies being implemented (Troupin *et al.*, 2016, pp. 3–4; Brunker *et al.*, 2018, p. 2; Gigante *et al.*, 2020, pp. 1-2). The application of HTS has recently helped expand the *Lyssavirus* genus with two newly added species including one isolated from a flying fox (*Pteropus medius*) from Asia (Amarasinghe *et al.*, 2018, p. 2283). WGS and subsequent phylogenetic and phylodynamic analysis pave the way for a better understanding of viral pathogenesis, trends in transmission and evolution, the impact of geography, and environmental factors on the transmission of RABV.

18.5 RABV Whole-Genome Sequencing (WGS)

In simple terms, a whole genome simply means 'all the genes' of an organism. Since the sequencing of the first human genome, the use of DNA sequencing technologies has revolutionized disease diagnosis and surveillance in human healthcare. The cost of DNA sequencing has drastically decreased since 2008 with the introduction of Illumina platforms that facilitated parallel sequencing of fragmented genomes and assembly for WGS of organisms. The third-generation sequencers have eliminated the need for specialized genetic laboratories, as these hand-held DNA sequencers are highly integrative of all components required for their portable use in disease surveillance (Gigante *et al.*, 2020, p. 2). In fact, the MinION Oxford Nanopore sequencers have facilitated mobile genetics laboratories that can transform the workflow of field researchers (Brunker *et al.*, 2020, p. 2). This also facilitates undertaking more field research studies using the genomics approach. WGS-based surveillance of microbes using modern sequencers has accelerated the identification of pathogen type, genotype tracking and people needing treatment. WGS has emerged as a popular technique that will revolutionize the delivery of real-time healthcare.

The first full genome sequencing of RABV was done using a 'mapping and primer walking' strategy which was a multistep approach. The WGS of the first RABV strain from India was performed by following sequential steps including polymerase chain reaction (PCR) to amplify genome fragments, cloning, and sequencing of plasmid DNA using vector and gene-derived primers (Nagaraja *et al.*, 2008, pp. 450–451). A similar approach was followed in the characterization of various *Lyssavirus* genotypes (Delmas *et al.*, 2008, pp. 1–2; Kuzmin *et al.*, 2008, p. 81). The availability of these complete genomes confirmed the conserved genomic organization, relevant mutation sites, and protein structures that are key to the sustenance and dissemination of RABV across host populations (Marston *et al.*, 2017, p. 1). The WGS of RABV from Asian regions and related information on sequencing platforms are listed in Table 18.1. Human and animal infection cases have been assessed through gene and genome sequencing of RABV in China, Jordan, Bhutan, India, Tibet, Laos, Iran and other countries. The modern DNA sequencers allow massive parallel sequencing in each run. These platforms are increasingly being used in the case of human and animal RABV infections, thus expanding the depth of RABV genetic information (Brunker *et al.*, 2020, p. 9; Gigante *et al.*, 2020, pp. 2–3).

A gene-sequence-based phylogenetic tree for RABV in Bangladesh was found to be homologous to the Arctic-like lineage, similar to the strain reported from Bhutan (Jamil *et al.*, 2012, p. 2021). This study highlighted the need for widening surveillance of RABV to neighbouring countries like Iran, Iraq, Afghanistan, India and Nepal to determine the origin and transmission pathways of the identified Arctic-like lineage RABV. In another study, WGS analysis of a cattle-derived strain (JSTZ190314) confirmed the transmission of canine rabies through animal bites. Phylogenetic analysis of the gene sequence from this study with RABV gene sequences derived from a public repository revealed that the JSTZ190314 strain was prevalent for 13 years and had infected six other species including deer, dogs and humans from 2006 to 2019 in China (Cheng *et al.*, 2020, pp. 1453–1456). It was possible to study the evolution of RABV and its prevalence across time in different animals owing to the availability of genetic data (for China) in the public databases. Similar reports from other Asian countries indicated the need to analyse the RABV strains among infected livestock and other animals which are in proximity to free-ranging dogs (Jamil *et al.*, 2012, p. 2023; Munnink *et al.*, 2020, p. 325). Nanopore technology-based sequencing platforms have been successfully used for gene sequencing of RABV in countries like Kenya, Vietnam and India (Brunker *et al.*, 2020, pp. 2–3). Yet, the studies providing phylogenetic insights into the evolution of RABV from countries such as Jordan and Bangladesh, and missing representation of countries like India and Bhutan, emphasize the scarcity of whole-genome data from Asian countries (Jamil *et al.*, 2012, pp. 2022–2023; Al-Eitan *et al.*, 2021, p. 1).

18.6 Applications of HTS in Understanding the Ecology and Evolution of RABV

Genetic characterization of (known and novel) pathogens in combination with meta-information holds great potential in disease epidemiology and outbreak response. With the technological advancements and decreasing cost of sequencers, phylogeographic and phylodynamic analyses have enhanced knowledge of the evolution and spread of RABV across countries and continents (Troupin *et al.*, 2016, pp. 13–14; Reddy *et al.*, 2018). Estimation of total viral heterogeneity is possible with the availability of the whole genome sequences where the identified mutation sites can serve as epidemiological markers of RABV. Moreover, variations in the RABV genome sequence are indicative of its contribution to establishing transmission within a new host (Al-Eitan *et al.*, 2021, pp. 5–7). Bayesian and frequentist methods are often used for understanding RABV genome divergence, substitutions, and classification of strains with respect to the geography of the study site to reveal the significance of environmental factors that contribute to the transmission of RABV (Troupin *et al.*, 2016, pp. 1–4; Nahata *et al.*, 2021, pp. 3–6, 15).

Currently, the National Center for Biotechnology Information (NCBI) contains a total of 42 Bioproject entries that include human (three), mammal (25) and virus (14) organism groups. A total of 43,424 entries are available for a 'rabies' search in the nucleotide database (inclusive of all partial gene sequences,

Table 18.1. Gene and whole-genome sequencing(WGS)-based studies of rabies virus (RABV) from human and animal samples reported from Asia and a few Middle Eastern countries. Table created by authors.

Sample source	Country	Year of sampling	Sequencing technique	Key findings	Reference
Human RABV vaccine strain CTN181	China	Not mentioned	PCR and sequencing	Mutations crucial to the function of leader region and L protein identified	Du *et al.* (2008)
HN10, recovered from brain tissue of a rabid patient	China	2009	Sequenced by outsourcing to service provider	RABV fully characterized for conservation of functional regions in genome	Ming *et al.* (2009)
Vaccine strain	China	2011	Positive clones sequenced	*N* gene was the most conserved indicating this gene is the most appropriate for quantitative genotype definition	Jiao *et al.* (2011)
Human	Sri Lanka	2008	ABI-3130 Genetic Analyser, Applied Biosystems	Independent lineage that did not cluster with other RABV lineages	Matsumoto *et al.* (2011)
Goat (dog bite to head)	Bangladesh	2010	ABI-3130 Genetic Analyser, Applied Biosystems	*N* gene sequence similarity indicated a common ancestor for RABVs circulating in Bangladesh and Bhutan	Jamil *et al.* (2012)
Dog	Laos	2004–2011	ABI-3130 Genetic sequencer, Applied Biosystems	Dogs were the main viral reservoir, three RABV lineages circulating in the country	Ahmed *et al.* (2015)
Dog and cow	China	2008–2012	PCR and sequencing	Six different China RABV lineages	Hao *et al.* (2016)
Wild carnivores and dogs	Taiwan	2013	ABI 3730xl DNA analyser, Applied Biosystems	RABV clustered with mountainous geographic segregation	Tsai *et al.* (2016)
Dog	Tanzania	2004–2013	Illumina MiSeq or NextSeq platform	Phylogeographic framework	Brunker *et al.* (2020)

Continued

Table 18.1. Continued

Sample source	Country	Year of sampling	Sequencing technique	Key findings	Reference
Patients died of rabies after receiving kidney transplants	China	2016	Sanger sequencing	Uncertain exposure history and misdiagnosis led to RABV spread	Chen *et al*. (2018)
Buffalo camel, cat, cattle, dog and horse	Different states of India	2013–2016	Paired-end Sanger's di-deoxy sequencing	Arctic-like and cosmopolitan lineage confirmed	Reddy *et al*. (2018)
Human and dog	Tibet	2012, 2015, 2017	PCR-based sequencing	Cross-border transmission confirmed	Tao *et al*. (2019)
Cattle	Eastern China	2019	Sanger sequencing	Dog lineage RABV detected from cattle sample	Cheng *et al*. (2020)
Fox, camel and human	Qatar (Nepalese patients diagnosed in Qatar)	2018–2019	Multiplexed metagenomic Nanopore sequencing using Nanopore sequencer	Need for epidemiological research on the human–animal interface	Munnink *et al*. (2020)

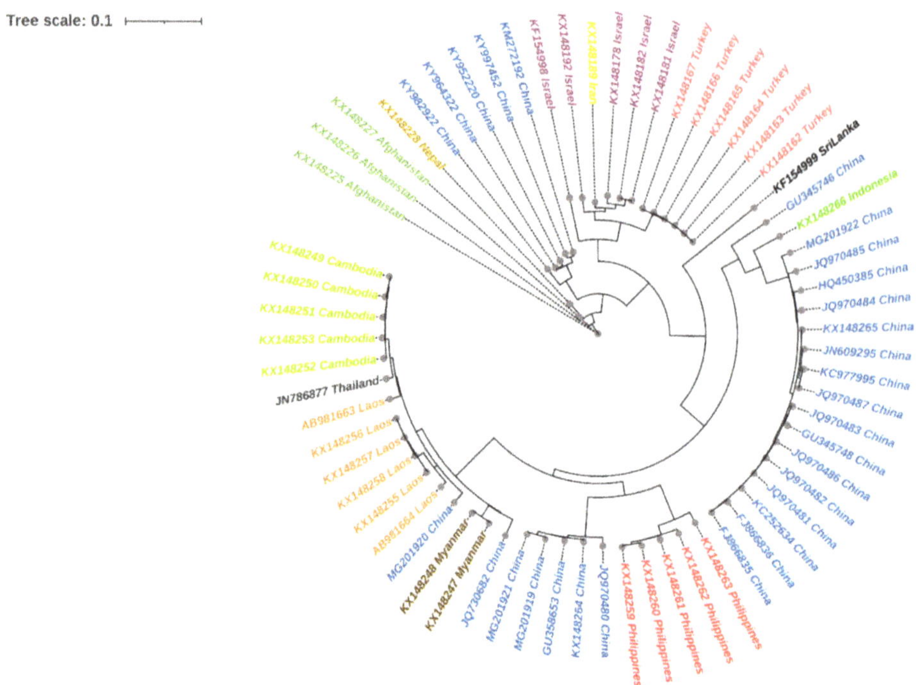

Fig. 18.2. Phylogenetic analysis of canine RABV whole-genome sequences from Asian and a few Middle Eastern countries (genome length >18,000 bp). The tree scale indicates the nucleotide substitutions per site. Created by authors.

complete gene sequences, and genomes of RABV as assessed on 15 October 2021). Of these total nucleotide entries, about 5% were submitted to the database from 2000 to 2005, 20% from 2005 to 2010, and 47% from 2010 to 2015, which indicates an exponential expansion of the RABV genes and genomes in the database. Canine rabies studies based on partial gene sequencing from various geographies have been recently reported and reviewed (Heydarabadi *et al.*, 2020; Nahata *et al.*, 2021, p. 1628). Canine RABV WGS from various countries, related strain information available in the NCBI nucleotide database, and their phylogenetic analysis are provided in Fig. 18.2. The RABV whole-genome sequences were retrieved from the NCBI nucleotide database and sequences with a length of at least 11,800 nucleotides and above were selected to perform phylogenetic analysis in raxmlGUI 2.0 (Silvestro and Michalak, 2012, pp. 335–337). The phylogenetic tree revealed distinct clades and branching patterns indicating a polyphyletic structure of the RABV population that confirms the presence of diverse lineages as reported in previous studies (Ahmed *et al.*, 2015, p. 1; Troupin *et al.*, 2016, p. 4). A few complete genomes of RABV from cattle and wildlife have been collectively assessed in a large-scale phylogeographic analysis and identification of diverse lineages of the virus in distinct animals (Troupin *et al.*, 2016, p. 4). The majority of the canine RABV whole-genome sequences present in the NCBI database represented various regions in China. Several countries remain under-represented as highlighted by fewer whole-genome sequences from Afghanistan, Nepal, Israel, Turkey, Sri Lanka, Indonesia and the Philippines, and no representation from countries like India in this phylogenetic reconstruction.

Knowledge gaps pertaining to viral genome evolution and its impact on host DNA, its impact on the efficacy of existing rabies vaccine, and the environmental factors driving the transmission of RABV in developing countries can be addressed by undertaking WGS and analysis. HTS holds great potential in genomic epidemiology as newer methods and models are being developed for genome sequencing, sampling, and metadata collection. HTS techniques offer effective methods in rabies epidemiology especially in developing countries, in ensuring safe organ transplants, and integrative surveillance of rabies at animal–animal and animal–human interfaces. These gaps can benefit from the application of cost-effective and accurate HTS-based genomic surveillance techniques.

The concept of human interaction with animals and the environment is not something that is a recent 'aha moment'. Hippocrates's treatise in *Air Water and Places* (as early as *c*.460–*c*.377 BC) mentioned how the environment, climate and water quality influence the physical condition of humans. The modern definition of One Health is: 'a collaborative, multisectoral, and transdisciplinary approach working at the local, regional, national, and global levels with the goal of achieving optimal health outcomes recognizing the interconnection between people, animals, plants, and their shared environment' (CDC, 2017). In the last century, major pandemic events have occurred with large death tolls, despite improvements in the medical capacity and public health infrastructure. With the increasing incidence of emerging and re-emerging viral infectious diseases, significant policy attention and lateral thinking have commenced for effective disease outbreak management (Wong *et al.*, 2020; Aborode *et al.*, 2021, p. 1; Cevik *et al.*, 2021, p. 170).

Surveillance is a crucial component in identifying viral outbreaks, evolution in zoonotic viruses, the emergence of viral variants, and transmission pathways associated with increased outbreaks (Desdouits *et al.*, 2020, pp. 1–2). We have witnessed the wide use of genomic analysis in addressing the recent SARS-CoV2 pandemic. A recent study applied the workflow developed for SARS-CoV2 nomenclature to the RABV genome sequences available in public databases (Campbell *et al.*, 2022, p. 1). More than 70 different lineages within the Cosmopolitan type RABV group could be classified based on an analysis of 650 rabies whole-genome sequences from 46 countries from 1950 to 2018. These genomic techniques could be extended to existing surveillance and response mechanisms of zoonotic diseases like rabies by integrating digital means of disease detection and the One Health approach (John *et al.*, 2021, p. 845).

Additionally, gene sequencing and whole-genome-based surveillance benefit the building of public databases, like NCBI's GenBank and the European Nucleotide Archive (Benson *et al.*, 1993, p. 2963; Leinonen *et al.*, 2010, p. 28). The curated databases are inclusive of standardized metadata (location and date of sampling, specimen information, host information, submitting authority and institute details) (Bernasconi *et al.*, 2021, p. 664). In addition to serving as an online data resource, the use of NCBI can facilitate specialized host–pathogen datasets for integrative surveillance of viral pandemics (COVID-19 Host Genetics Initiative, 2020). With the increased availability of cost-effective, rapid and accurate sequencing platforms, genome sequencing will provide deeper insights into the evolution of the RABV in animals and host–pathogen interaction (Brunker *et al.*, 2020, pp. 1–6; Gigante *et al.*, 2020, pp. 1–2).

18.7 HTS in a One Health Approach for Rabies Elimination

Elimination of rabies in developing countries is largely hindered by critical knowledge gaps in the identification of rabies hotspots, knowledge of high-risk condition areas, frequency of rabies cases in free-ranging dogs, and transmittance to other animals. These gaps highlight the need for an integrative approach to address the social and ecological drivers of genetic congruence in RABV. Serological assays, molecular tools and gene detection techniques have been extensively used, especially in ante-mortem testing of human and animal rabies. The negative-sense RABV genome exhibits unusual genomic plasticity as it lacks proofreading activity owing to which this virus often dodges immune surveillance and the virus remains highly adaptative in new host animals. Optimization of HTS techniques can facilitate advanced diagnostic assays to address drawbacks of current rabies diagnostics. The portable sequencing machines offer quick and cost-effective means of on-field surveillance (Brunker *et al.*, 2020).

Population-scale genomics in combination with temporal and spatial analysis provides valuable insights for designing effective dog vaccination programmes. However, adopting HTS in rabies surveillance will necessitate the optimization of methods and expertise in molecular and *in silico* analysis. This will require the collaboration of experts from interdisciplinary domains including veterinary, wildlife, epidemiology and social science. The fragmented attempts at rabies elimination in India have been addressed in the newly introduced National Action Plan for Dog Mediated Rabies Elimination in India which relies on the broad framework of the One Health approach and is based on key components of political will, intersectoral planning, sustained funding, community planning, coordination and review and operational research (NHP, 2021). This joint inter-sectoral programme will involve government officials and the private sector, veterinarians, epidemiologists, wildlife ecologists, policy makers and social scientists. While technological advancements like HTS can help in understanding rabies transmission, disease surveillance and diagnostics, only coordinated efforts can translate plans to actions.

18.8 Conclusion

WGS-based studies have facilitated the detection of trends in rabies transmission across temporal and spatial scales thus indicating the emergency and scale of interventions required. It is possible to recognize demographic factors driving RABV transmission through the combination of genome sequencing of the pathogen in animals, animal ecology, and environmental parameters impact assessment. A strong genetic database of RABV can benefit the forecasting of the spatial spread and emergence of rabies in new regions. Whole-genome-based RABV analysis has provided information on nucleotide substitutions in RABV street strains and their impact on the efficacies of existing RABV.

Comparative analysis of RABV genomic elements has indicated that the increasing divergence (or nucleotide substitutions) in street strains could render the vaccines ineffective (Faber *et al.*, 2004, p. 16328). The use of WGS can also help in understanding fine-scale dynamics of spread, including testing hypotheses related to the role of multi-species interactions, domestic-sylvatic cycles, and the spread of rabies from urban to rural areas or vice versa (Belsare and Gomper, 2010). Ultimately, such information can help in designing more effective mitigation strategies that target these pathways of transmission or deploy scarce resources in a more organized fashion to achieve local and ultimately global rabies elimination targets.

Acknowledgements

This study was funded by DBT/Wellcome Trust India Alliance (Grant No. IA/CPHI/15/1/502028 to Abi Tamim Vanak).

Authors' Declaration

All authors declare they have no conflict of interest.

All authors have approved this manuscript, agree with its submission, and share collective responsibility and accountability.

This manuscript has not been published or is not under review elsewhere.

References

Aborode, A.T., Hasan, M.M., Jain, S., Okereke, M., Adedeji, O.J. *et al.* (2021) Impact of poor disease surveillance system on COVID-19 response in Africa: time to rethink and rebuilt. *Clinical Epidemiology and Global Health* 12, 100841. DOI: 10.1016/j.cegh.2021.100841.

Ahmed, K., Phommachanh, P., Vorachith, P., Matsumoto, T., Lamaningao, P. *et al.* (2015) Molecular epidemiology of rabies viruses circulating in two rabies endemic provinces of Laos, 2011-2012: regional diversity in Southeast Asia. *PLoS Neglected Tropical Diseases* 9(3), e0003645. DOI: 10.1371/journal.pntd.0003645.

Al-Eitan, L.N., Wu, G., Golding, M., Tang, Y., Goharriz, H. *et al.* (2021) Whole-genome sequencing and phylogenetic analysis of rabies viruses from Jordan. *PLoS Neglected Tropical Diseases* 15(5), e0009431. DOI: 10.1371/journal.pntd.0009431.

Amarasinghe, G.K., Aréchiga Ceballos, N.G., Banyard, A.C., Basler, C.F., Bavari, S. *et al.* (2018) Taxonomy of the order Mononegavirales: update 2018. *Archives of Virology* 163(8), 2283–2294. DOI: 10.1007/s00705-018-3814-x.

Araújo, D.B., Langoni, H., Almeida, M.F. and Megid, J. (2008) Heminested reverse-transcriptase polymerase chain reaction (hnRT-PCR) as a tool for rabies virus detection in stored and decomposed samples. *BMC Research Notes* 1(1), 1–6. DOI: 10.1186/1756-0500-1-17.

Aravindhbabu, R.P., Manoharan, S., Ramadass, P. and Chandran, N.D.J. (2011) Rabies in South Indian cows: an evidence of Sri Lankan rabies virus variant infection based on the analysis of partial nucleoprotein gene. *Indian Journal of Virology* 22(2), 138–141. DOI: 10.1007/s13337-011-0050-3.

Belsare, A.V. and Gomper, M.E. (2010) Rabies control in developing countries: A model-based approach to contrast mass vaccination strategies. Poster presentation for Council of Graduate Students. Missouri State University, Rolla, Missouri at Rolla, Missouri. DOI: 10.13140/RG.2.1.3956.5207.

Benson, D., Lipman, D.J. and Ostell, J. (1993) GenBank. *Nucleic Acids Research* 21(13), 2963–2965. DOI: 10.1093/nar/21.13.2963.

Bernasconi, A., Canakoglu, A., Masseroli, M. and Ceri, S. (2021) The road towards data integration in human genomics: players, steps and interactions. *Briefings in Bioinformatics* 22(1), 30–44. DOI: 10.1093/bib/bbaa080.

Bharti, O.K., Sharma, U.K., Kumar, A. and Phull, A. (2018) Exploring the feasibility of a new low cost intra-dermal pre & post exposure rabies prophylaxis protocol in domestic bovine in Jawali Veterinary Hospital, District Kangra, Himachal Pradesh, India. *World Journal of Vaccines* 08(1), 8–20. DOI: 10.4236/wjv.2018.81002.

Brunker, K., Lemey, P., Marston, D.A., Fooks, A.R., Lugelo, A. *et al.* (2018) Landscape attributes governing local transmission of an endemic zoonosis: rabies virus in domestic dogs. *Molecular Ecology* 27(3), 773–788. DOI: 10.1111/mec.14470.

Brunker, K., Jaswant, G., Thumbi, S.M., Lushasi, K., Lugelo, A. *et al.* (2020) Rapid in-country sequencing of whole virus genomes to inform rabies elimination programmes. *Wellcome Open Research* 5, 3. DOI: 10.12688/wellcomeopenres.15518.2.

Campbell, K., Gifford, R.J., Singer, J., Hill, V., O'Toole, A. *et al.* (2022) Making genomic surveillance deliver: a lineage classification and nomenclature system to inform rabies elimination. *PLoS Pathogens* 18(5), e1010023. DOI: 10.1371/journal.ppat.1010023.

CDC (Centers for Disease Control and Prevention) (2017) One Health basics. Available at: https://www.cdc.gov/onehealth/basics/index.html (accessed 15 October 2021).

Cevik, M., Marcus, J.L., Buckee, C. and Smith, T.C. (2021) Severe Acute Respiratory Syndrome Coronavirus 2 (SARS-CoV-2) transmission dynamics should inform policy. *Clinical Infectious Diseases* 73(Suppl 2), S170–S176. DOI: 10.1093/cid/ciaa1442.

Chen, J., Liu, G., Jin, T., Zhang, R., Ou, X. *et al.* (2018) Epidemiological and genetic characteristics of rabies virus transmitted through organ transplantation. *Frontiers in Cellular and Infection Microbiology* 8, 86. DOI: 10.3389/fcimb.2018.00086.

Cheng, K., Chu, H., Ren, Y., Xie, X., Yu, Z. *et al.* (2020) Phylogenetic analysis of the whole genome sequence of a dog lineage rabies virus detected from cattle in eastern China, 2019. *Brazilian Journal of Microbiology* 51(3), 1453–1458. DOI: 10.1007/s42770-020-00264-7.

Cherian, S., Singh, R., Singh, K.P., Reddy, G.M., Kumar, G.R. *et al.* (2015) Phylogenetic analysis of Indian rabies virus isolates targeting the complete glycoprotein gene. *Infection, Genetics and Evolution* 36, 333–338. DOI: 10.1016/j.meegid.2015.09.024.

COVID-19 Host Genetics Initiative (2020) The COVID-19 Host Genetics Initiative, a global initiative to elucidate the role of host genetic factors in susceptibility and severity of the SARS-cov-2 virus pandemic. *European Journal of Human Genetics* 28(6), 715–718.

Dacheux, L., Wacharapluesadee, S., Hemachudha, T., Meslin, F.-X., Buchy, P. *et al.* (2010) More accurate insight into the incidence of human rabies in developing countries through validated laboratory techniques. *PLoS Neglected Tropical Diseases* 4(11), e765. DOI: 10.1371/journal.pntd.0000765.

Delmas, O., Holmes, E.C., Talbi, C., Larrous, F., Dacheux, L. *et al.* (2008) Genomic diversity and evolution of the lyssaviruses. *PloS ONE* 3(4), e2057. DOI: 10.1371/journal.pone.0002057.

Desdouits, M., de Graaf, M., Strubbia, S., Oude Munnink, B.B., Kroneman, A. *et al.* (2020) Novel opportunities for NGS-based one health surveillance of foodborne viruses. *One Health Outlook* 2(1), 14. DOI: 10.1186/s42522-020-00015-6.

Du, J., Zhang, Q., Tang, Q., Li, H., Tao, X. *et al.* (2008) Characterization of human rabies virus vaccine strain in China. *Virus Research* 135(2), 260–266. DOI: 10.1016/j.virusres.2008.04.002.

Faber, M., Pulmanausahakul, R., Nagao, K., Prosniak, M., Rice, A.B. *et al.* (2004) Identification of viral genomic elements responsible for rabies virus neuroinvasiveness. *Proceedings of the National Academy of Sciences of the United States of America* 101(46), 16328–16332. DOI: 10.1073/pnas.0407289101.

Fisher, C.R., Streicker, D.G. and Schnell, M.J. (2018) The spread and evolution of rabies virus: conquering new frontiers. *Nature Reviews. Microbiology* 16(4), 241–255. DOI: 10.1038/nrmicro.2018.11.

Fooks, A.R., Johnson, N., Freuling, C.M., Wakeley, P.R., Banyard, A.C. *et al.* (2009) Emerging technologies for the detection of rabies virus: challenges and hopes in the 21st century. *PLoS Neglected Tropical Diseases* 3(9), e530. DOI: 10.1371/journal.pntd.0000530.

Gardy, J.L. and Loman, N.J. (2018) Towards a genomics-informed, real-time, global pathogen surveillance system. *Nature Reviews Genetics* 19(1), 9–20. DOI: 10.1038/nrg.2017.88.

Garg, S.R. (2014) Rabies manifestations and diagnosis. In: *Rabies in Man and Animals*. Springer, New Delhi, pp. 37–49.

Gigante, C.M., Dettinger, L., Powell, J.W., Seiders, M., Condori, R.E.C. *et al.* (2018) Multi-site evaluation of the LN34 pan-lyssavirus real-time RT-PCR assay for post-mortem rabies diagnostics. *PLoS ONE* 13(5), e0197074. DOI: 10.1371/journal.pone.0197074.

Gigante, C.M., Yale, G., Condori, R.E., Costa, N.C., Long, N.V. *et al.* (2020) Portable rabies virus sequencing in canine rabies endemic countries using the Oxford Nanopore MinION. *Viruses* 12(11), 1255. DOI: 10.3390/v12111255.

Gold, S., Donnelly, C.A., Nouvellet, P. and Woodroffe, R. (2020) Rabies virus-neutralising antibodies in healthy, unvaccinated individuals: What do they mean for rabies epidemiology? *PLoS Neglected Tropical Diseases* 14(2), e0007933. DOI: 10.1371/journal.pntd.0007933.

Hao, L., Guo, Z.Y., Zhang, J., Tao, X.Y., Zhu, W.Y. *et al.* (2016) Whole genome sequencing and comparisons of different Chinese rabies virus lineages including the first complete genome of an Arctic-like strain in China. *Biomedical and Environmental Sciences* 29(5), 340–346. DOI: 10.3967/bes2016.044.

Heydarabadi, F.H., Baessi, K., Bashar, R., Fazeli, M. and Sheikholeslami, F. (2020) A phylogenetic study of new rabies virus strains in different regions of Iran. *Virus Genes* 56(3), 361–368. DOI: 10.1007/s11262-020-01752-6.

Jamil, K.M., Ahmed, K., Hossain, M., Matsumoto, T., Ali, M.A. *et al.* (2012) Arctic-like rabies virus, Bangladesh. *Emerging Infectious Diseases* 18(12), 2021–2024. DOI: 10.3201/eid1812.120061.

Jiao, W., Yin, X., Li, Z., Lan, X., Li, X. *et al.* (2011) Molecular characterization of China rabies virus vaccine strain. *Virology Journal* 8(1), 1–11. DOI: 10.1186/1743-422X-8-521.

John, G., Sahajpal, N.S., Mondal, A.K., Ananth, S., Williams, C. *et al.* (2021) Next-Generation Sequencing (NGS) in COVID-19: a tool for SARS-CoV-2 diagnosis, monitoring new strains and phylodynamic modeling in molecular epidemiology. *Current Issues in Molecular Biology* 43(2), 845–867. DOI: 10.3390/cimb43020061.

Kuzmin, I.V., Wu, X., Tordo, N. and Rupprecht, C.E. (2008) Complete genomes of Aravan, Khujand, Irkut and West Caucasian bat viruses, with special attention to the polymerase gene and non-coding regions. *Virus Research* 136(1–2), 81–90. DOI: 10.1016/j.virusres.2008.04.021.

Lafon, M. (2005) Rabies virus receptors. *Journal of Neurovirology* 11(1), 82–87. DOI: 10.1080/13550280590900427.

Leinonen, R., Akhtar, R., Birney, E., Bower, L., Cerdeno-Tárraga, A. *et al.* (2010) The European nucleotide archive. *Nucleic Acids Research* 39(Database issue), D28–D31. DOI: 10.1093/nar/gkq967.

Luo, J., Zhang, B., Wu, Y. and Guo, X. (2020) Amino acid mutation in position 349 of glycoprotein affect the pathogenicity of rabies virus. *Frontiers in Microbiology* 11, 481. DOI: 10.3389/fmicb.2020.00481.

Mani, R.S. and Madhusudana, S.N. (2013) Laboratory diagnosis of human rabies: recent advances. *The Scientific World Journal* 2013, 569712. DOI: 10.1155/2013/569712.

Mani, R.S., Madhusudana, S.N., Mahadevan, A., Reddy, V., Belludi, A.Y. *et al.* (2014) Utility of real-time Taqman PCR for antemortem and postmortem diagnosis of human rabies. *Journal of Medical Virology* 86(10), 1804–1812. DOI: 10.1002/jmv.23814.

Marston, D.A., Horton, D.L., Nunez, J., Ellis, R.J., Orton, R.J. *et al.* (2017) Genetic analysis of a rabies virus host shift event reveals within-host viral dynamics in a new host. *Virus Evolution* 3(2), vex038. DOI: 10.1093/ve/vex038.

Matsumoto, T., Ahmed, K., Wimalaratne, O., Yamada, K., Nanayakkara, S. *et al.* (2011) Whole-genome analysis of a human rabies virus from Sri Lanka. *Archives of Virology* 156(4), 659–669. DOI: 10.1007/s00705-010-0905-8.

Ming, P., Du, J., Tang, Q., Yan, J., Nadin-Davis, S.A. *et al.* (2009) Molecular characterization of the complete genome of a street rabies virus isolated in China. *Virus Research* 143(1), 6–14. DOI: 10.1016/j.virusres.2009.02.014.

Moore, S.M., Gordon, C.R. and Hanlon, C.A. (2013) Measures of rabies immunity. In: Jackson, A.C. (ed.) *Rabies Scientific Basis of the Disease and its Management*, 2nd edn. Elsevier, Amsterdam, pp. 449–455.

Muleya, W., Namangala, B., Mweene, A., Zulu, L., Fandamu, P. *et al.* (2012) Molecular epidemiology and a loop-mediated isothermal amplification method for diagnosis of infection with rabies virus in Zambia. *Virus Research* 163(1), 160–168. DOI: 10.1016/j.virusres.2011.09.010.

Munnink, B.B.O., Abd Farag, E.A.B., GeurtsvanKessel, C., Schapendonk, C., van der Linden, A. *et al.* (2020) First molecular analysis of rabies virus in Qatar and clinical cases imported into Qatar, a case report. *International Journal of Infectious Diseases* 96, 323–326. DOI: 10.1016/j.ijid.2020.04.070.

Nagaraja, T., Madhusudana, S. and Desai, A. (2008) Molecular characterization of the full-length genome of a rabies virus isolate from India. *Virus Genes* 36(3), 449–459. DOI: 10.1007/s11262-008-0223-3.

Nagarajan, T., Mohanasubramanian, B., Seshagiri, E.V., Nagendrakumar, S.B., Saseendranath, M.R. *et al.* (2006) Molecular epidemiology of rabies virus isolates in India. *Journal of Clinical Microbiology* 44(9), 3218–3224. DOI: 10.1128/JCM.00801-06.

Nahata, K.D., Bollen, N., Gill, M.S., Layan, M., Bourhy, H. *et al.* (2021) On the use of phylogeographic inference to infer the dispersal history of rabies virus: a review study. *Viruses* 13(8), 1628. DOI: 10.3390/v13081628.

NHP (National Health Portal) India (2021) National action plan for dog mediated rabies elimination. Available at: http://nhp.gov.in/healthprogramme/national-action-plan-rabies-elimination (accessed 15 October 2021).

Realegeno, S., Niezgoda, M., Yager, P.A., Kumar, A., Hoque, L. *et al.* (2018) An ELISA-based method for detection of rabies virus nucleoprotein-specific antibodies in human antemortem samples. *PLoS ONE* 13(11), e0207009. DOI: 10.1371/journal.pone.0207009.

Reddy, G.B.M., Krishnappa, S., Vinayagamurthy, B., Singh, R., Singh, K.P. *et al.* (2018) Molecular epidemiology of rabies virus circulating in domestic animals in India. *Virus Disease* 29(3), 362–368. DOI: 10.1007/s13337-018-0478-9.

Rupprecht, C.E., Fooks, A.R. and Abela-Ridder, B. (eds) (2018) *Laboratory Techniques in Rabies*, 5th edn. World Health Organization, Geneva, Switzerland. Available at: https://apps.who.int/iris/bitstream/handle/10665/310836/9789241515153-eng.pdf (accessed 15 October 2021).

Silvestro, D. and Michalak, I. (2012) raxmlGUI: a graphical front-end for RAxML. *Organisms Diversity & Evolution* 12(4), 335–337. DOI: 10.1007/s13127-011-0056-0.

Tao, X.-Y., Li, M.-L., Wang, Q., Baima, C., Hong, M. *et al.* (2019) The reemergence of human rabies and emergence of an Indian subcontinent lineage in Tibet, China. *PLoS Neglected Tropical Diseases* 13(1), e0007036. DOI: 10.1371/journal.pntd.0007036.

Troupin, C., Dacheux, L., Tanguy, M., Sabeta, C., Blanc, H. *et al.* (2016) Large-scale phylogenomic analysis reveals the complex evolutionary history of rabies virus in multiple carnivore hosts. *PLoS Pathogens* 12(12), e1006041. DOI: 10.1371/journal.ppat.1006041.

Tsai, K.J., Hsu, W.C., Chuang, W.C., Chang, J.C., Tu, Y.C. *et al.* (2016) Emergence of a sylvatic enzootic formosan ferret badger-associated rabies in Taiwan and the geographical separation of two phylogenetic groups of rabies viruses. *Veterinary Microbiology* 182, 28–34. DOI: 10.1016/j.vetmic.2015.10.030.

Wacharapluesadee, S. and Hemachudha, T. (2001) Nucleic-acid sequence based amplification in the rapid diagnosis of rabies. *The Lancet* 358(9285), 892–893. DOI: 10.1016/S0140-6736(01)06041-X.

Walker, P.J., Blasdell, K.R., Calisher, C.H., Dietzgen, R.G., Kondo, H. *et al.* (2018) ICTV virus taxonomy profile: Rhabdoviridae. *Journal of General Virology* 99(4), 447–448. DOI: 10.1099/jgv.0.001020.

WHO (World Health Organization) (2021a) Rabies. Available at: https://www.who.int/news-room/fact-sheets/detail/rabies (accessed 15 October 2021).

WHO (World Health Organization) (2021b) WHO collaborating centres for rabies. *WHO Rabies Bulletin Europe* 2021. Available at: https://www.who-rabies-bulletin.org/resource/who-collaborating-centres-rabies (accessed 15 October 2021).

WOAH (World Organisation for Animal Health) (2021) Infection with rabies and other lyssaviruses [Chapter 2.1.17]. In: *Manual of Diagnostic Tests and Vaccines for Terrestrial Animals*. WOAH, Paris.

Wong, J.E., Leo, Y.S. and Tan, C.C. (2020) COVID-19 in Singapore–current experience: critical global issues that require attention and action. *Journal of the American Medical Association* 323(13), 1243–1244. DOI: 10.1001/jama.2020.2467.

19 Maintenance of Rabies-free Status in Japan for 65 Years and Application of Lessons Learned to Other Countries Working Towards Zero Human Deaths

Koji Kanda[1*], Ananda Jayasinghe[2], Chandrika Jayasinghe[2] and Takahiko Yoshida[1]

[1]Asahikawa Medical University, Asahikawa, Japan; [2]University of Peradeniya, Peradeniya, Sri Lanka

Abstract

A few countries in Asia have achieved rabies-free status, including Japan. In Japan, various countermeasures have been implemented for eliminating indigenous rabies, as well as preventing rabies reintroduction for more than half a century. These include: (i) dog registrations; (ii) annual rabies vaccination; (iii) surveillance; and (iv) strict quarantine on imported animals. These activities have been put into practice through multisectoral cooperation. However, maintaining a rabies-free country requires enormous expense and effort. In addition, there is a concern for small but potential risks of rabies re-outbreak due to the decline of rabies awareness among the general public, subsequent stagnation of dog vaccination rate, and an increasing number of unregistered dogs. This chapter provides an overview of the success and challenges of maintaining rabies-free status in Japan and discusses how such experiences can apply to other countries working to eliminate canine rabies and zero human deaths by 2030.

19.1 Introduction

The World Health Organization (WHO) depicts the progress towards rabies elimination as having five different phases from endemicity to eradication: (i) endemic; (ii) control; (iii) zero human deaths; (iv) elimination; and (v) maintenance (WHO, 2018). In Asia, only a few countries, including Japan, the Maldives and Singapore, have enjoyed the 'maintenance' status for more than half a century. Among them, Japan has maintained no indigenous dog-mediated rabies since 1957 due to strict enforcement of pet dog registration, mandatory annual canine vaccination, and stray dog confinements with the enactment of rabies-related laws such as the Rabies Control Act of 1950. With continuous efforts of canine rabies control after nationwide elimination and following several amendments of relevant laws and regulations during the 'maintenance' phase, only four imported human rabies cases have been reported until now. This unique 65-year experience of 'rabies-free' maintenance would be beneficial and may apply to other nations fighting rabies. Particularly, it would help boost

*Corresponding author: kkanda@asahikawa-med.ac.jp

DOI: 10.1079/9781800622975.0019

elimination activities among countries close to zero human deaths, such as Sri Lanka. In this chapter, we will review the success stories and challenges of rabies control and prevention in Japan and propose effective recommendations towards rabies elimination and during the maintenance stage for the countries currently close to zero human deaths.

19.2 Japan's Experience with Rabies Elimination and Maintenance of the 'Rabies-Free' Status

19.2.1 History towards elimination

The first record of 'mad-dog' control activity in Japan was found in the eighth century when the basic law of the ancient Japanese political system called 'Yoro Law' mentioned the slaughter deposition of dogs (Taniguchi, 2012; Kanda *et al.*, 2020). Although there is no way to find out whether 'mad-dog' in the law indicates rabies or not, it was suggestive that rabies might have already existed in the country. Since then, several dog bites and their subsequent human deaths were sporadically reported in the modern era or mid-18th century. When the Epizootics Prevention Act of 1896 and the Infectious Diseases Prevention Act of 1897 were introduced, the first rabies statistics were released, identifying 71 canine rabies cases (Fig. 19.1). The number of canine rabies cases gradually increased, and the first human rabies cases were reported in 1912, which were 83 human deaths with 696 canine cases (Iwabuchi, 1970, p. 368). Both statistics peaked after a year of the Great Kanto Earthquake in 1923, when the public health infrastructure in the Tokyo Metropolitan Area was destroyed (Iwabuchi, 1970, p. 368). More than 3000 canine and 235 human rabies deaths occurred in 1924. Bovine and equine rabies were also sporadically reported during this time, with a maximum of 31 and 25 annual cases, respectively (Metropolitan Police Department Division of Hygiene, 1938). Due to the rapid increase of animal rabies, the first Domestic Animal Infectious Diseases Control Act of 1922 was enacted to allow the killing of all livestock infected with rabies and the control of stray dogs. The number of human and canine rabies cases dropped in the 1930–1940s, perhaps due to incomplete surveillance activities and unfavourable social situations such as war and natural disasters. It was said that many pet dogs were destroyed because of criticism, such as

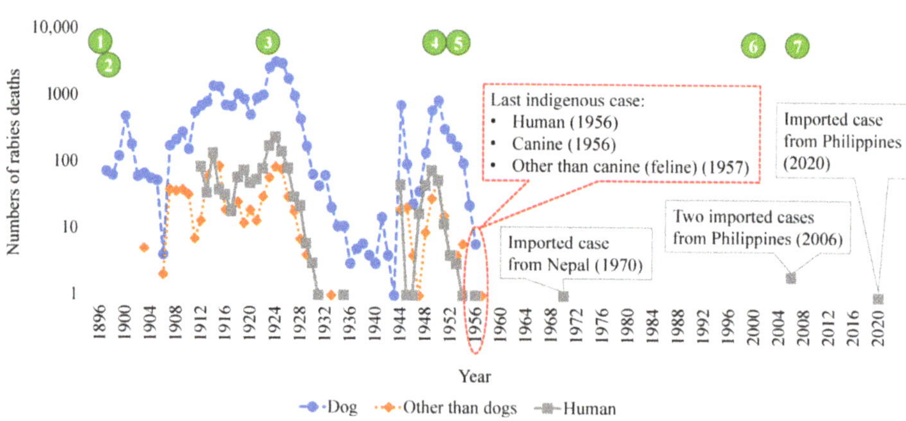

Fig. 19.1. Human and canine rabies deaths in Japan since 1896. (1) Epizootics Prevention Act (1896); (2) Infectious Diseases Prevention Act (1897); (3) Domestic Animal Infectious Diseseases Control Act (1922); (4) Rabies Prevention Act (1950); (5) New Domestic Animal Infectious Diseases Control Act (1951); (6) Infectious Diseases Control Act (1999); (7) New Quarantine Program for Importation and Exportation of Animals and Pets (2005). Data from Iwabuchi (1970); MHLW, n.d.a. Figure created by Koji Kanda.

owning dogs being considered a luxury during the war period (Ueki, 2007, p. 3). Almost all of the annual canine rabies cases in the country during 1937–1944 were reported only in the Tokyo Metropolitan Area (Ueki, 2007, pp. 4–5).

After World War II, rabies prevention and control activities were greatly improved through the nation's reconstruction of the public health system and regulations, including the enactment of the Rabies Prevention Act of 1950 and the New Domestic Animal Infectious Diseases Control Act of 1951. In particular, the Rabies Prevention Act of 1950 mandated pet owners to follow their pet dog registration and twice-a-year anti-rabies vaccinations and enforced local governments to confine stray dogs. Civil servants with veterinary licences were appointed rabies prevention officers to manage relevant activities. These powerful strategies with strong political commitment successfully achieved the rapid reduction of indigenous rabies for a short period of time. The last indigenous case in humans and dogs occurred in 1956, and the last animal case (domestic feline) was recorded in 1957 (MHLW, n.d.a; Tojinbara, 2011) (Fig. 19.1).

After World War II, more detailed statistics were recorded in the Tokyo Metropolitan Area, where the country's highest number of rabid dogs were observed. Among 793 rabid dogs identified in 1949–1955, nearly half of them were unregistered dogs (49.0%), followed by stray dogs (33.4%) (Ueki, 2007, p. 11). The proportion of registered rabid dogs gradually decreased as the number of vaccinations increased, particularly after 1952 (Table 19.1). The annual vaccination rate among registered dogs extended from 37.2% in 1949 to 82.6% in 1955 (Ueki, 2007, p. 220). The canine population estimate was founded on 1.5 times the number of registered dogs, which generated 165,383 dogs in 1951 and 226,085 in 1955 (Ueki, 2007, p. 34). Based on these figures, the human population and number of households in contact with each estimated dog during 1951–1956 were 32.5–38.8 and 7.8–9.4, respectively.

Behind the achievement of rabies elimination in a short period, there were also several difficulties at the beginning of the enforcement of the Rabies Prevention Act of 1950. For example the different ministries in charge of administration (Ministry of Health and Welfare) and implementation (Ministry of Agriculture and Forestry) of the Act and no regulations regarding the entity responsible for administering anti-rabies vaccinations (Tokutomi, 1950). Also, many dog owners were not cooperative in dog registration due to the requirement for twice-a-year vaccinations which had a reasonably high cost (Ueki, 2007, pp. 11–12). There were some cases where dog bites in a community were treated as court issues (Ueki, 2007, p. 12). However, these problems were quickly resolved through intersectoral cooperation, which aligns with the current concept of One Health.

19.2.2 Maintenance of 'rabies-free' status after elimination

Current strategies of keeping zero rabies are based on the Infectious Diseases Control Act of 1999 and the Rabies Prevention Act of 1950. To monitor the rabies outbreaks in the country, rabies is listed as a Category IV Infectious Disease under the Infectious Diseases Control Act of 1999, meaning that all cases have to be reported immediately to the governor of residence by the doctors who diagnosed rabies patients. Rabies viruses must be detected by the isolation and identification of pathogens, using the fluorescent antibody test (FAT), polymerase chain reaction (PCR) or enzyme-linked immunosorbent assay (ELISA) before the diagnosis (MHLW, n.d.a) Also doctors have to report immediately to the authority if they are suspected of rabies infection after a post-mortem examination of a body with clinical features of rabies. Likewise, a veterinarian who diagnoses or examines a rabid or suspected rabid dog must report the case to the director of the nearest public health centre under the Rabies Prevention Act of 1950. Also the local government prepares for practical responses to possible rabies outbreaks among both humans and dogs in the community based on two guidelines: (i) 'Rabies Response Guideline 2001 in case of a rabies outbreak in Japan' (MHLW, 2001); and (ii) 'Rabies Response Guideline 2013 for the risk management of rabid dogs found in Japan' (MHLW, 2013).

Table 19.1. Rabies situation and canine-related statistics in Tokyo, 1950–1956. Data from: (1) Iwabuchi, 1970; (2) (MHLW, n.d.a); (3) Ueki, 2007; (4) calculated based on Ueki, 2007. Table created by Koji Kanda.

Rabies cases and canine-related statistics	1950	1951	1952	1953	1954	1955	1956
Human rabies cases nationwide[1,2]	54	12	4	3	1	0	1
Canine rabies cases[3]	259	112	70	122	43	3	0
Registered dogs (%)[3]	56 (21.6)	21 (18.8)	6 (8.6)	10 (8.2)	4 (9.3)	0 (0.0)	0 (0.0)
Unregistered dogs (%)[3]	128 (49.4)	53 (47.3)	37 (52.9)	63 (51.6)	23 (53.5)	1 (33.3)	0 (0.0)
Stray dogs (%)[3]	75 (29.0)	38 (33.9)	27 (38.6)	49 (40.2)	16 (37.2)	2 (66.7)	0 (0.0)
Number of estimated dogs (registered dogs × 1.5)[3]	59,129	165,383	191,480	220,004	228,542	226,085	218,976
Size of human population (no.) per estimated dog[4]	106.2	38.8	35.6	32.5	32.8	34.4	36.7
No. of households per estimated dog[4]	23.9	9.4	8.6	7.8	7.9	8.3	8.9
Area (1000 m^2) per estimated dog[4]	34.4	10.5	9.0	7.9	7.6	7.7	7.9
Number of registered dogs[3]	39,419	110,255	127,653	146,669	152,361	150,723	145,983
Vaccination rate among registered dogs[3]	133.8%	74.4%	78.1%	81.9%	82.1%	82.6%	81.3%

Fig. 19.2. Examples of rabies awareness leaflets (left: dog vaccination; right: rabies prevention during travelling abroad). (1) 'Protecting people and dogs from rabies'; (2) Rabies vaccination months (April to June); (3) Pet dog registration, vaccination, registration and vaccination tags to dog collar as a mandatory to pet owners under the Rabies Prevention Act; (4) About rabies; (5) Countermeasures; (6) No unnecessary approach to animals and no feeding; (7) Rabies epidemiology; (8) When bitten, wash wounds carefully, go to hospital for post-exposure prophylaxis (PEP), and consult with a quarantine officer after returning to Japan. From MHLW (n.d.a). Only numbers in green circles have been added by Koji Kanda.

19.2.2.1 Pet owners' responsibility: registration and vaccination

It is the pet owner's responsibility that pet dogs aged 91 days and older have to be registered to their municipality of residence under the Rabies Prevention Act of 1950. After the registration, the owner is required to attach the registration tag to a collar worn by their pets. The registration fee is dependent on each municipality, but a cost of approximately ¥3000 (US$26; US$1 = ¥114.92 (Japanese yen), as of 7 March 2022) per pet dog is borne by the owner. Annual anti-rabies vaccination is also mandatory. The owner has to pay an additional ¥3000 (US$26) per vaccination. Until March 1985, vaccinations were required twice a year. Currently, every April–June is set as 'rabies vaccination months' to encourage all pet owners to complete their pet's annual vaccination at the nearest animal clinic or mass dog vaccination site (Fig. 19.2). According to the Ministry of Health, Labour and Welfare of Japan (MHLW) report, there are approximately 6.2 million registered dogs with 71.3% of vaccination coverage in 2019 (MHLW, n.d.b). However, it is estimated that there are 8.8 million pet dogs, including unregistered ones (Japan Pet Food Association, 2020), meaning that the vaccination coverage is only 49.7%, which is much lower than the WHO recommendation of at least 70% coverage (which, anyway, mainly applies to endemic countries). The Act mentions the penalty for those failing to comply with pet dog registration or annual vaccination, with a maximum fine of up to ¥200,000 (approximately US$1740). However, this penalty has not been strictly applied so that vaccination coverage among registered dogs dropped from nearly 100% in 1985 to the current level of 71.3%.

19.2.2.2 Prevention of imported animal cases

On the other hand, a national quarantine system has been improved and strictly enforced at the port of entry. Under the new quarantine regulation of importation and exportation of pets and animals with the Rabies Prevention Act of

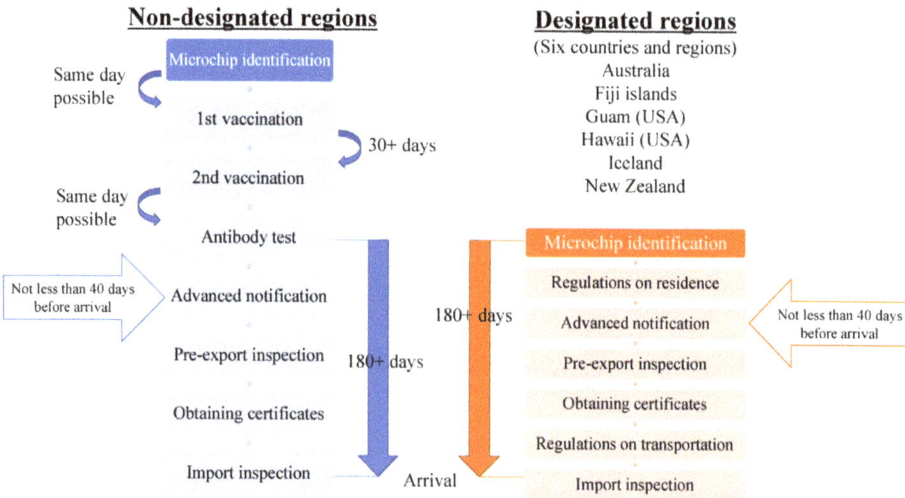

Fig. 19.3. Animal quarantine procedures. From Animal Quarantine Service of the Ministry of Agriculture, Forestry and Fisheries, 2021a, b. Figure created by Koji Kanda.

1950 and the New Domestic Animal Infectious Diseases Control Act of 1951, dogs, cats, foxes, racoons and skunks are required to undergo import quarantine. The quarantine procedures differ according to the type and origin of animals. For example, the importation of dogs and/or cats requires eight steps until the release of the imported animals from a quarantine office at the port of entry. These are: (i) microchip identification; (ii) two separate rabies vaccinations; (iii) rabies antibody test; (iv) observation of 180 days and more; (v) advance notification form to a quarantine office of entry via mail, fax or e-mail at least 40 days before arrival; (vi) pre-export inspection; (vii) obtaining certificates; and (viii) import inspection (Fig. 19.3). The total length of procedures for importation is 210 days and more. Six countries and regions (Australia, Fiji, Guam (USA), Hawaii (USA), Iceland and New Zealand) are designated as rabies free by the Minister of Agriculture, Forestry and Fisheries as of July 2013 so that they are exempt from vaccination and antibody test. Prior rabies vaccinations are not required for foxes, racoons and skunks, but the animals are stationed at a quarantine entry office for 12 hours if they are from the above six designated countries and regions, or 180 days if they are from the rest of the world. To qualify for the 12-hour quarantine of these species, an importer must meet all the requirements before entry to Japan, which are microchip implantation, 40-day advanced notification to the port of entry, approval of import inspection of animal, clinical inspection before departure, and obtaining a health certificate from the responsible exporting country government agency (Animal Quarantine Service of the Ministry of Agriculture, Forestry and Fisheries, 2013, pp. 3–4). Under this regulation, 4155 dogs, 1681 cats and 11 foxes were successfully screened and imported into Japan in 2020 (Animal Quarantine Office of Japan, 2022).

Therefore, as long as the current quarantine system is maintained or deregulated by only 20%, the mathematical modelling estimates that the risk of rabies reintroduction into Japan from abroad is infinitely low, indicating the importance of full compliance of current regulations for animal quarantine (Kwan *et al.*, 2017, p. 1168; Yamada *et al.*, 2019, p. 204).

19.2.2.3 Raising awareness for rabies prevention

No new indigenous rabies cases for over 60 years in Japan has made people's awareness of rabies prevention very low. In particular, some pet owners have questioned the need for annual dog vaccination due to Japan's rabies-free status for more than half a century, so the vaccination rate, including non-registered dogs, is currently limited to about 50%. The awareness among stakeholders involved in rabies control and prevention is also not high due to the extremely small number of suspected rabies cases. Periodical training for rabies personnel has helped enhance their awareness and understanding of appropriate responses towards rabies control, but opportunities for such training are limited (Kigami *et al.*, 2016, pp. 18–21). In addition, many Japanese are unaware of the worldwide rabies epidemic, resulting in very few of them receiving pre-exposure vaccinations before travelling unless they are required to do so. Several reasons for limited pre-exposure vaccinations among travellers are considered. One is that an anti-rabies vaccine is usually only available at travel clinics, many of which are located in major cities. The cost of pre-exposure vaccination is prohibitively high at ¥15,000–20,000 (US$130–174) per shot. Three-time vaccinations are also burdensome until obtaining the required antibodies. As a result, it is very unlikely to be vaccinated on short trips abroad, and regretfully four people have died from rabies after being bitten by a dog during travel after the nationwide elimination in 1957. Several journals have recommended rabies awareness raising and pre-travel vaccination for Japanese travellers (Basnyat *et al.*, 2000, p. 37; Kashino *et al.*, 2014, p. 246). A public awareness campaign on World Rabies Day seems not to be well activated in the country. Therefore, it is of great concern how to raise and keep rabies awareness in a country rabies free for several decades, where rabies is considered 'a forgotten disease'.

19.2.2.4 Economic benefits of rabies control

Huge financial resources and efforts are spent on implementing Japan's current rabies control measures. Of those, annual mandatory vaccination among pet dogs is a unique practice that major rabies-free countries and territories, including Australia, France, Hawaii, Hong Kong and the UK have not adopted (Yamada *et al.*, 2019, p. 206-207). As previously mentioned, Japan's average cost of vaccination is approximately ¥3000 (US$26) per shot. If all of the 6.3 million registered dogs in 2019 were vaccinated, the total cost would have been ¥18.9 billion (US$163.8 million) annually. The cost would be estimated to reach ¥26.7 billion (US$231.4 million) if 2.6 million unregistered dogs are included in the calculation. Besides this, a one-time registration cost of approximately ¥12 billion (US$10.4 million) in total and unknown indirect costs would be added for pet dog management for rabies control. The indirect cost would be twice as much as the total vaccination expenditure (Koba *et al.*, 2009, p. 71). However, the economic burden of a canine rabies outbreak has been estimated to be only about US$1.7 million with the current vaccination schedule and US$5 million without the schedule, concluding that the current vaccination strategy in Japan is poorly cost-effective (Yamada *et al.*, 2019, p. 205). Therefore, the World Organisation for Animal Health (WOAH, founded as OIE) raised the concern that the current annual vaccination schedule may need to be reviewed by undertaking a cost–benefit analysis of the mandatory rabies vaccination of dogs using a risk assessment approach and considering international 'best practices' (Weaver *et al.*, 2017, p. 7).

On the other hand, sylvatic rabies occurred among ferret badgers in Taiwan in 2013 for the first time in 52 years since 1961, and 816 positive cases were confirmed by February 2020 (Wu *et al.*, 2014; WOAH, 2020, p. 2). This resulted from surveillance of wildlife animals after a long period of infection and incubation among them. After this outbreak, the Taiwanese government strengthened preventive measures mainly by vaccinating domestic dogs and cats and animal quarantine personnel. Thus, the Japan Veterinary Medical Association (2019, p. 192) has expressed that the current review of loosening rabies vaccination is questionable, as sylvatic rabies surveillance is unavailable in Japan. Costs for rabies elimination and maintenance of zero rabies cases are huge but may be required at some point. Japan has eliminated rabies by stray dog confinement and biannual vaccinations with strong political commitment. Implementing sustainable measures based on regular post-elimination monitoring is required to maintain a rabies-free country.

19.3 Japan's Lessons for Countries Working Towards Zero Human Deaths: The Case of Sri Lanka

Takayama (2000) summarized the reasons for achieving rabies elimination in Japan in the 1950s into four points:

- an island country that could carry out quarantine activities efficiently;
- no sylvatic rabies throughout the country;
- no overlap in the living areas between bats and dogs; and
- the people's highly cooperative attitude in rabies control activities.

Regarding practical aspects of rabies control, the following three issues are worth mentioning: (i) robust data collection; (ii) pet dog registrations and stray dog confinement; and (iii) maintenance of pet dog vaccination rate. We will briefly discuss the above practical issues to provide further clues to reaching rabies elimination for the countries with zero human deaths.

19.3.1 Robust data collection

Robust data collection enhanced and promoted rabies control activities in the Tokyo Metropolitan Area, where the rabies epidemic was scarce at that time. During the years in the stage of achieving zero human deaths, the Tokyo Metropolitan Government regularly collected and published detailed rabies data such as both human and animal rabies cases (especially dogs), results of the canine rabies examination, estimated dog population by the type of their attributes (pet dogs, stray dogs, roaming dogs) and locations, the number of stray dog confinements, and vaccination rates (e.g. Shimada *et al.*, 1954; Kitaura *et al.*, 1955, pp. 249–250; Ueki, 2007, pp. 1–229). These are still accessible today, even though rabies elimination was achieved 65 years ago. However, rabies surveillance data in many countries is not recorded accurately and analysed systematically. Particularly, it is usually handled by different government organizations due to its zoonotic characteristics; therefore, data sharing is difficult.

For example, in Sri Lanka, where rabies elimination is about to be achieved, rabies data collection has been conducted independently by three different government organizations. Assessing the data trends was often difficult due to the lack of sharing information on animal population sizes, unquantified sampling biases owing to inequities in access to diagnostic capacities, regulatory and administrative barriers, and continued reliance on clinical means to establish a diagnosis (Nihal *et al.*, 2019, p. 1). As a result, the reduction in human rabies deaths has stagnated since 2015, and inadequate data management has become a barrier to the development and implementation of effective rabies control strategies, despite the extensive efforts of canine vaccination and animal birth control (Kanda *et al.*, 2021, pp. 8–9).

The concept of One Health is now widely recognized, and collaboration among relevant departments of rabies control is much easier than when Japan achieved rabies elimination. Léchenne *et al.* (2021) also pointed out the advantage of sharing information between the health and veterinary sectors through close communication. Fortunately, rabies data collection does not require special technology and can be easily achieved with continuous fieldwork and the application of systematic surveillance systems. Therefore, it is essential to establish and operate a sector-wide data collection system, and to promptly and periodically visualize the collected information, while working towards rabies elimination and maintenance of the 'rabies-free' status at the earliest opportunity.

19.3.2 Pet dog registrations and stray dog confinement

The registration of pet dogs is critical for understanding the canine population and estimating the vaccination rate. It is not technically difficult, as it requires a one-time notification to the nearest local authority when pet owners gain possession of their pet. Rather, sufficient awareness raising of the need for registration is required. For example, there are a certain number of unregistered dogs in Japan, while registration is mandatory. Fortunately, unregistered dogs are not a major problem because there are now virtually no stray dogs in the country; however, the vaccination rate based on the registered and estimated unregistered dogs is well below the maintenance target of 70%.

Although it is only available in Japanese, MHLW releases the number of registered dogs as part of annual statistics for rabies control (MHLW, n.d.c). Japan Pet Food Association (2020) also conducts a pet dog survey annually to estimate the number of registered and unregistered dogs using economic statistics such as dog food consumption. It is, however, merely possible to observe dog registration statistics in endemic rabies countries, including Sri Lanka. Among rabies control experts, a conventional dog to human ratio of 1:8 has been used for implementing the anti-rabies measures (Perera *et al.*, 2007, p. 69), but a systematic review indicated there was a wide range of dog population estimates due to location and timing of the study and methods applied (Kanda *et al.*, 2021, p. 5). In other words, the denominator of vaccination coverage is not always calculable accurately, negatively affecting the current and future vaccination strategy and birth control activities. Therefore, if such estimates are available, it will be easier to estimate the total number of dogs, including non-registered dogs. It is true that if an unregistered dog is fed human waste, it will be excluded from the statistics. However, if the number of registered dogs is not expected to increase, such estimates may be useful in understanding the current situation of domesticated dogs.

On the other hand, applying Japan's technical experience in terms of stray dog confinement might be unrealistic. Stray dog control during the stage of achieving zero human deaths in Japan focused on the capture and detention of roaming dogs and registration and vaccination. In Tokyo, where the rabies epidemic was particularly serious, the detention period of captured dogs was shortened (or they were destroyed) and roaming dogs on the street were fed poisonous food. In particular, the latter method was carried out during a campaign period with public awareness. As a result, roaming dogs were destroyed and the number of stray pet dogs themselves was reduced (Kitaura *et al.*, 1955, p. 250). However, it is prohibitive in current society to carry out physical removals of stray dogs for religious reasons and/or increased animal welfare. Community surveys in Sri Lanka revealed that stray dogs were an obstacle in people's daily life and that their removal was banned (Matibag *et al.*, 2007, p. 85; 2009, p. 58; Muthunuwan *et al.*, 2017). To compensate for such situations, sheltering stray dogs is voluntary (Jayasinghe, 2009). In addition, a simple method to count the number of stray dogs in the field was demonstrated (Conan *et al.*, 2015, p. 179) to help manage the dog population in total.

19.3.3 Maintenance of pet dog vaccination rate

To meet the worldwide target of 70% vaccination coverage, it is essential to comprehend the accurate size of the dog population. Japan successfully determined the precise number of dogs during the pre- and post-elimination stage of rabies control. In Tokyo, for example, an estimated dog population during 1950–1966 determined by the local government more or less matched the figure established by multiplying the number of registered dogs by 1.5 when the existing data were analysed in 1967 (Ueki, 2007, pp. 33–35). As a result, the maintenance of the required vaccination coverage and the prevention of indigenous rabies outbreaks have been achieved by accurate estimation of the size of the canine population or the denominator of the vaccination rate. Nowadays, it is not impossible to determine the number of stray dogs more easily and accurately by applying recently developed techniques, such as readily available livestock marking methods (Conan *et al.*, 2015, p. 179). To identify stray dogs, rabies controllers need to put their time and effort into field surveys; they are not technically difficult and are a necessary step towards early elimination. Therefore, a strong political commitment to their implementation is highly desirable.

Regular vaccinations for domesticated dogs are required until and after rabies elimination. Although it is not cost-effective to continue annual vaccination for decades after elimination, as is the case in Japan, efforts must be made to maintain a vaccination coverage of 70% to hold a zero rabies status. To do this, it is essential to develop the necessary infrastructure during the phase of achieving zero human deaths, including the provision of human resources and supplies for vaccination. Public awareness is also important during and after the elimination stages. The general public needs to be educated about first aid and post-exposure prophylaxis, as well as about the obligation of dog owners to register and vaccinate their dogs. In Japan's early days of rabies elimination, public awareness was frequently raised using radio, posters and street advertising. Today, the Internet and social networking sites are also effective in raising awareness. In Sri Lanka, rabies awareness activities as part of health education in primary schools were introduced as a pilot study, resulting in close collaboration between health and educational sectors (Kanda *et al.*, 2015). For local community people, the distribution of simple awareness leaflets was effective (Matibag *et al.*, 2009, p. 61). These interventions can be more powerful if social epidemiological surveys grasp the actual situation beforehand. Therefore, it is widely expected that awareness-raising activities will be promoted by government officials and researchers with various backgrounds, including preventive medicine, veterinary medicine, sociology and anthropology.

19.4 Conclusion

This chapter provides an overview of the successes and challenges of eliminating rabies and maintaining rabies-free status in Japan. We discussed how such experiences would be applied to countries with zero human deaths and contribute to worldwide rabies elimination by 2030. Japan strictly enforced robust data collection for human and animal rabies cases, stray dog confinement, pet dog registration and vaccination. All measures except stray dog confinement are still implemented to prevent rabies reintroduction, with an enhancement of the quarantine required for imported pets and wild animals. Such countermeasures require huge costs and efforts to maintain the services, particularly in annual vaccination. In addition, there is a concern for the decline of rabies awareness among the general public, causing the stagnation of the dog vaccination rate and the increasing number of unregistered dogs, which may be a small but potential risk of the reintroduction of the disease. However, such practical experiences would be highly applicable to countries with nearly zero human deaths. In particular, detailed data collection as part of sector-wide surveillance among relevant departments and estimating the dog population through commercial statistics towards the target vaccination coverage of 70% would help accelerate nationwide elimination of canine rabies and ultimately achieve the Zero by 30 target.

Acknowledgements

This study was funded by JSPS KAKENHI (grant numbers 18H06337 and 19K21421).

Authors' Declaration

All authors declare they have no conflict of interest.

All authors have approved this manuscript, agree with its submission, and share collective responsibility and accountability.

This manuscript has not been published or is not under review elsewhere.

References

Animal Quarantine Office of Japan (2022) Annual bulletin of animal quarantine [written in Japanese]. Available at: https://www.maff.go.jp/aqs/tokei/attach/pdf/toukeinen-40.pdf (accessed 14 February 2022).

Animal Quarantine Service of the Ministry of Agriculture, Forestry and Fisheries (2013) Guide to importing dogs, cat and other animals subject to animal quarantine from designated regions into Japan (Final revision: July 2013). Available at: https://www.maff.go.jp/aqs/animal/pdf/guidefree-e.pdf (accessed 17 September 2021).

Animal Quarantine Service of the Ministry of Agriculture, Forestry and Fisheries (2021a) Guide to importing dogs and cats into Japan from non-designated regions (Final revision: March 2021). Available at: https://www.maff.go.jp/aqs/animal/dog/attach/pdf/import-other-42.pdf (accessed 17 September 2021).

Animal Quarantine Service of the Ministry of Agriculture, Forestry and Fisheries (2021b) Guide to importing dogs and cats into Japan from the designated regions (Final revision: March 2021). Available at: https://www.maff.go.jp/aqs/animal/dog/attach/pdf/import-free-23.pdf (accessed 17 September 2021).

Basnyat, B., Pokhrel, G. and Cohen, Y. (2000) The Japanese need travel vaccinations. *Journal of Travel Medicine* 7(1), 37. DOI: 10.2310/7060.2000.00011.

Conan, A., Kent, A., Koman, K., Konink, S. and Knobel, D. (2015) Evaluation of methods for short-term marking of domestic dogs for rabies control. *Preventive Veterinary Medicine* 121(1–2), 179–182. DOI: 10.1016/j.prevetmed.2015.05.008.

Iwabuchi, H. (1970) History of rabies epidemic and prevention [written in Japanese]. *Journal of Japanese Veterinary Medical Association* 23(6), 367–376. DOI: 10.12935/jvma1951.23.367.

Japan Pet Food Association (2020) A national survey of dog and cat ownership 2020 [written in Japanese]. Available at: https://petfood.or.jp/data/chart2020/index.html (accessed 2 April 2021).

Japan Veterinary Medical Association (2019) Statement of the Japan veterinary medical association towards the opinions of the revision of rabies vaccination schedule. *Journal of the Japanese Veterinary Medical Association* 72(4), 192–195. Available at: http://nichiju.lin.gr.jp/mag/07204/a3.pdf (accessed 15 September 2021).

Jayasinghe, A. (2009) *Rabies: The Dreadful Disease*. Tha-Ro publishers, Kandy, Sri Lanka.

Kanda, K., Obayashi, Y., Jayasinghe, A., Gunawardena, G.S.P. de S., Delpitiya, N.Y. *et al.* (2015) Outcomes of a school-based intervention on rabies prevention among school children in rural Sri Lanka. *International Health* 7(5), 348–353. DOI: 10.1093/inthealth/ihu098.

Kanda, K., Jayasinghe, A., Jayasinghe, C. and Yoshida, T. (2020) Toward rabies elimination: Japan's experience, Sri Lanka's goal by 2020, and its application to Myanmar context. *Journal of Myanmar Academy of Arts and Science* 2020(Sub-theme IV), 19–26. Available at: https://www.maas.edu.mm/sca19/Sub%20Theme%20IV.pdf (accessed 18 August 2020).

Kanda, K., Jayasinghe, A., Jayasinghe, C. and Yoshida, T. (2021) Public health implication towards rabies elimination in Sri Lanka: a systematic review. *Acta Tropica* 223, 106080. DOI: 10.1016/j.actatropica.2021.106080.

Kashino, W., Piyaphanee, W., Kittitrakul, C., Tangpukdee, N., Sibunruang, S. *et al.* (2014) Incidence of potential rabies exposure among Japanese expatriates and travelers in Thailand. *Journal of Travel Medicine* 21(4), 240–247. DOI: 10.1111/jtm.12124.

Kigami, T., Koyama, M., Kitano, R., Nakayama, J., Torii, J. *et al.* (2016) Practical training of the sampling method for rabies diagnosis given for rabies-preventive staffs. *Annual Report of Kyoto Prefectural Institute of Public Health and Environment* 61, 18–21. Available at: https://www.pref.kyoto.jp/hokanken/documents/nenpou61-04.pdf (accessed 5 December 2022).

Kitaura, Y., Tanaka, J., Miki, H. and Murata, M. (1955) A study on the poisoning removal of roaming dogs in Tokyo. *Journal of the Japan Veterinary Medical Association* 8(5), 249–250. DOI: 10.12935/jvma1951.8.249.

Koba, N., Hidano, A., Kawasaki, R., Obayashi, Y. and Tamashiro, H. (2009) Rabies control after world war II and future challenges in Japan – a systematic review. *Hokkaido Journal of Public Health* 23(2), 67–72.

Kwan, N.C.L., Sugiura, K., Hosoi, Y., Yamada, A. and Snary, E.L. (2017) Quantitative risk assessment of the introduction of rabies into Japan through the importation of dogs and cats worldwide. *Epidemiology and Infection* 145(6), 1168–1182. DOI: 10.1017/S0950268816002995.

Léchenne, M., Miranda, M.E., Mauti, S., Mbilo, C. and Zinsstag, J. (2021) Integrated rabies control. In: Zinsstag, J., Schelling, E., Crump, L., Whittaker, M., Tanner, M. *et al.* (eds) *One Health: The Theory and Practice of Integrated Health Approaches*, 2nd edn. CAB International, Wallingford, UK, pp. 234–249.

Matibag, G.C., Kamigaki, T., Kumarasiri, P.V.R., Wijewardana, T.G., Kalupahana, A.W. *et al.* (2007) Knowledge, attitudes, and practices survey of rabies in a community in Sri Lanka. *Environmental Health and Preventive Medicine* 12(2), 84–89. DOI: 10.1007/BF02898154.

Matibag, G.C., Ohbayashi, Y., Kanda, K., Yamashina, H., Kumara, W.R.B. *et al.* (2009) A pilot study on the usefulness of information and education campaign materials in enhancing the knowledge, attitude and practice on rabies in rural Sri Lanka. *Journal of Infection in Developing Countries* 3(1), 55–64. DOI: 10.3855/jidc.106.

Metropolitan Police Department Division of Hygiene (1938) *Tokyo Rabies Epidemic Journal*. Metropolitan Police Department, Tokyo [written in Japanese].

MHLW (Ministry of Health, Labour and Welfare of Japan) (n.d.a) Rabies [written in Japanese]. Available at: https://www.mhlw.go.jp/bunya/kenkou/kekkaku-kansenshou10/ (accessed 14 September 2021).

MHLW (Ministry of Health, Labour and Welfare of Japan) (n.d.b) Number of registered and vaccinated dogs by prefecture, FY2012-2017 [written in Japanese]. Available at: https://www.mhlw.go.jp/bunya/kenkou/kekkaku-kansenshou10/01.html (accessed 14 September 2021).

MHLW (Ministry of Health, Labour and Welfare of Japan) (n.d.c) Annual trend of the number of registered and vaccinated dogs in 1955–2019 [written in Japanese]. Available at: https://www.mhlw.go.jp/bunya/kenkou/kekkaku-kansenshou10/02.html (accessed 14 September 2021).

MHLW (Ministry of Health, Labour and Welfare of Japan (2001) Rabies response guideline 2001 in case of a rabies outbreak in Japan [written in Japanese]. Available at: https://www.mhlw.go.jp/bunya/kenkou/kekkaku-kansenshou18/pdf/05-01.pdf (accessed 22 September 2021).

MHLW (Ministry of Health, Labour and Welfare of Japan (2013) Rabies response guideline 2013 for the risk management of rabid dogs found in Japan [written in Japanese]. Available at: https://www.mhlw.go.jp/bunya/kenkou/kekkaku-kansenshou18/pdf/guideline2013.pdf (accessed 22 September 2021).

Muthunuwan, J.T., Ganhewa, A.G.K.H., Perera, H.D.S.G., Hishaam, M., Bandara, W.M.M.S. *et al.* (2017) Preliminary survey on knowledge, attitudes and practices regarding rabies. *Sri Lankan Journal of Infectious Diseases* 7(1), 38. DOI: 10.4038/sljid.v7i1.8133.

Nihal, P.D.B., Dangolla, A., Hettiarachchi, R., Abeynayake, P. and Stephen, C. (2019) Surveillance opportunities and the need for intersectoral collaboration on rabies in Sri Lanka. *Journal of Veterinary Medicine* 7808517. DOI: 10.1155/2019/7808517.

Perera, I.N.G., Pinidiyapathirage, M.J. and Wickremasinghe, A.R. (2007) Rabies vaccination of domesticated dogs in the Central Province of Sri Lanka. *Ceylon Medical Journal* 52(2), 69–70. DOI: 10.4038/cmj.v52i2.1031.

Shimada, K., Kato, T., Ueki, H., Ohishi, J. and Murakami, H. (1954) Results of the rabies examinations in Tokyo in 1953 [written in Japanese]. *Journal of the Japan Veterinary Medical Association* 7(7), 333–337. DOI: 10.12935/jvma1951.7.333.

Takayama, N. (2000) *Human Rabies: Forgotten, Deadly Disease*. Jiku Publishers, Tokyo.

Taniguchi, K. (2012) Dog History in Japan: A 10,000-Year Story With the Human Being [written in Japanese]. Yoshikawa Ko-bun Kan, Tokyo.

Tojinbara, K. (2011) Rabies history and epidemic in Kanagawa prefecture [written in Japanese]. *Japanese Journal of Veterinary History* 48, 22–34. Available at: https://square.umin.ac.jp/jsvh/archives/pdf/48/048022034.pdf

Tokutomi, G. (1950) Enforcement of the rabies prevention act and the role of private veterinarians [written in Japanese]. *Journal of the Japan Veterinary Medical Association* 3(10), 25–26. DOI: 10.14837/jvma1948.3.335.

Ueki, H. (2007) *Tokyo Rabies Epidemic Journal*, Reprint edn. Jiku Publishers, Tokyo [written in Japanese].

Weaver, J., Punderson, J. and Stratton, J. (2017) OIE PVS evaluation report of the veterinary services of Japan. World Organisation for Animal Health, Paris. Available at: https://www.oie.int/app/uploads/2021/03/20180727-final-oie-japan-pvs-report.pdf (accessed 15 September 2021).

WHO (World Health Organization) (2018) WHO expert consultation on rabies – Third report. *World Health Organization Technical Report Series* 1012, 1–182. Available at: https://apps.who.int/iris/bitstream/handle/10665/272364/9789241210218-eng.pdf (accessed 27 July 2019).

WOAH (World Organisation for Animal Health) (2020) Current status of the rabies and prevention strategy in Taiwan. Available at: https://rr-asia.oie.int/wp-content/uploads/2020/05/status_of_report-taiwan.pdf (accessed 26 September 2021).

Wu, H., Chang, S.S., Tsai, H.J., Wallace, R.M., Recuenco, S.E. *et al.* (2014) Notes from the field. Wildlife rabies on an island free from canine rabies for 52 years – Taiwan, 2013. *Morbidity and Mortality Weekly Report* 63(8), 178. Available at: https://www.ncbi.nlm.nih.gov/pmc/articles/PMC4584525/pdf/178.pdf (accessed 5 December 2022).

Yamada, A., Makita, K., Kadowaki, H., Ito, N., Sugiyama, M. *et al.* (2019) A comparative review of prevention of rabies incursion between Japan and other rabies-free countries or regions. *Japanese Journal of Infectious Diseases* 72(4), 203–210. DOI: 10.7883/yoken.JJID.2018.431.

20 Conclusions

Deborah Nadal[1,2,3]* **and Vanessa Slack**[4]
[1]Ca' Foscari University of Venice, Italy; [2]University of Glasgow, UK; [3]World Health Organization, Geneva, Switzerland; [4]Texas Department of State Health Services, Zoonosis Control Branch, Texas, USA

The aim of this book was to bring together practical, real-world experiences with dog rabies control and dog-mediated human rabies elimination at the ground level, with the purpose of sharing ideas, suggestions and lessons learnt on how to accelerate towards dramatically reducing the burden of this lethal disease in animals and humans. It is unlikely that the goal of reaching zero human rabies deaths by 2030 will be achieved in all rabies-endemic countries, but crucial progress can be made in the next few years to set the groundwork for the 'last mile' in the decade 2030–2040. With this timeline in mind, we hope that this collection will be of inspiration for current but especially prospective rabies practitioners, decision makers, scholars and any stakeholders across any sector, discipline, community and country – within and outside Asia.

This concluding section builds on the take-home message that we, as editors, asked each chapter author to provide, as well as the messages that we personally took home from two international meetings that we attended in autumn 2022 when we were writing the conclusions of this book. These two events are the Rabies in the Americas conference (Mexico, October 2022) and the World One Health Congress (Singapore, November 2022), where rabies was among the most represented diseases. These take-home messages can be recapitulated as follows.

The international rabies community requires collaboration from social scientists, communication specialists, and economics and policy experts in order to understand what obstacles are hampering a fast and smooth transition from the theory to the practice of rabies elimination. Particular attention must be paid to mass dog vaccination and the need to implement it wherever necessary in a coordinated, systematic and sustained way. The future of rabies elimination lies in persistence. While mass dog vaccination is widely advocated, it is the lack of sustained implementation that is causing this control method to fall short of its yearly target goals, as well as the 2030 goal.

To do this, the adoption of a collaborative, intersectoral One Health approach is integral to success. The Association of South-east Asian Nations (ASEAN) should aim to advance competencies in One Health science and implementation, by developing more capable leaders and advocates who fully embrace the One Health approach. ASEAN member states should strive to narrow the One Health development gaps,

*Corresponding author: nadal.deborah@gmail.com

 One Health for Dog-mediated Rabies Elimination in Asia (eds V. Slack *et al.*)
DOI: 10.1079/9781800622975.0020

commit to establishing formal One Health multisectoral coordination mechanisms, and look for incentives to collaborate between stakeholders to break down siloed approaches. This is crucial not only to co-design and co-implement rabies control strategies, but also to strengthen the surveillance of human, dog and wildlife rabies. Moreover, because of the crucial importance of cross-border mass dog vaccination, regional networks such as ASEAN can play a key role in ensuring that not only single countries, but entire areas of the world, move fast towards rabies elimination.

Structured conversational processes such as the World Café are very helpful to create a safe, welcoming environment where it is possible to increase awareness on why and how to eliminate rabies, connect and share multiple ideas and perspectives, generate novel solutions, and encourage collective action. The assumption is that collective discussion can trigger the paradigm shift that a One Health approach to rabies control requires.

One Health collaboration needs to include rabies stakeholders at all levels of society, from the international to the local one. At the local level, the network should include not only communities and traditional stakeholders (within and outside the health sector) but non-traditional stakeholders as well, such as religious leaders and traditional healers. This is necessary because rabies control strategies need to be tailored to the local social, cultural and religious context, and to build on local networks, resources and communication channels.

Rabies One Health networks also need to consider animal welfare activists, to ensure, if possible, that all rabies stakeholders are working towards the same goal – although, sometimes, in different ways. For example, animal welfare activists can be pivotal in convincing government bodies and civil society that dog culling is not only ineffective, but counterproductive, in controlling dog rabies.

We need to make a strong investment case to finance mass dog vaccination as the most cost-effective control measure considering the high costs of post-exposure prophylaxis (PEP) and overall rabies control. Financial resources for mass dog vaccination should come from both the human and the animal health sectors. Examples of this One Health co-investment model built on the principle of preventing rabies at the source (i.e. dogs) – and not just treating dog bites in humans – are available.

The mass vaccination of puppies, using rabies vaccines of high immunogenic capacity, can make it easier to achieve and maintain 70% vaccination coverage. Additionally, the introduction of oral dog vaccination can be a game changer in complementing parenteral vaccination, by enhancing vaccination coverage in hard-to-catch free-roaming dogs. Evidence exists to demonstrate the feasibility and scalability of oral dog vaccination and its potential contribution to the elimination of dog-mediated human rabies. Coordination and assistance among countries, development partners and the private sector are urgently needed to make the best use of this tool.

Dog ecology influences the efficacy of mass dog vaccination. The collection of field data is essential to improve knowledge of local aspects such as population size and roaming ranges, habitat utilization, pack sizes, contacts, and dog population demographics and dynamics, and to form the basis of evidence-based animal and public health policy for the control of dog rabies.

The contribution of dog population management to rabies control is in increasing the accessibility of free-roaming dogs, through measures such as the promotion of responsible behaviour, veterinary care, dog vaccination, dog population control, dog bites and rabies awareness, and law enforcement. The Asian continent has a high free-roaming dog population and relatively little investment in dog population management systems and associated small-animal veterinary services. Dog population management should provide an opportunity to bring additional funding, political will, and community support based on its wider aims reaching beyond only rabies control. Nevertheless, it should not divert resources away from vaccination. Rather it should support greater vaccination coverage and drive down the cost of vaccination campaigns and PEP.

Lacking or poor waste management increases the quantity of food available to free-roaming dogs and supports the growth of their population, which in turn decreases the fraction of vaccinated dogs. Efficient waste management is necessary to complement dog population management and support mass dog vaccination efforts.

Accurate epidemiologic data both for humans and for animals is significant in streamlining rabies prevention and control and evaluating the success of intervention strategies. To have accurate data, laboratory confirmation is of utmost importance. The straw and hook methods are techniques for brain specimen collection that are reliable, cheap, fast and easy to perform, and require fewer biosafety requirements compared to opening the skull (the standard method of collecting brain specimens for rabies diagnosis).

Advancements in high throughput DNA sequencing tools and genomic techniques can help to better understand the transmission of rabies across geographies, the emergence of new lineages, and host-specific evolutionary and transmission dynamics of the virus. Whole-genome sequences can help classify circulating rabies strains for the identification of high-risk areas and targeted control programmes for rabies elimination in Asia. With further optimization of protocols, training of professionals, and infrastructure, these advanced techniques can facilitate rabies surveillance.

Resource-limited countries should implement cost-effective, PEP-saving measures, to benefit both national health systems and at-risk communities. Vaccine-pooling strategies that adopt the intradermal administration of vaccines are cost-effective options designed to administer fractional doses of rabies biologicals, without compromising efficacy and by making sure that PEP is always available and free of charge for high-risk populations.

Animal Bite Treatment Centres provide an efficient system for widespread PEP delivery at the local level, but this infrastructure needs to be complemented with an animal investigation for a more holistic approach to bite management and for making sure that expensive PEP is used only when necessary. Collaboration between animal and human health workers should be strengthened, for example through the setting up of an Integrated Bite Case Management system.

Based on Japan's experience in the pre-elimination stage of rabies, robust and cross-sectoral surveillance, mandatory pet dog registration, the maintenance of high levels of dog vaccination coverage, free-roaming dog population control, and continuous awareness-raising are indispensable tools to achieving and maintaining rabies elimination.

If an impact is to be made by 2030, it is crucial to implement and scale up mass dog vaccination and PEP delivery to target those people and dogs living in socially, economically and marginalized areas. Actual rabies prioritization (beyond formal disease prioritization workshops) and long-term political and financial commitment to a One Health approach to rabies elimination are now all that is necessary to reach the final goal as soon as possible. When advocating for rabies elimination, it is useful to remember that:

- We have the tools and the knowledge to start making an impact now, and improving interventions as we go.
- Rabies is 99.9% fatal but 100% preventable: let's make that 0.1% count.
- Rabies elimination contributes to the achievement of several Sustainable Development Goals: (1) no poverty, (3) good health and well-being, (10) reduced inequalities and (17) partnerships for the goals.
- Rabies elimination strengthens health systems to the benefit of neglected and non-neglected diseases.

There is no single, simple solution to eliminate dog rabies from endemic areas nor to achieve zero human rabies deaths. Every aspect of rabies control carries a set of challenges to overcome in order to be successful. A multi-pronged approach is hence needed, centring around a One Health collaborative effort. We must emphasize the need for sustained mass dog vaccination, continuous communication and open information sharing to learn from one another's successes.

Authors' Declaration

All authors declare that they have no conflict of interest.

All authors have approved this manuscript, agree with its submission, and share collective responsibility and accountability.

This manuscript has not been published or is not under review elsewhere.

Index

Letters '*b*', '*f*' and '*t*' after page numbers indicate box, figures and tables, respectively

www.ingramcontent.com/pod-product-compliance
Lightning Source LLC
LaVergne TN
LVHW070513160926
848142LV00009B/42